FERTILITY AND MORTALITY

Fertility and Mortality

Theory, methodology and empirical issues

EDITED BY

K. MAHADEVAN

with

P.J. REDDY & D.A. NAIDU

SAGE PUBLICATIONS
New Delhi/Beverly Hills/London

First published in 1986 by

Sage Publications India Pvt. Ltd.
32 M Block Market, Greater Kailash I
New Delhi 110 048

Sage Publications Inc
275 South Beverly Drive
Beverly Hills, California 90212

Sage Publications Ltd
28 Banner Street
London EC1Y 8QE, England

Published by Tejeshwar Singh for Sage Publications India Pvt Ltd, phototypeset by South End Typographics, and printed by Pearl Offset Press.

Library of Congress Cataloging in Publication Data

Fertility and mortality.

1. Fertility, Human—Addresses, essays, lectures. 2. Fertility, Human—India—Case studies—Addresses, essays lectures. 3. Mortality—Addresses, essays, lectures. I. *Mahadevan*, K., 1940– II. Reddy, P.J. III. Naidu, D.A.
HB901.F468 1986 304.6'3 85–22287
ISBN 0–8039–9497–4

To

felicitate Dr C Chandrasekaran for his demographic contributions

Contents

Preface

This book is an outcome of a seminar on 'Society and Population Dynamics' held at the Department of Population Studies, S.V. University, Tirupati, India in January 1984. The major aim of the Seminar was to pool the latest knowledge on fertility and mortality and to bring out a book on the dynamics of fertility and mortality. On this occasion Dr C. Chandrasekaran was felicitated for his contributions in the field of Population Studies and this book is dedicated to him.

The papers included in this volume were prepared for the seminar and for this book. However, a few papers that were received after the seminar were also included here. The major theme of the book focuses on theory, methodology and issues of fertility and mortality. The nineteen papers included in this volume have been divided into five sections: Theories of fertility, Culture and fertility, Social change and fertility, Analytical framework, models and mortality, and A profile of Dr C. Chandrasekaran. Since this book focused only on themes related to fertility and mortality, we could not include several other valuable papers presented at the seminar because of their diversity from the central theme. I am sure that the authors of these papers will bear with us the omission, though these papers merit publication.

The coverage of topics and their discussion relate to sex roles, theory, socio-cultural determinants, social change, value and cost of children and fertility, theoretical expositions, methodological aspects and determinants of mortality. The critical discussion of these topics is the core foundation of the growing and expanding field of population science. They have universal relevance as well. Several case studies focusing on different cultures and regions provide a cross-cultural insight into the diverse determinants of fertility behaviour. These case studies along with other empirical findings on the socio-economic determinants of fertility and its regulation are valuable themes for formulating decentralised policies and successfully implementing population programmes. The three chapters on demographic transition and determinants of rapid demographic and social change in Kerala, India and Indonesia are the outcome of successful experiments in the field of population control worthy of replication elsewhere. The chapters on the significance of modernisation along theoretical and empirical lines would enhance our understanding of socio-demographic change. The fourth section dealing with analytical framework, conceptual models, socialisation, cultural and nutritional determinants of mortality may form the basis for future research and for formulating policies and planning effective mortality control programmes in developing and developed countries. Hence, this book will be a valuable addition to the existing body of knowledge in the field of population sciences. The chapters included in this volume may be suitable either independently or in combination with other existing literature to meet the growing demands for advanced literature in the field of population and to cater to the requirements of population scientists.

The seminar and the publication of this book was possible because of the sincere efforts of several contributors and colleagues who volunteered to write different chapters. Special credit for the success of the seminar must be given to Dr P.J. Reddy and Dr D.A. Naidu. We are indeed grateful to the S.V. University administration, faculty, staff and research scholars of the department of Population Studies for their administrative support and to the University Grants Commission of India for their financial assistance. We also acknowledge the financial assistance offered by the Dorabji Tata Endowment Trust for the publication of this volume. We would also like to thank our research students and the typists for reading the manuscript as well as for the secretarial and typing assistance rendered by them.

A couple of articles included in this volume were originally presented at different conferences and published in various journals. However, they were modified or expanded for the present publication. At the request of Dr Kingsley Davis and a few other authors, we thank the editor of PDR and the other organisers of these conferences for permitting us to include these articles in this volume.

Once again we thank the S.V. University administration for the facilities provided by them which enabled us to compile this volume.

K. MAHADEVAN

I
Theories of Fertility

1

Demographic Foundations of New Sex Roles

KINGSLEY DAVIS AND PIETRONELLA VAN DEN OEVER

A fundamental dilemma faced by the advanced societies today is whether all ascription of status on the basis of sex will disappear or whether some division of labour between men and women will be retained. Hardly anyone believes that the traditional system will, or can, be brought back completely, but many people doubt whether all assignment of rights and obligations on the basis of gender can be abolished. They feel either that selected traditional elements must persist or that a new utopian structure of sex roles based on modern technology must be worked out; and in either case they base their argument on the biological differences between the sexes that cannot be totally ignored in the division of labour. If they are right, then the debate shifts from the question of whether men and women shall be treated exactly alike in all assignment of tasks to the question of what would be an equitable division of labour between the sexes.

The present paper does not address these issues in their entirety; rather, it considers an important but somewhat neglected aspect—the effect on sex roles of a few important demographic changes in advanced industrial societies—changes such as, increased longevity, widening sex differences in mortality, aging populations, and low fertility. The argument is that such changes give rise to new circumstances between men and women that force alterations in the sex roles. This, in turn, brings forth ideological developments such as the feminist movement. It is not that the demographic trends are perceived by each individual and thus consciously adapted to, but rather that they arise spontaneously in advanced societies and alter the conditions of life. Ironically, they strengthen the public demand for equal treatment of men and women while widening the differences between them.

The Demography of Age and Sex

The basic demographic changes in modern times are familiar. In addition to those already mentioned, they include decreases in household size, unbalanced sex ratios, shortened reproductive activity, and near universal survival to old age. What is less familiar is the tendency of these developments to affect the sexes differentially and to alter the traditional organisation of sex roles.

In analysing the effects, we can start with any demographic change, since the changes are all interrelated. Let us begin with mortality.

New Patterns of Mortality

Everyone is aware about the dramatic fall in human mortality during the last two centuries. Many are also aware about the greater fall in female than in male mortality, but few realise the pace and scope of that change. In Japan, for instance, the male-female difference in the expectation of life at birth increased from six months at the beginning of this century to over five years recently.[1] Elsewhere the shift has been comparable. As a result, female longevity now substantially exceeds male longevity virtually throughout the industrialised world. In the white population of the United States in 1980, the difference in life expectancy was 7.6 years; in France in 1976 it was 8.0 years; and in the USSR in 1971–72 it was 10.0. As Table 1 shows, the female advantage characterises developing as well as developed countries, but to a lesser degree.

TABLE 1 Difference between Male and Female Life Expectancy in Developing and Developed Countries in the 1970s

	Life Expectancy at Birth (years)[a]		*Difference (years)*	*Difference as Per cent of Male Life Expectancy*
	Male	*Female*		
16 developing countries	60.5	63.9	3.4	5.7
20 developed countries	68.4	74.8	6.4	9.4

[a] Unweighted averages of individual country figures
SOURCE: Calculated from United Nations Demographic Yearbook 1978, Table 22

The difference in mortality shows up not only in life expectancy but also in crude death rates. However, since the female population is older than the male population, the difference in crude rates grossly understates the real difference in mortality. Thus, in the Netherlands in 1977, the male crude death rate was 27 per cent higher than the female rate, but if the female population had the same age structure as the male population, the male rate would have been 72 per cent higher.

[1] Differences in life expectancy in Japan at the start of the century and in modern times were as follows:

	Male %	*Female%*	*Difference*
1902–13	44.2	44.7	0.48
1976	72.2	77.4	5.20

SOURCE: Institute of Population Problems. The 22nd Abridged Life Tables, Research Series No. 194, 15 July 1970, Appendix Tables; United Nations Demographic Yearbook 1978, p. 408

A Demographic Illusion

A vague awareness of such facts has led to a widespread illusion—namely, that the male-female difference in mortality has unbalanced the sex ratio of whole populations. Intuitively, one thinks that if more men than women were to die, there would be fewer men than women; extrapolating the trend, one can easily envisage a disconcerting feminisation of the population and conjure up radical changes in the sex roles. To be sure, if the situation was that there were four women to every man, sex roles would be different from those seen today.

Although basically false, this idea of progressive feminisation contains a grain of truth. Worldwide data show that there is, on an average, a surplus of females, especially in the industrial nations. For dates in the 1970s, the mean number of males per 100 females in twenty-one developing countries was 99.3; in nineteen more advanced countries it was 96.1.

In view of the pronounced sex differences in mortality, however, the striking feature about the overall sex ratios is how balanced they are. A sex ratio in the 90s is, after all, fairly close to parity. Evidently, in the popular notion of overall feminisation, a point is being overlooked. The point that is being overlooked is, first, the fact that mortality is not the only factor that affects the sex ratio, and second, that the influence of mortality varies strikingly according to the age group being considered. As a result, the significant distortions in the sex ratio occur in particular age groups, and in different directions. It follows that to discuss the sex ratio without reference to the age structure has little meaning. For instance, the fact that the entire population of the Netherlands in 1977 had 98.9 males per 100 females meant little to those aged 70 and over, among whom the proportion was 67.7 per 100.

Historically, as mortality declined, sex ratio became more feminine at advancing ages, while it became more masculine at the younger ages. This is illustrated by the white population of the United States. If we exclude the effects of migration and population growth, the sex ratio changed as shown in Table 2.[2]

TABLE 2 Changes in Sex Ratio Excluding the Effects of Migration and Population Growth

	Males per 100 Females	
	1910	*1980*
Under age 50	101.9	103.7
Age 50 and over	94.2	84.9

At age levels below 50, sex ratio became more masculine despite the fact that female survival in those ages improved as rapidly as male survival. Over the age of 50, with the survival of women rising twice as fast as that of men, sex ratio became drastically more feminine.

To understand this movement of sex ratio in opposite directions according to age, and to explain why overall feminisation is so slight, we must examine the nature and interaction of the determinants of sex ratio by age.

[2] These are the ratios in a sex-adjusted life table population.

Determinants of Sex Ratio by Age

In a population long subject to constant vital rates and no migration—that is, a stable population—the sex ratio at any given age is determined by two factors: the sex ratio at birth and the survival of males and females up to the given age. Significantly, the second factor can be further divided into two parts that affect the sex ratio in contrary ways: first, the difference in survivorship between men and women, and second, the mean level of survivorship for both. The lower the mean level, the more important a given difference will be.

At exact age O, there is only one influence—the sex ratio at birth. If we suppose that no other factor entered at the later ages, the sex ratio at every age would be identical with the ratio at birth, which means it would be masculine. However, from the moment of birth the difference in mortality between males and females exercises an influence, and this influence is cumulative. Since the difference normally favours females, it eventually overcomes the masculine advantage at birth and, at the older ages, makes the sex ratio feminine.

The operation of these factors is shown in Table 3, which uses data from Norway.[3] The first column gives the actual sex ratios by age. These ratios are the result of all the factors combined, including migration and the absolute level of mortality. The second column shows the constant sex ratio that would result if the natal ratio was the only factor involved. The third column gives the results of mortality alone; since, in the absence of infanticide, from the moment of birth boys die more frequently than girls, the sex ratio from mortality alone cannot be masculine at any age. The last column shows the combined influence of the sex ratio at birth and subsequent mortality. It resembles the first column precisely because in Norway migration and population growth are minor factors, leaving the natal sex ratio and mortality as the main determinants.

How important the sex ratio at birth is in keeping the population masculine through at least the reproductive ages can be seen by contrasting the third and fourth columns. These demonstrate that in the stationary population the surplus of male infants causes all age groups to be masculine up to about 50 years of age, after which the cumulative effect of greater male mortality is so tremendous that it makes all the older age groups feminine.

We can now understand why, historically, the population below the age of 50 years has become more masculine rather than more feminine. The fact was that a high proportion of deaths occurred at the younger ages. This meant that a pronounced difference in mortality between males and females could distort the sex ratio even at the younger ages. Now death rates have fallen so far that nearly everybody survives till the age of 50 (89.8 per cent of white males and 94.7 per cent of white females, according to the 1980 US life tables). This being true, there is no longer room at the younger ages for differential mortality to make an impact, and such improvements as have taken place must be mainly among the males. If we imagine improvement until nobody dies below the age of 50, then (as comparison of the second and fourth columns in Table 3 shows)

[3] Norway has had a minimal influence from migration in recent decades, and its data on age, sex, and marital status are excellent.

the sex ratio of people under 50 would be entirely determined by the sex ratio at birth and, therefore, be even more masculine than it is now.

TABLE 3 Determinants of the Sex Ratio by Age, Norway, 1979

Age	*Sex Ratio in Actual Population*[a]	*Sex Ratio in Population if Influenced by*		
		Sex Ratio at Birth Only[b]	*Sex Difference in Mortality Only*[c]	*Sex Ratio at Birth and Mortality*[d]
0–19	105.1	105.9	99.6	105.5
20–29	105.1	105.9	99.0	104.8
30–39	106.4	105.9	98.3	104.1
40–49	102.0	105.9	97.2	102.9
50–59	98.3	105.9	94.0	99.6
60–69	86.1	105.9	83.5	88.4
75+	62.4	105.9	59.0	62.5
All ages	98.3	105.9	90.8	96.2

[a] Norway, Central Bureau of Statistics, Population by Age and Marital Status, 31 December 1979, Table 5, pp. 32–35

[b] This is the sex ratio at each age if the sex ratio at birth was the sole determining factor

[c] These are the sex ratios that would result from the influence of mortality alone, obtained by dividing L_x values in the male life table by the corresponding L_x values in the female life table. Since both life tables have a radix of 100,000, the sex ratio at birth exercises no influence. The life table for 1977–78 is used. Norway, Central Bureau of Statistics, Vital Statistics and Migration Statistics, 1979, Table 39, pp. 58–69.

[d] These are the sex ratios that would result from the combined influence of the sex ratio at birth and, at each age, the sex difference in mortality. Obtained by calculating a sex-adjusted life table—that is, one in which the ratio between the male and female radixes equals the sex ratio at birth. (Arithmetically this equals the product of the corresponding figures in columns 2 and 3, divided by 100)

It follows that the mortality gap between men and women is overwhelmingly due to the older ages. In fact, of the 7.6 year difference in life expectancy between white American males and females in the United States in 1980, only 2.2 years, or 29 per cent, comes from lower female mortality under age 50, leaving 71 per cent to be accounted for by differential mortality above the age of 50. In Norway, where improvement has gone a bit further than in the United States, the proportions are even more dramatic—of the 6.3 year difference in life expectancy at birth, 22 per cent comes from lower female mortality under age 50, and 78 per cent from differences at the older ages.

'Surpluses'

Instead of using the sex ratio to describe the numerical relation between the sexes, one can speak of 'surplus' males or females—meaning the number or proportion above a 50–50 ratio. One can easily convert Table 3 to these terms. For instance, a sex ratio of 105.9 (in the second column of Table 3) means that the surplus of males is 5.9 per cent, and this is what it would be at all ages if the sex ratio at birth were the sole determinant. Similarly, in the third column, a sex ratio of 59.0 at ages 75 and above means that, on the basis of mortality alone, only 59 per cent of the women would have a male in the same age group, leaving 41 per cent of the women without such a man and therefore 'surplus'. If mortality alone was the determinant, there would be an excess of females at all ages, but the excess would be only 1 per cent at ages 20–29 reaching high percentages only at the age of 60 years and above. It is the sex ratio at birth that makes the surpluses masculine up to the age of 45 or 50 years, after which the cumulative force of mortality makes the surplus feminine. This means that it is the older ages—the ones in which mortality is high—that show the greatest imbalances of the sexes, and these imbalances have increased as the male-female gap in mortality has widened.

Clearly, then, the remarkable extension of life in the advanced societies has not been an unmixed blessing. Not only has it achieved a reduction in fertility and, therefore, an aging of the population, but by favouring females more than males, it has produced a revolutionary imbalance of the sexes at the older ages. As the population aged, this imbalance affected an ever greater proportion of the population. Although at the younger ages a rise in the male surplus has occurred, at age levels above 55 years both the proportion and the number of women in the 'surplus' category have become very large.

Marriage and the Imbalance of the Sexes

So far we have considered only the imbalances in the sex ratio by age. Now we must analyse the imbalances by marital status as well. One naturally tends to assume that in any adult age group the sex ratio influences marital chances—that is, a 'surplus' of one sex or the other reduces the probability of marriage for that sex. This observation is roughly correct, but it understates the case. Since each person already has a marital status—either married or unmarried—his or her chances of getting married depend less on the sex ratio of his or her age group as a whole than on the sex ratio of those in the group who are unmarried.

Age Selection in Marriage

It turns out that the sex ratio imbalances among the unmarried are greater than can be explained by the age-sex distribution alone. For instance, among people in the age group 60–64 in Norway in 1979, there was a moderate surplus of women—8.2 per cent. However, among the unmarried in this age group, the surplus of women was 45.7 per cent. Obviously, some factor other than the sex ratio at birth and differential mortality is at work.

This some other factor is well-known. Let us call it 'age selection,' or 'age choice,' in marriage. It is the fact that people do not necessarily choose mates in their own age group. In general, not only are grooms somewhat older than their brides at the time of the first marriage, but after divorce or widowhood a larger proportion of men than women remarry and, when they do, they tend to marry women younger than themselves by a margin much wider than was the case in their first marriage. With this source of distortion added to the sources already described, the imbalances in the sex ratio of the married and unmarried in some age groups become spectacular.

Table 4 shows the data for two groups of countries, industrial and developing. Clearly, the married and unmarried populations complement each other. In the married population the sex ratios are highly feminine at the younger ages and masculine at the older ages, whereas in the unmarried population the reverse is true.

TABLE 4 Sex Ratios by Age and Marital Status, Developed and Developing Countries, ca. 1970

Age	*Males per 100Females*			
	9 Industrial Countries[a]		*13 Developing Countries*[b]	
	Married	*Unmarried*	*Married*	*Unmarried*
15–19	17.6	114.3	17.0	119.4
20–24	53.5	168.1	46.3	194.3
25–29	86.5	177.6	74.7	215.0
30–34	96.5	133.2	92.4	134.6
35–39	100.7	103.2	102.2	89.6
40–49	100.4	55.7	115.4	53.3
50–59	107.2	39.5	142.2	37.6
60–74	133.4	32.4	194.4	34.2
75+	200.1	30.2	259.8	41.3

[a] Australia, Austria, Belgium, Czechoslovakia, Finland, Hungary, Japan, Norway, and Yugoslavia
[b] Brazil, Fiji, Malaysia, Mexico, Morocco, Panama, Puerto Rico, Singapore, South Korea, Syria, Thailand, Tunisia, and Turkey
SOURCE: United Nations Demographic Yearbook, 1978, Table 22. The figures are unweighted averages, each country counting as one regardless of size

Among the married, the reason for the female surplus at the younger ages is that many young women are married to older men, thus excluding a substantial portion of men of their own age from the married column. At the advanced ages the sex ratio of the married population is masculine—for the same reason—many men are married to younger women. In the unmarried population, by contrast, males predominate at the younger ages. To a minor extent this is because there are more males than females at those age levels. The main reason, however, is that many males either choose to remain unmarried during their youth or lose out in competition with older men and thus are forced to do so. At the older ages, on the other hand, females predominate both because they are numerous at those age levels compared to males and because a sizable portion of males at those age levels have married women considerably younger than themselves. In the industrial countries (Table 4), for example, at ages 20–24 there are, on an average, nearly 60 unmarried women for every 100 unmarried men, while at ages 50–59 the ratio is more than 250 for every 100.

A look at Table 4 shows that non-industrial countries have age- and marital-status-specific sex ratios roughly similar to those seen in the industrial nations. Indeed, by comparison it appears that in the developing countries older men are somewhat more successful in competing for younger women, and this compensates for the lesser male-female differential in mortality. In any case, we are dealing with a phenomenon which is so fundamental that it is independent of economic development.

There is, however, one difference between the economically advanced and the less advanced countries. Due to low birth rates, the advanced nations have much older populations. This means that the excess of unmarried women in the older age levels is much larger as a proportion of the total population. Table 5 clearly highlights the difference. Unmarried females aged 50 years and above, in proportion to all females aged 15 years and above, are two-and-a-half times more in the industrialised than in the non-industrialised countries. On the other hand, as Table 5 reveals, the surplus of married men at the older ages, in ratio to the total adult male population, is less in the developed countries. In other words, in the developed countries the imposing number of elderly unmarried women is not due to some aberrant behaviour of men but to the much greater proportion of the elderly in these countries. In fact, in the industrial countries the propensity of men to marry younger women is less than it is in traditional societies.

TABLE 5 Surplus Married Men and Unmarried Women Aged 50 and Over as a Per Cent of the Adult Population, ca. 1970

	Weighted Averages		*Unweighted Averages*	
	9 Industrial Countries	*13 Developing Countries*	*9 Industrial Countries*	*13 Developing Countries*
Surplus married men age 50+ as per cent of all men age 15+	4.9	5.6	4.5	5.8
Surplus unmarried women age 50+ as per cent of all women age 15+	17.3	6.5	14.5	6.0

NOTE: Identity of the countries in each group is the same as in Table 4
SOURCE: See Table 4

THE CONTRIBUTION OF AGE CHOICE TO SEX IMBALANCES

In discussions of sex imbalances, people tend to view the involuntary causes—mainly the sex ratio at birth and the male-female mortality differential—in a different light from the voluntary cause, age choice. The latter seems to be more socially determined, and more subject to control. From a policy point of view, we should measure its separate influence.

Before we can do this, we must first get an expected, or theoretical age-sex marital distribution derived solely from the sex ratio at birth and mortality. This can then be compared with a distribution based on age selection as well. The difference between the two models will measure the influence of age selection alone.

We begin with the stationary population of a sex-adjusted life table (Table 3). To obtain the first model, we incorporate marital status by assuming that in each age group the married and unmarried populations have the same sex ratio as found in the stationary population.[4] To obtain the second model, we assign the married and unmarried populations the sex ratios that they have in the actual population. Since the sex ratios in the first model are derived solely from the sex ratio at birth and mortality, while those in the second are due to age choice as well, the difference between the two models stems from the latter factor.

The results of the calculations are given in Tables 6 and 7, which show that the age choice factor exercises tremendous influence, especially at either end of the age scale. As Table 6 shows, if the factor of age choice was absent, the sex ratio of the married and unmarried alike would be predominantly masculine at the younger age levels and feminine at the older age levels. This is actually the case among the unmarried, as can be seen from the last column, but the imbalances are greatly exaggerated. Among the married the distortion runs the other way: the sex ratio is very feminine at the younger age levels and masculine at the older age levels. The figures in the table indicate that older males compete with younger males for young women, leading to an excess of unmarried men in the younger age groups and of unmarried women in the older age groups. Between the ages of 35 and 55, a rough balance is struck. The competition for younger women by men of all ages, together with the masculinity of the sex ratio below the age of 45, maximises the possibility of mating for women during their reproductive span.

The same information can be expressed in terms of surpluses.[5] In the married population the surpluses in any age group run counter to the natural sex ratio. There is a surplus of married women at the younger age levels, where the sex ratio is masculine, while married men are in surplus at the older age levels, where the natural sex ratio is feminine. The surpluses are almost entirely attributable to age selection. With the unmarried, however, the natural sex ratio and age choice both work in the same direction, and thus the surpluses are explained by both age choice and involuntary factors, though age choice is the larger factor.

For this reason, and because the sex imbalances among the unmarried are greater than among the married, one can understand why discussions of marital chances focus on the unmarried. A more important reason, however, lies in the logic of monogamous marriage. If the entire married population is taken independently of age, its sex ratio in a monogamous society has to be 100; and, at any

[4] It must be remembered that the question at issue is how age choice affects the sex ratio of the married and unmarried population in each age group. In the absence of such choice, the sex ratio in these marital categories would be the same as in the life table population.

[5] 'Surplus' here is defined as the percentage by which a given sex exceeds the other sex in a given age and marital status group.

TABLE 6 Sex Ratios by Age and Marital Status, with and without Age Selection, Norway, 1979

Age	*Males per 100 Females*			
	Married Population		*Unmarried Population*	
	Without Age Selection[a]	*With Age Selection*[b]	*Without Age Selection*[a]	*With Age Selection*[b]
15–19	105.2	9.3	105.2	108.0
20–24	104.8	46.1	104.8	141.5
25–29	104.6	82.6	104.6	165.8
30–34	104.1	97.6	104.1	151.6
35–39	103.8	101.0	103.8	130.7
40–44	103.2	98.2	103.2	123.8
45–49	102.3	99.0	102.3	117.5
50–54	100.7	100.9	100.7	97.4
55–59	98.2	103.7	98.2	74.9
60–64	94.3	110.1	94.3	54.3
65–69	88.7	122.4	88.7	43.3
70–74	80.8	133.6	80.8	35.6
75+	62.4	171.2	62.4	33.8

[a] Sex-adjusted life table population. From the Central Bureau of Statistics (see note c, Table 3 above), Table 39, pp. 58–59. The ratios are applied to the married and unmarried populations alike; hence the first and third columns are identical

[b] Sex ratios of the actual population in 1979. From the Central Bureau of Statistics (see note a, Table 3 above), Table 5, pp. 32–35

particular age, a surplus of married males or females means only that at least the numbers indicated by the size of the surpluses are married to persons outside their own age group. Within the unmarried population, however, surpluses affect the probability of getting married. Insofar as people deem marriage to be desirable, the unmarried who are surplus are viewed as being handicapped in reaching this state.

In view of the association of marriage with young adulthood and reproduction, it is odd that people apparently regard being unmarried at an advanced age as worse than being unmarried at a younger age. For instance, the surplus of unmarried males in the age group 20–39 years is large, but it is not generally thought of as a social problem. The reason, doubtless, is that most of the surplus men eventually end up getting married; their problem is temporary. The surplus of elderly unmarried women attracts more attention for several reasons. First, the surplus is larger than that of unmarried young males (in Norway in 1979 there was a surplus of 1,62,000 unmarried females in the age group 55 years and older, compared with a surplus of 88,000 unmarried males between the ages of 20–44). Second, the likelihood of these women getting either married or remarried seems remote. Last, their unmarried status often imposes on them economic as well as other personal disadvantages. Their situation, therefore, seems more likely than that of young men to generate proposals for social reform.

Social Responses to Sex Imbalances

Upon learning of unbalanced sex ratios such as those just described, for example, many people intuitively jump to certain policy recommendations. They say that men should or should be compelled to marry women older than themselves, or that a legal limit should be placed on the extent to which the husband's age can exceed the wife's age. Such proposals are worthy of study, but in addition to the evasions and unforeseen consequences they seem likely, if enacted, to produce, they address themselves to only one of the causes of the problem—age choice. They leave untouched the other two factors—mortality and the natal sex ratio. Also, deliberate intervention is not the only way social adaptation comes about. The view adopted here is that major changes in marital patterns and sex roles result less from deliberate policy than from unconscious adaptation to demographic changes. Finally, deliberate measures to control the factor of age choice in marriage would have to be quite dramatic to succeed.

Marriage 'Rules' and their Consequences

Consider three hypothetical age choice rules: (*a*) the groom cannot exceed his bride's age by more than five years; (*b*) men must marry women in their own five-year age group; (*c*) men must marry women in the next older five-year age group. How would each of these rules affect marital chances? Rule (*a*) would have little effect, because most marriages in industrial societies already involve an age gap of less than five years. In the United States Marriage Registration Area in 1977, for instance, the difference in the mean ages of brides and grooms was only 2.8 years (median: 2.3 years).[6] In Norway in 1979 the mean difference was 2.4 years for first marriages and 2.7 years for all marriages.[7] A five-year limit would, therefore, do little to raise the marital chances of either young men or older women.

Under Rule (*b*), although the ages of the partners could be as much as five years apart, the average difference would approach zero. As the middle columns in Table 6 show, this rule would reduce but would not eliminate the surpluses of both unmarried young men and unmarried older women. It would do more for young men because their unmarried surplus is due overwhelmingly to age choice, whereas for women over 65, their surplus is substantially due to the cumulative effect of mortality.

Rule (*c*) would virtually eliminate all the surpluses. If random choice within the older five-year age group is assumed, the bride would, on an average, be five years older than the groom, although she may be as much as ten years or even as little as one day older. As the last column of Table 7 shows, this rule

[6] National Center for Health Statistics, Vital Statistics of the United States, 1977, Vol. 3, 'Marriage and divorce', pp. 1–10, 1–21.

[7] Central Bureau of Statistics, Norway. Population by age and marital status, 31 December 1979, pp. 32–35. See also *The development in marriage and deaths 1911–1976* (Norway, Oslo, 1978), p. 36.

would especially affect the marital chances of women over 60. Their surpluses would all but disappear. However, it would not help the men in the age group 35–44 years because they were mainly born before World War II, whereas the women who are five years older were born during the war and were consequently few in number. If it were not for this special circumstance, Rule (*c*) would produce very small unmarried surpluses throughout the age scale. Other rules can be imagined, some of them quite complex; but in general, given a normal age distribution, any rule of marital selection under which men marry older women will offer a higher number of potential mates for women.

If the factor of age choice was purely habitual or customary, there would be little difficulty in moving toward Rule (*c*) or some other similar measure; and if it was purely economic, there would be a tendency for women, as they achieve financial success, to marry men younger than themselves. However, the three observations suggest that age choice is not determined solely by culture or economics. First, the pattern of age hypergamy is too universal to be attributable to arbitrary custom. Second, there is little evidence that women of high economic status tend to marry younger men. Third, marriage normally involves a biological component—sexual attraction—that is linked with age. To the extent that marriage is artificial, as it has been in aberrant groups, such as, royalty or in tribes with the levirate, or ghost marriages or woman-woman marriages, it need not involve sexual attraction. But since the most fundamental obligation of marriage is sexual access, sexual attraction is normally an element in mate selection. Thus, it is likely that the pattern of age selectivity in marriage is, at least in part, an evolutionary product. It concentrates the attention of men, regardless of age, on women in age groups in which reproduction is possible.

This latter interpretation fits our data on marital choice. For instance, the main sources of distortion due to age selection are second marriages and male competition for younger women. At the younger ages, the two sexes strongly attract each other, and this is when most marriages are consummated. In 1977 in the United States, 97 per cent of the brides and 93 per cent of the grooms in first marriages were below the age of 30. In such marriages the difference in age is usually small. The large surpluses among the unmarried are produced by widowhood, divorce and remarriage, which allow older men to compete for younger women. In marriages with one or both partners previously married, according to data from the United States, the proportion of women under 25 marrying men aged 35 and above was 7.4 per cent, whereas in the case of first marriage for both partners the proportion was only 0.5 per cent.[8]

Our conclusion, then, is that age hypergamy is too deeply rooted in the species to be extirpated by legislation. No law will prevent the sexual competition of men of all ages for younger women, nor is any law likely to cancel the sex difference in mortality. Control of the sex ratio at birth may be achieved, but if it is, changes in fashion may cause greater distortions than those found at present. If preference was given to male offsprings, it would make the sex

[8] Calculated from National Center for Health Statistics, Vital Statistics of the United States, 1977, Vol. 3, 'Marriage and divorce', p. 21.

TABLE 7 Unmarried Surpluses in Norway, 1979, Under Actual Patterns of Age Choice and Under Two Hypothetical Restrictive Patterns

Age	*Actual Unmarried Surplus*[a]			*Surplus if Married in Same 5-year Age Group*[b]			*Surplus if Men Married Women in Next Older Age Group*[c]	
	Sex	*Per cent of Pop. of that Sex*	*Per cent of Unmarried Pop. of that Sex*	*Sex*	*Per cent of Pop. of that Sex*	*Per cent of Unmarried Pop. of that Sex*	*Sex*	*Per cent of Pop. of that Sex*
15–19	M	7.4	7.4	M	5.0	5.1	–	–[d]
20–24	M	24.4	29.3	M	4.4	5.5	M	5.2
25–29	M	17.2	39.7	M	5.4	17.0	M	7.1
30–34	M	8.3	34.1	M	6.5	28.7	M	1.7
35–39	M	4.5	23.5	M	5.3	38.2	M	27.1
40–44	M	3.6	19.2	M	2.0	12.0	M	21.4
45–49	M	2.8	14.9	M	1.9	10.9	M	2.6
50–54	F	0.5	2.6	M	0.2	1.1	F	7.4
55–59	F	6.2	25.1	F	3.4	15.6	F	13.5
60–64	F	15.0	45.7	F	8.2	31.6	M	4.8
65–69	F	25.0	56.7	F	14.4	39.7	F	0.7
70–74	F	37.1	64.4	F	22.8	52.7	M	2.2
75–79	F	38.6	72.4	F	33.0	59.0	F	1.7
80–84	F	42.3	67.0	F	39.4	58.4	M	4.5
85+	F	44.2	60.8	F	46.2	57.3	F	3.2

[a] The population in Norway in December 1979. From the Central Bureau of Statistics (see note a, Table 3 above), pp. 32–35. Surplus refers to the excess in the unmarried population of one sex or another above an equal number, and this surplus is taken first as a percentage of the total population of that sex in the age group and second as a percentage of the total unmarried population of that sex in the age group

[b] The hypothetical surpluses are those that would arise if all people marry within their own age group. For instance, at ages 75–79, there were 11,169 more married males than females. If they had married in the age group, they would have reduced the number of unmarried females by that amount. This gives us a reduced surplus of unmarried females aged 75–79, and a new total unmarried female population in terms of which the percentage unmarried can be calculated

[c] The hypothetical surpluses are those that would arise if all women married a man in the next younger age group as long as there was a man in that age group to marry. For instance, at ages 30–34 there were 1,52,838 women in the population, and in the next younger five-year group (25–29), there were 1,55,496 males. Under the rule, all the women could marry, but there would be 2,658 males aged 30–34 who could not. This surplus divided by the total population of males in the 25–29 group yields a ratio of 1.7 per cent. Since we assume universal marriage whenever a mate of suitable age is available, there is no unmarried population against which to calculate the second percentage. In other words, the surplus unmarried and the unmarried population are identical

[d] It is assumed that women between 15–19 years do not marry

imbalance worse for young men while alleviating it for older women; if a switch were made favouring female offsprings, there would ultimately be a generation of women suffering a mate famine. It seems more probable that new social patterns will gradually emerge—in fact, are already emerging—as unintended and largely unconscious adaptations to underlying demographic changes. The question arises: what are these patterns?

New Social Institutions

Perhaps the main structural adaptation is that marriage—in the sense of a permanent sexual and reproductive union—is falling out of fashion. The manner in which demographic forces are bringing this about is clear. In the past, when a woman married she could normally expect to spend most of her remaining life in wedlock or at least with the children born out of the union. That, however, was because her life was shorter and the period of childbearing more prolonged. Assuming that she married at the age of 22, even as late at 1900 she could expect to live only till the age of 64. True, she would be widowed by the age of 60, but the chances were that she would die before her last child left home. At present, however, according to the 1980 life tables for the United States, a white woman marrying at the age of 22 would live to the age of 79.4 years. Yet she would stop having children at around 30 years of age, much earlier than her grandmother did. Assuming that the last child would leave home at the age of 18, the mother (at 48) would still have 32.5 years more to live free from the responsibilities of caring for her children.

What is she to do during this period, which comprises more than half of her lifetime after marriage? Some of this time she can spend with her husband: he would survive till she is 69 years. That is long enough, but at 69 she still has an expected 16 years of life remaining, which is 43 per cent of what is now (since she has survived to the age of 69) her postchild lifetime. And the chances are that the marriage would have broken anyway, long before she reaches the age of 69. The 1975 risk tables for divorce in the United States indicate that 47 per cent of first marriages ultimately end in divorce, and that a quarter of them last only seven years or even less.[9] On an average, the wife would have become a widow or divorcee by the time she is 44 years old. If she does not remarry, she would retain that status for the next 36 years. Her motivation to remarry may therefore be great, and the chances of remarriage are fair because, in the mid-forties, the sex ratio of the unmarried is still reasonably favourable. If she does remarry, say at 45 years of age, she would still have another 35 years to live. A part of this time would be spent with her second husband, but the difference in their ages would be more as compared to that in the first marriage, and the marriage would be more susceptible to divorce or widowhood. She would stand an even chance of being divorced or widowed a second time within 11 years, when she would still have more than 26 years of life remaining.

The underlying demographic changes diminish the part of marriage and children in women's lives, and, by doing so, lead women to attach less importance to their domestic role. Women's prospects are that around two-thirds of their adult years would be spent without children in the household, and possibly between half to two-thirds without a husband. For long periods they would probably have to depend on their own resources and would therefore, need employment. This means that, in planning their lives, women must look to their own careers as separate individuals. For best results, they must enter

[9] James A. Weed, *National estimates of marriage dissolution and survivorship: United States* (Washington, D.C., National Center for Health Statistics, November 1980), pp. 5, 16 and 19.

employment while still young and remain employed throughout in order to build up experience, seniority, reputation, and whatever other cumulative benefits accrue from occupational commitment.

Once under way, the system of change exhibits a dynamic of its own. Insofar as demographic trends lead women to downgrade marriage and stress employment, they also lead them to reduce not only their own dependence on their husbands but also their service to them. Men, in turn, are induced to reconsider the costs and benefits of marriage. They sense that, at the older ages, men are increasingly scarce compared to women, that they do not have to marry to enjoy female company, and that if they do marry, their role as father and family head has somehow been eroded. Not surprisingly, the divorce rate has risen to unprecedented levels, making the institution of marriage less secure and, therefore, less valuable for both sexes. Marriage undergoes attrition in two ways: it is either postponed or not undertaken at all, and when it is undertaken, it is increasingly brittle.

In such terms one can understand why, in recent years, the proportion of unmarried population has risen steeply, especially at ages below 45 years (Table 8). Doubtless, as household surveys indicate, many couples now live together without marrying, but this fact underscores the decline of marriage. Consensual unions are less stable and less conducive to childbearing than regular unions, a circumstance that means low fertility and, therefore, still more freedom for women to pursue outside interests and contacts.

TABLE 8 Proportion of Women not Married, By Age: United States and Norway, Recent Dates

Age	*United States White Women*		*Norwegian Women*	
	1967	*1980*	*1972*	*1980*
15–19	89.1	91.2	94.5	97.7
20–24	31.2	54.0	48.4	63.8
25–29	9.4	30.2	19.7	29.8
30–34	6.9	23.1	13.5	18.2
35–44	7.9	18.6	12.1	16.2
45–54	13.8	18.9	17.8	17.8
55–64	29.2	30.0	30.8	28.4

SOURCE: Bureau of the Census, United States. Marital status and family status: March 1967, *Current Population Reports*, Series P–20, No. 170, 23 February 1968: 12; Marital status and living arrangements: March 1980, *Current Population Reports*, Series P–20, No. 365, October 1981: 8. (In 1967 the youngest age group in the US panel was 14–19 instead of 15–19). Central Bureau of Statistics, Norway. *Social Survey* 1974, p. 8; Population by age and marital status, 31 December 1980, pp. 36–37.

The Feminist Movement

Seen in this light, the rise of the women's movement in recent decades in industrial societies becomes comprehensible. It seems to be an inevitable response to the basic demographic changes that we have described. Like other movements, it is an ideological reaction to alterations in the underlying conditions of life. Such alterations make old patterns and ideals unworkable and thus push individuals into new modes of behaviour that, because they are new, have no ethical justification or support in the received social order. Yet people need to feel 'right' about what they do. When they are forced into new paths, they welcome a movement that justifies their new behaviour and thus morally reassures them. The feminist movement normally supports new female behaviour patterns by showing that these patterns are widespread and hence not aberrant, that they are justified by the highest moral principles, and that they are opposed only by people who are misguided. Thus, under the banners of very broad principles such as 'equality,' 'privacy,' and women's 'rights,' the feminist movement legitimates a variety of innovations that are in women's interest under the changed circumstances of contemporary society. These include bread-and-butter programmes like day-care centres and paid maternity leave, as well as radical measures like the approval of free abortion, sexual freedom, childlessness, and affirmative action.

If we try to distil the most extreme or radical goal of the feminist movement, it appears to be the rejection of all division of labour based on sex. Underlying the specific programmes, then, is the general idea that women should have the same opportunities as men, the same occupations, and the same privileges. According to this doctrine, any difference in rights and obligations among people should be based on merit, not on sex. Indeed, gender should be expunged from the language.

In the light of the present analysis, this goal is understandable, because it extrapolates to its end the actual diminution of marriage and childbearing in women's lives that we have documented. However, for that very reason it is a questionable goal, for it encounters the stubborn obstacle that it is still women who become pregnant and bear children. Until advanced technology frees them from this specialisation, a modicum of sex-based division of labour is unavoidable if there is to be a birth rate.

Recognising this difficulty, the feminist movement generally pursues a less extreme goal. It acknowledges women's special reproductive role but seeks to maximise the rewards and minimise the costs of that role. In practice, however, it has done less to maximise the rewards, evidently because it wishes to de-emphasise women's reproductive specialisation and because in the past women were rewarded for reproduction, but in ways that reinforced their ascribed status and their dependence on the family. Instead, feminist action has concentrated on lessening the costs of reproduction for women, especially the career costs. Thus, women are not to be penalised in the workplace for pregnancy or parenthood, and men, through domestic and child care duties, are to make an investment in reproduction that equals the investment made by women.

While this strategy is in keeping with the underlying trends, it nevertheless encounters an ancient problem. Whenever there is a division of labour, a troublesome issue arises—the issue of fairness. Is what each side gives and gets a square deal? In the occupational sphere, the issue is resolved by the forces of the economic and political market-place, but between men and women reciprocal obligations and rewards have traditionally not been determined that way. Instead, they have been governed by an institutional structure, the very genius of which was to create solidarity and identity in the family and, thus, to protect its members from raw economic and political competition among themselves. Now, however, the mutual rights and obligations of men and women are the subject of dispute, and the question of what is fair or unfair seems virtually insoluble. As demographic and social change has upset the conditions supporting the old system, women have turned more and more to earning their rewards in the market-place, not in return for reproduction but in return for activities similar to those of men. The result, in most industrial societies, is a large increase in female labour force participation and a drastic decline in fertility. How to reward men and women for reproduction, at least sufficiently to provide replacement fertility, is a moot question.

It would be odd if the feminist movement did not overstate its case, because that is what ideological movements generally do. We must ask, therefore, whether from a demographic point of view the idea that except for reproduction men and women are to be treated exactly alike goes too far, possibly against the self-interest of women—like saying that a fox and a rabbit are just alike except that one is carnivorous. Fairness is not necessarily achieved by treating people who are different as though they were just alike. One complication is that it is hard to set a close boundary on women's reproductive role. After a child is born, it needs a great amount of attention including (according to current medical opinion) breastfeeding. Some economic activities are more easily combined with this attention than others. Thus reproductive specialisation tends to ramify. Another complication is that there may be biological differences between men and women other than sheer pregnancy and childbirth. We cannot argue that question here, but we can say that the very demographic forces that underlie the feminist movement and give women more opportunity to be like men have, in other ways, widened the differences. The two sexes have become increasingly distinct in such important matters as longevity, age structure, and marital chances. To ignore these differences is in some ways unfair to men and in some ways unfair to women.

A viable division of labour cannot simply be fair; it must also be efficient. Insofar as women in industrial societies today are not motivated to achieve replacement fertility but instead are rewarded for non-familial activities, they are curtailing the one activity that most distinguishes them from males and are acting out the ideology of feminism. In the long run, however, such an arrangement is self-defeating. If it persists, the social order that gave rise to it will be replaced by another—either one that supports traditional sex roles or some new order that rewards women adequately for reproduction. The feminist movement may undertake to construct such a new order. If it does, or if such a new order evolves by trial and error, it will have to be in keeping with the new demography found in advanced societies.

2

Theories of Fertility Decline

RONALD FREEDMAN

Despite decades of intensive and sophisticated work, we do not know a good deal about the determinants of fertility. The work by Bongaarts and others[1] on the proximate determinants of fertility is a major accomplishment in quantifying the effect of the Davis and Blake intermediate variables on fertility. It now provides the challenge of specifying the determinants of the proximate determinants. We know that the large variations between cultures in pre-industrial fertility is immediately due to the variations in breastfeeding practices and the age at marriage, but what determines the variation in those is still unknown. One may speculate that these proximate variables evolved over time to fit normative differences in desired fertility, but that would be merely speculation.

The socio-economic variables are the determinants on which most analysts have relied for decades. Education, urbanisation, occupation, labour force status, income and the like are still the primary variables studied, if only because they are most often available in censuses and surveys. But this reliance on socio-economic variables has proved weak for several reasons. They did not have the expected consistent explanatory power in the Princeton historical studies of the European demographic transition. This has, fortunately, led to a consideration of other kinds of variables which may supplement the socio-economic variables. Ironically, disappointment with the socio-economic variables in Europe was expressed at a time when LDC political statements stressed the socio-economic development variables as primary for fertility change in LDCs.

In fact, significant but unexpected fertility declines occurred without the expected higher levels of socio-economic development in—for example, Indonesia, Thailand and the PRC. Other forces were also at work.

More recently, the World Fertility Survey has provided new insights into the less consistent patterns of the socio-economic variables. In comparisons of the relation of similar socio-economic variables to fertility across many countries, we find that the relationships vary considerably in degree and even

[1] John Bongaarts, 'The fertility inhibition effects of the intermediate fertility variab les,' *Studies in Family Planning*, Vol. 13 (617), June/July 1982; J. Hobcraft and R.J. Little, 'Individual level analyses for assessing the contribution of proximate determinants to socio-economic differences in fertility', WFS, Tech 17 September 1982.

in direction. Something appears to be wrong with the straightforward conventional approach.

A promising approach by Hermalin and Mason[2] views the variation in the micro-relationship of the socio-economic variables and fertility as the fact to be explained and uses macro variables characterising the countries as the independent variables. Apart from a possible solution to the particular problem, this offers a more general approach in which the macro country level context is searched for explanations for the micro level relationships.

In at least one version of the classical form, the social structural variables were believed to influence fertility by their effect on the family.[3] Modernisation was supposed to move functions from the family to other institutions and thus change the costs and benefits of children. But almost all the studies related the socio-economic variables directly to fertility without any data on the crucial familial relationships and structure. That may be one reason why the fit of the relations between the socio-economic and fertility variables is so loose. There is, of course, now a renaissance of interest in changing family relations as the rest of the society changes and this has become a major focus of research.

All this is not to suggest that we discard the socio-economic variables. First, they still explain more than the other variables in some situations although this does not say a great deal. Second, they have a useful if incomplete rationale. Third, their limitations which are partly in their definition and operationalisation can now be improved. We are often limited by official and conventional definitions. It may be helpful, for example, to define women's labour force activity to indicate how work affects their familial roles. In LDCs variables defining risks and risk avoidance may be more helpful than other income variables.

One reason for our failure to identify and explain the determinants of fertility more successfully may be our concentration on micro level variables to explain phenomena that are social in character. The problem, too, is social. Many years ago I made a statement to this effect: the important question is not why couples in a particular country have two, four or six children but why the frequency distribution and mean of family size is different as between countries A, B, C or D. For social policy, the social rather than the individual differences are important for most purposes. On another occasion, I stated that it would be a sociological anomaly if there were not social norms governing a problem so prevalent, central, and recurrent as how many children to have.

There has been a growth of interest in macro variables characterising the community, its subgroups, the region, province, state, or the nation. The WFS tried out a set of community-level variables without great success in the first

[2] A.I. Hermalin and W.M. Mason, 'A strategy for the comparative analysis of WFS data, with illustrative examples,' in United Nations Fund for Population Activities. *The United Nations Programme for Comparative Analysis of World Fertility Survey Data.* (New York, 1980), pp. 90–168.

[3] F.W. Notestein, 'Economic problems of population change,' *Proceedings of the Eighth International Conference of Agricultural Economists* (London: Oxford University Press, 1953).

empirical trials that I have seen. These followed more or less closely a model I had prepared for the WFS. I do not apologise for these results, because before the fact it was plausible that these community variables should be related to fertility, so even negative results are useful. Beyond that, however, the basic idea of relating the social structure of communities to demographic behaviour appears to me to be theoretically right, so I look forward to a better selection, conceptualisation, and tests of such variables.

Certainly, one of the most important macro variables is the changing character of the economy and how it affects authority and other relations within the family. For a sociologist, it is an act of faith that social determinants and improved definitions of social norms for the desired family size will help to determine fertility levels and distributions.

The role of ideas as independent determinants of fertility has only recently attracted attention again as a supplement to the socio-economic variables. Most demographers are inclined to believe that ideas are reflexive epiphenomena of the basic social and economic structure. But attempts to explain fertility changes mainly by social and economic changes have not been very successful. There is, at least, a great deal to be explained after such variables are considered. Lesthaeghe and Livi Bacci have published interesting empirical results using indicators of a secular mentality in relation to fertility decline for Europe.[4] Francine and Etienne van de Walle have called my attention to data on similar relationships for eighteenth century France and Switzerland. It is, at least, plausible that secularism is relevant for such deviant cases as the Geneva Bourgeoisie, the Italian Jews, and the European nobility. Secularism and rationality may have influenced both the process of fertility decision-making and the specific substantive benefits and costs of children under the influence of specific new ideas and frames of reference.

Secularism implies that nothing is sacred in considering alternatives. This brings us close to the idea of rationality when that is defined as the self-conscious calculated choice of alternatives toward a desired end. Rejection of a traditional possible course of action is then possible but not necessary. One way of defining a traditional society is that it is one in which socially inherited norms offer reasonably satisfactory solutions for recurring problems so that consideration of alternative solutions is less necessary or frequent. Rational actions are not necessarily right. They are the result of deliberate self-conscious choice. There are reports that in some African and other surveys the question 'how many children do you want to have?' is greeted with stunned silence or incredulous laughter, presumably because everyone knows that the number of children should be as many as possible—not a subject for rational calculation.

It is well known that in the European Demographic Transition, fertility tended to have distinctive values unexplainable by customary structural

[4] R. Lesthaeghe, 'A century of demographic and cultural change in Western Europe,' Manuscript (Brussels, Vrije Univesiteit, 1982); R. Lesthaeghe, *The history of the Belgian fertility decline, 1800–1970* (Princeton, Princeton University Press, 1977); M. Livi Bacci, *A century of Portuguese fertility* (Princeton, Princeton University Press, 1971); M. Livi Bacci, *A history of Italian fertility decline in the last two centuries* (Princeton, Princeton University Press, 1977).

variables in local areas defined by language, religion, and ethnicity. It also appears that the region is more important than the customary social, economic, and demographic values in relation to contraceptive use rates in Thailand and Indonesia. Something beyond the usual social and economic variables, perhaps culturally defined local frames of reference or ideas, are at work here. We need to know more about such situations.

A specific substantive point of relevance is that the idea of family limitation has an independent role in the acceptance of fertility control even after more children are born than are wanted. John Knodel[5] has made a good case for the relevance of this idea both in the Demographic Transition and in contemporary LDCs. Legitimising the idea of family limitation may occur through undirected social processes. However, it is also true that an important function of a family planning programme is to try to speed up this process so that prospective clients perceive family planning as accepted by significant others and regarded as legitimate by the accepted authority.

At least in LDCs an important influence on fertility which involves changing ways of thinking is the extent to which the local leaders and the population are linked to a national and world system of transportation and communication. In this way people are related to new models, ideas, and things which transcend the local and familial levels. Since this involves comparison and doubt, it creates the basis for secularism and rationality. Such styles of thought may be more important than the new substantive ideas. John Knodel[6] has suggested that such an expanded system forms the basis for social change and fertility decline in Thailand. One can make a case for such an influence in Indonesia. Even in the PRC where the bridges to the rest of the world are limited, there is an internal national system which has such influences, albeit highly directed. At least it transcends the commune and other local units.

There is little doubt that contraception and abortion are the major immediate causes of fertility decline in the modern world. What is in controversy is whether the availability of the idea, information, and the means of fertility control have an active independent role or whether once the motivation to limit family size is present the idea of, and the means for, fertility control will inevitably become available. The fact that withdrawal and illegal abortion were the principal means of fertility control leading to fertility decline in Europe is often cited as evidence of the potential for fertility decline (given the motivation) without any organised family planning programme, without modern contraception, and in the face of the opposition of the state, the church, and the medical profession. These facts are correct and important, but we do not know whether with improved methods, with more information, and with the support of legitimising institutions the decline would have been more rapid, especially for the disadvantaged strata of the population.

A reasonable prevalence of the desire to limit family size is generally essential

[5] J. Knodel, 'Family limitation and the fertility transition: evidence from the age patterns of fertility in Europe and Asia', *Population Studies*, 31 July, 219–49.

[6] J. Knodel and N. Debavaiya, 'Thailand's reproductive revolution,' *International Family Planning Perspectives*, 4 (Summer), pp. 34–39.

for a fertility decline, but there must also be an acceptance of the idea that the practice of family limitation is socially legitimate, as already indicated.

Even where there is fairly widespread motivation to limit family size, I believe that the provision of safe, effective, and inexpensive contraceptives under legitimate and convenient auspices will increase the adoption of family limitation practices, depending on the extent of doubts about the legitimacy and safety that are likely in a period of social change from traditional to modern society. Thailand, Indonesia, Taiwan and Korea are examples with which I have some familiarity. In Korea and Taiwan, for example, it is likely that contraceptive practice would have eventually reached the present very high levels without the organised programme but it seems unlikely that it would have happened as quickly, especially among those from the disadvantaged strata. Contraceptive use differentials have virtually disappeared in Taiwan, for example, in a relatively short period. In all these cases, serious questions were raised before the campaigns were launched about the possibility of such a rapid rate of fertility decline, especially among the disadvantaged strata which in fact have adopted fertility control in large proportions.

Indonesia and the PRC are both examples of family planning programmes in countries where the socio-economic levels were lower than those required for fertility decline by the demographic transition criteria. While these programmes are very different in many ways, both countries have a very important element in common. That is, the capacity to transmit influences from the centre to the local units through a political bureaucratic system and to send to and receive from the local units information about the programme. This involves the capacity to mobilise the population and resources towards national purposes. Both the extent of such control and its purposes differ substantially as between the PRC and Indonesia.

The organised social pressures in the PRC working previously for the two-child family and then for the one-child family are so tremendous and awesome, and so reinforced with sanctions that in our terms the situation is close to compulsion. As a result, the costs to the individual of having a child beyond the one permitted are so great that while the continuing elements of the traditional Chinese family make one child grossly unsatisfactory to many couples, large numbers apparently conform.

In evaluating the effects of family planning programmes, it is apparently necessary to go beyond the programme narrowly defined as a marketing system and to consider also the nature of the social system which affects both the operation of the programme and the incentives and disincentives attached to the acceptance of family limitation.

To assert that organised family planning programmes have an independent role once motivation is present is not to assert they have such a role everywhere and without reference to motivation. India, Bangladesh, and Pakistan are illustrative of situations where organised programmes appear to have had relatively little impact, except in a few local areas, presumably because of the absence of motivation. It is possible to argue that the reason for this is that the programme in these cases were poorly organised and executed. While that is possible, it is hard to maintain that position after several decades of large scale effort and considerable financial input. Something more substantial would have been expected by now if motivation were reasonably prevalent.

Another indication of the role of organised programme intervention is indicated by the data on the decrease in the incidence of unwanted fertility in the United States during the sixties and the seventies. There was initially a large percentage of women with unwanted fertility following poor contraceptive practice—especially among the blacks, the poor, and the uneducated. Following a significant programme to introduce family limitation among women from these strata and to produce a significant increase in the availability of legal abortion, there was a significant decline in the percentage of women with unwanted children and a substantial increase in the effective use of contraception.

If fertility control is only a reflexive aspect of motivation, and a substitute for what would happen anyway without a programme, then discontinuation of an organised family planning programme should have no lasting effect. The well-known Romanian example is a good test case. When the Romanian government suddenly made abortion unavailable in 1966, the birth rate shot up from fourteen to twenty-seven in one year. The consensus of published expertise then and since has been that after an initial temporary effect fertility would return to its former low level as couples found substitutes for legal abortion. While the birth rate has gone much of the way towards its former level, sixteen years later fertility has still not reached its former low and Romania has about the highest fertility level in Eastern Europe rather than one of the lowest. Barney Berelson[7] pointed out some years ago that Romanian officials, whose goal was to increase the population and growth rate of Romania, may have had some satisfaction that their intervention had made a difference in the desired direction.

I began with the statement that we do not know a great deal about the determinants of fertility. True, we are still learning a lot about the proximate determinants, but what we really need to know is the prior determinants of both proximate variables and of desired family size, which together constitute the primary determinants of fertility.

The classical list of determinants, such as education, urbanisation and labour force status, have turned out to vary in their relation to fertility from country to country. This may turn out to be more useful information than if they had followed a uniform pattern, especially if we can establish that there are macro determinants that explain their varying relationships. While the failure of these variables to explain the course of the Demographic Transition is disappointing, it is consistent with the similar lack of relationships for the LDCs. Further, there are interesting and promising approaches both in Europe and the LDCs. There are interesting potentials in the role of ideas and the mental framework, in changing familial structure, in macro variables at the community and other collective levels, especially with reference to the changing economy, in family

[7] B. Berelson, 'Romania's 1966 Anti-Abortion decree: the demographic experience of the first decade,' *Population Studies*, 33, No. 2, pp. 209–22.

planning programme, especially as affected by the central capabilities for mobilising the population and resources.

Perhaps my initial pessimism was unjustified. There are promising new directions for further work and even the conventional socio-economic variables have taken a new turn which may lead to a greater understanding. And, of course, even negative findings can help us if they guide us to avoid paths leading in the wrong direction. The glass which is half empty is also half full.

3

Societal Development and Population Change—A Reappraisal of Selected Theories

N. Krishnan Namboodiri

More than three decades ago, Rupert Vance (1952) quoted Colin Clark as saying:

> There is room for only two or three economic theorists in each generation, no more. Only men of transcendental powers of reasoning can be candidates for these positions. The rest of us should be economic scientists content steadily to lay stone on stone in building the structure of ordered knowledge. Instead, it seems to be the ambition of nearly every teacher of economics to put his name to a new formulation of economic theory. The result is a vast output of literature of which it is safe to say, scarcely one syllable will be read in fifty years' time.

The idea Vance wanted to convey was that Colin Clark's characterisation of economists did not apply to the masons, carpenters, and coolie workers of population studies. In fact, he felt, when he wrote the paper, that the field of population studies could use some theorists. Have demographers made much progress in respect to theory building since then? I propose to examine this question, starting with a brief look at the now-disfavoured demographic transition theory and some of the recent attempts to modify it, followed by an examination of two opposing orientations to theory building. In the end I shall point out that the progress in respect to theory building in demography has been rather modest since the early 1950s when Rupert Vance expressed the optimistic view that the demographic transition theory would soon provide the needed 'binder' for the 'diverse and particularized findings' of population studies (Vance, 1952).

I

The Demographic Transition Theory

Coale and Hoover (1958) provide the following outline of the demographic transition theory:

> The agrarian low-income economy is characterized by high birth and death rates—the birth rates relatively stable, and the death rates fluctuating in response to varying fortunes. Then as the economy changes its form to a more interdependent and specialized, market-oriented economy, the average death rate declines. It continues to decline under the impact of better organization and improving medical knowledge and care. Somewhat later the birth rates begin to fall.
>
> The two rates pursue a more or less parallel downward course with the decline of the birth rate lagging behind. Finally, as further reduction of the death rate becomes harder to attain, the birth rate again approaches equality with death rate and a more gradual rate of growth is re-established, with, however, low risks of mortality, and small families as the typical pattern. Mortality rates are now relatively stable from year to year and birth rate—now responsive to voluntary decisions rather than to deeply imbedded customs—may fluctuate from year to year. This short description fits the experience of most countries whose economies have undergone the kind of reorganization we have been calling economic development (Coale and Hoover, 1958: 12–13).

In the classical formulation of the model, three types of societies have been distinguished: pre-transitional, transitional and post-transitional. The first type of societies are said to be in a state of high growth potential, and the last type in a state of incipient decline. Thus, one may say that according to the classical formulation, the demographic transition theory visualises societies moving from an initial state of high growth potential to an end state of incipient decline, through an intermediate state of high growth. But the states thus described (high growth potential; high growth; and incipient decline) are demographic states. From a theoretical standpoint, what is needed is a description of the various states in terms of the determinants of demographic characteristics. Labels such as 'low-income economy' and 'interdependent and specialized, market-oriented economy' (Coale and Hoover, 1958) are illustrative of how scholars have attempted to meet this requirement. There is a consensus among scholars, however, that it is virtually impossible to be specific about the particular material improvement, structural changes, or shifts in attitudes or beliefs responsible for the demographic transition. It is, therefore, fair to say that we do not have today a clear statement which is specific with respect to the non-demographic characteristics of populations before, during, or after the demographic transition. A few selected writings have been reviewed here to illustrate the directions in which scholars have elected to move in their examination of the demographic transition theory.

Kingsley Davis' Theory of Multiphasic Response

As an adjunct to the demographic transition theory, Kingsley Davis (1963) suggested nearly twenty years ago that a sustained natural increase resulting from continued declines in mortality levels in the context of economic development produces a multiphasic demographic response, involving postponement of marriage, increased celibacy, resort to abortion, use of contraception, and migration.

> Although generally overlooked because of our preoccupation with the contraceptive issue, the fact is that every country in northwest Europe reacted to its persistent excess of births over deaths with virtually the entire range of possible responses. Regardless of nationality, language, and religion, each industrializing nation tended to postpone marriage, to increase celibacy, to resort to abortion, to practice contraception in some form, and to emigrate overseas. The timing and relative importance of the reactions were not identical in the various countries, and of course methods could not be used that were not then technically feasible for the public at large (e.g., harmless sterilization); but the remarkable thing is that all of the northwest European countries reacted, that they did so in each case with the reappearance of the whole range of responses, and that virtually the entire panorama was later repeated in Japan (Davis, 1963: 350–51).

Thus, persistent excess of births over deaths was the stimulus, and a combination of control of nuptiality, limitation of marital fertility, and migration was the response. How were the stimulus and response connected?

> My own view is that no society has been geared to a sustained high rate of natural increase except by conquest. Under a prolonged drop in mortality with industrialization, people in northwest Europe and Japan found that their accustomed demographic behavior was handicapping them in their effort to take advantage of the opportunities being provided by the emerging economy. They accordingly began changing their behavior. Thus it was in a sense the rising prosperity itself, viewed from the standpoint of the individual's desire to get ahead and appear respectable, that forced a modification of his reproductive behavior.
>
> Mortality decline impinged on the individual by enlarging his family. Unless something were done to offset this effect, it gave him, as a child, more siblings with whom to share whatever derived from his parents as well as more likelihood of reckoning with his parents for a longer period of life; and, as an adult, it gave him a more fragmented and more delayed share of the patrimony with which to get married and found his own family, while at the same time it saddled him, in founding the family, with the task of providing for more children—for rearing them, educating them, endowing their marriages, etc.—in a manner assuring them a status no lower than his . . . . The parent needed to conserve some means for himself, because of longer life-expectancy and because of the importance of capital for seizing opportunities or staving off disaster in the fluid situation of the times (Davis, 1963: 352).

In a nutshell, sustained mortality decline, in the context of economic development, produces an incompatibility between people's accustomed demographic behaviour and their ability to attain their mobility aspirations, which, in turn, causes a multiphasic response involving postponement of marriage, non-marriage, marital fertility limitation, and migration.

The theory relies on the assumption that individuals make everyday decisions on the basis of their perceptions regarding changes in their environment (people will respond to mortality reduction only if they perceive it). It also seems to assume that the individuals' solutions to their level-of-living problems provide, in the aggregate, the needed solution to the societal problem of disequilibrium between the population structure and the socio-economic organisation. But the question arises whether it is possible to harmonise what is 'good' for the individual with what is 'good' for the society. If 'yes', what is the mechanism that produces the harmony? If 'no', what are the consequences of disharmony? How much disharmony can be tolerated? How long can the disharmony last without causing catastrophic outcomes? We need answers to these and similar questions.

Another problematic is the absence of any clue in the theory as to why couples decide to have any children at all, given that children handicap their parents' efforts to take advantage of the opportunities offered by the changing economy. (The notion of incompatibility between the accustomed demographic behaviour and mobility chances invoked in the theory raises the possibility that people in the developed countries may opt to have no children at all.)

Yet another problem seen in the theory is that it does not specify how to predict the composition of the multiphasic response in different populations or subpopulations. The emphasis on the fact that 'all of the northwest European countries reacted . . . with the . . . whole range of responses, and that virtually the entire panorama was later repeated in Japan' (p. 352), gives one the impression that the entire range of responses is to be expected in almost all instances. But nowadays we know international migration is a very negligible option; the response of some populations may be predominantly in the form of non-marriage and delayed marriage; in some, birth control may be the predominant response; and so on. The question is—why the variation in the mixture of responses? What specific mixtures of responses are linked to given specific factor combinations?

Then there are the usual problems seen in nearly all theories. For instance, no time schedule is proposed for the various events (mortality decline, economic development, and multiphasic response); the causal language is informal; the evidence adduced is historical; and the competing models are not analysed. (There is, with respect to this last point, a reference to the possibility that population growth may be drastically curtailed to avoid misery and starvation, but this possibility is dismissed with the assertion that such was not the mechanism in any of the important historical cases known.)

CALDWELL'S 'WEALTH-FLOW' THEORY

Caldwell (1976) believes that the demographic transition theory is faulty because its major premises are not valid:

> In general, in societies of every type and stage of development, fertility behavior is rational . . . . Whether high or low fertility is economically rational is determined . . . by the direction of the intergenerational wealth flow. This flow has been from younger to older generations in all traditional societies; and it is apparently impossible . . . for a reversal . . . to occur before the family is largely nucleated . . . .
>
> From the demographic viewpoint, the most important social exports [from Western Europe] have been the concept of the predominance of the nuclear family with its strong conjugal tie and the concept of concentrating concern and expenditure on one's children. The latter does not automatically follow from the former, although it is likely to follow continuing Westernization; but the latter must be preceded by the former. There probably is no close relationship in timing between economic modernization and fertility . . . . [F]amily nucleation and the reversal of the intergenerational wealth flow are likely to penetrate deeply into the Third World in the next half century, almost independently of the success of industrialization, and almost inevitably, they will guarantee slower global population growth (pp. 355–56).

'Generations' within the family and 'wealth' are two basic concepts in this theory. Caldwell's definition of these are:

> [*Generations*] In some traditional societies the older generations [within the family] may merely consist of parents and possibly grandparents, as well as parents-in-law; in others, uncles, aunts, and even great uncles or older cousins may play a role. Sometimes—and admittedly this places some strain on a strictly intergenerational approach—a rather similar role may be played by older siblings or even siblings of the opposite sex (or occasionally siblings with greater education or power in the new society). In many societies the relation of wives to their husbands is similar to that of children to fathers . . . . (Caldwell, 1982: 173–74).
>
> [*Wealth*] This includes all sort of goods, services, and guarantees or power to command these. Examples are (1) the ability of a mother-in-law to order about her daughter-in-law even if it does not make housekeeping more efficient, and (2) guarantees of help in years of scarcity and devastation, of assistance if one's neighbor raises his hand against one, . . . of support in old age . . . . (Caldwell, 1982: 174).

Three major propositions can be discerned in the theory:

(*a*) only if the family becomes nucleated will fertility drop;

(*b*) industrialisation is not a necessary condition for fertility decline; and
(*c*) fertility decline starts with the reversal of the direction of intergenerational 'wealth flow,' which in all traditional societies has been from the younger to the older generation.

One noteworthy feature of these propositions is that they are stated in a falsifiable form. The comments by Freedman (1979) amount to a rejection of the validity of the first proposition: [A]dopting of the western nuclear model and abandonment of traditional familial values is not necessarily a prior condition for large-scale adoption of contraception and a fertility decline (Freedman, 1979: 6).

If the nucleation of the family is not a necessary condition, neither is it a sufficient condition, because according to Caldwell's own thesis, the nucleation of the family does not guarantee that the intergenerational 'wealth flow' would change direction (Caldwell, 1976: 356).

Similarly, industrialisation is not regarded as a necessary condition for the onset of fertility decline or for sustained low fertility (Caldwell's theory; also see Freedman, 1979). Some scholars believe that industrialisation is not a sufficient condition either. [This is based on the finding that pockets of industrialised countries have had sustained high fertility long after industrialisation (Coale, et al., 1979).]

Another noteworthy feature of Caldwell's theory is that, unlike Davis' (1963), it addresses directly the issue as to why fertility does not go down to zero in any population.

> Once the direction of the wealth flow has reversed, most couples will continue to have children but they will manage to do so by receiving all the parental pleasure they desire from a smaller and more concentrated source . . . . They will have that number in the full knowledge that having children is not economic, but that one's own children provide a unique form of pleasure which is not substitutable and that they can afford that expenditure for such unique pleasure. They do not have fewer because such a number would create a family which in some way would be of incomplete size and the pleasure would be somehow eroded. They do not have more because the family already has sufficient completeness and the additional expenditure would make the life of the existing family or parents marked by less satisfaction. Clearly what constitutes a complete family is a concept subject to change in social attitudes and is affected by comparison with the size of other families (1982: 175).

Thus, the explanation for non-zero fertility is couched in terms of the notion that there is a family size that optimises parental pleasure, at each level of the parents' ability to afford children. Such notions make the theory appear similar to the decision-making models discussed later.

While the notion of parental pleasure is invoked in connection with the discussion of why fertility does not go down to zero, Caldwell carefully avoids using the notion in his discussion of why fertility starts to decline, or remains high. While elaborating on the concept of 'wealth flow,' he emphasises that

> We are talking of 'economic' flows—even if only the fetching of father's slippers or the guarantee of help against starvation or attack—as activities, or promised activities, or goods or money that have a material value and most of which are to some extent capable of substitution. We are not talking about *pleasure*, although the action would probably not be required and would not have a positive economic value if it did not give rise to some pleasure. The point that I am making is that I do not believe in the validity of the approach to explaining high fertility which argues that fertility will be high if the value of children, judged by *both* their economic contribution over a lifetime and the pleasure they give, is positive (Caldwell, 1982: 174).

The emphasis is on the material value of activities. Caldwell recognises that the material value of an activity may be often exclusively a function of the pleasure the activity gives someone. An example may be cited to illustrate this—Cinderella's wicked stepmother orders Cinderella to mop the floor. Cinderella politely points out that the floor is clean. Thereupon she is ordered to dirty the floor first and then mop it clean. [Caldwell's own example is that of a mother-in-law ordering about a daughter-in-law 'even if it does not make housekeeping more efficient' (1982: 174).] If it were possibe to attach a material value to the pleasure Cinderella's stepmother gets in watching her stepdaughter dirty the floor and then mop it clean, there is no need to be concerned with the pleasure as such—the material value created by it will do just as well. The problem is that there are no simple formulas to impute material values to pleasures. Take, for instance, playing with someone else's infant. Presumably both the grown-up person and the infant derive pleasure from the activity. From the little we know about child development, play activities do contribute to the infant's growth, even though the participants may not be aware of it. Unfortunately, our knowledge about child development is not deep enough to enable us to impute a material value to the play activity. The crucial question is whether there is any pleasure- or pain-causing activity that does not have a material consequence. I do not think so. It does not seem to matter, therefore, whether one focuses on the pleasure or pain aspect of an activity or on the material value of it, until one becomes concerned with quantification. I believe that were we to consider parental pleasure derived from playing with an infant or from watching one's children grow as returns to their investments, what may change over time in any society is the composition of the investments (time, goods, money) and that of the returns therefrom (pleasure, pain, old age support) rather than the direction of intergenerational transfers.

There is yet another point worth noting in this connection. Imagine a landlord's family, in which none of the members engage in any manual labour, except that the female members of the household are responsible for cooking food. The older and the younger generations live together. The younger generation would inherit the family property from the older generation. It is possible that in such a family net the 'flow of wealth' is *not* from the younger to the older generation. Caldwell covers such situations by suggesting that members of the younger generation run all sorts of errands (for example, fetch slippers) for the older generation and that one should focus on the 'rules about who can get what from whom (and how much) and who can tell whom to do what' (Caldwell, 1982: 175). Obviously, in the family just referred to, members of each generation interact with members of each other generation in accordance with certain rules. In a typical South Indian landlord family of the olden days, for example, a grandfather could have asked a grandson to be a messenger boy; but not vice versa. However, a grandchild could have asked a grandparent to tell a story, to answer a question as to, say, why the sun is so high, to teach him/her to do this or that (for example, how to snap his/her fingers). Grandparents take it as their duty to answer questions raised by their grandchildren, explain to them things as best as they can, and in general socialise and train them. Children can demand and get a lot of attention from the older generation. Obviously, whether the net 'wealth flow' is from the younger generation to the older or vice versa is an empirical question. I do not think that we have enough information on this score to say that in pre-industrial societies the net transfer is invariably from the younger to the older generation.

Finally, one wonders whether the focus should be on the familial unit or the society as a whole when intergenerational transfers are considered. Caldwell's focus is on the familial unit. But the bookkeeping can very well be viewed from a societal perspective. Imagine, for example, that there are three age classes—the very young, the working people, and the old. The working people support themselves and those in the other two age classes, either directly or through an intermediary such as the government. In doing this, they repay the debt they incurred while they were growing up (in the first age class) and earn credit for supporting those in the old age class, a credit that entitles them for support when they themselves become old. Those in the working age group also contribute toward economic development through savings and investments, in addition to bringing up the very young and taking care of the old. The contribution towards economic growth obviously involves deferred gratification (sacrificing present consumption for a better tomorrow). It is possible that the fruits of such investments, because of long gestation periods of particular development schemes, may be enjoyed more by subsequent generations than by those who are responsible for the investments. As a consequence of this, the net flow of benefits is more likely to be from the older cohorts (generations) to the younger ones, in growing economies. In other words, economic growth often necessitates net intercohort (intergenerational) transfers from the older to the younger ones.

Integration of macro perspectives such as this with the micro perspective of the kind Caldwell suggests awaits attention (see Willis, 1982).

Freedman's Modifications to the Transition Theory

Ronald Freedman (1979) concluded a recent appraisal of the theories of fertility decline by saying that 'the cut and dried transition theory . . . does not seem to be enough' (p. 16). He advances several hypotheses, among which are:

(*i*) A small subset of objective developmental changes is sufficient to make parents perceive the advantage of having fewer children than has been customary in the past, and this perception can and does trigger motivation for lower fertility.

(*ii*) The exposure to the world warehouse of ideas, models, and other cultural items concerning different life styles affects the desire for, and the belief in the possibility of attaining and in the legitimacy of adopting, a way of life different from one's own and these in turn develop motivation for lower fertility.

(*iii*) If motivation for lower fertility develops, then (*a*) the specific methods of family limitation available, and (*b*) the very adoption of the idea that family limitation within marriage is legitimate affect the timing of the widespread adoption of birth control and the rate of fertility decline.

Freedman agrees with Caldwell that the mass media is a powerful medium to make converts out of large-family-oriented people. However, as mentioned earlier, he rejects Caldwell's notion that the nucleation of the family is a necessary condition for people to become motivated for lower fertility. He recognises that the antecedents in proposition (*a*) are not specified: we do not 'know just how much change in which subset of conditions is sufficient to motivate fertility declines. Probably more than one combination will turn out to be sufficient' (Freedman, 1979: 3).

Theories Based on the Institutional Context of Reproduction

Several scholars have recently adopted the position that the institutional context of reproduction deserves more attention than has hitherto been given to it. To know just what is meant by institutional context in this connection, let us examine the writings of Adnan (1982), and McNicoll (1982).

With particular reference to Bangladesh, Adnan (1982) developed the thesis that the stratification system in a country may perpetuate high fertility. There are essentially three broad classes in Bangladesh: Class I primarily consists of owners of large landholdings; Class II consists of sharecropping tenant farmers and owners of small landholdings; and Class III includes most others, mainly landless labourers. The relationships between these classes and the power structure in control of the relationships tend to perpetuate high fertility in all three classes.

The members of Class I dominate the power structure. They thrive on the rent from their land and from the interest or rent they charge on whatever is borrowed from them (for example, money, farm equipment). It is in their self-interest, as landlord-cum-money-lenders, to prevent the introduction of better agricultural technology to increase productivity, because increased productivity may improve the status of tenants, reduce their dependence on the landlord-cum-money-lender, and consequently prevent the maximisation of the landlord-cum-money-lender's earnings from rent and interest. The sharecropper tenants have no incentive to improve productivity either, because under the existing arrangements, the more they produce the less will be their share of the incremental product. Moreover, the tenant is afraid that the landlord may lease the land to someone else the next season lest some new law declares that continued possession by one person is tantamount to ownership. In short, the existing class structure inhibits technological innovation and improved productivity, and this, in turn, perpetuates the existing class structure.

As for class-linked fertility orientation, each class stands to gain by having large rather than small families. For the members of Class I, both sons and daughters are assets, in that they help perpetuate the dominance of the class. Daughters do this by helping to build a network of 'influential connections' through matrimonial ties. Sons, in addition to helping in this way, receive education and join one or another prestigious profession such as engineering or medicine, or become part of the powerful bureaucracy. Thus, for couples in Class I, there are obvious advantages in having large families.

For the sharecropper tenants and small farmers (Class II), the need for workers on their own farms and for the labour exchange between families serves to promote high fertility. Furthermore, possibilities of additional income through local hired labour, small-scale self-employment opportunities, and migration to areas of labour shortage put a premium on having many sons. To the degree that more earners means more savings, which implies better chances of buying land and accomplishing upward mobility, a high fertility orientation is consistent with the position of Class II members. Of course, partition of the family property resulting in fragmentation of landholdings may cause downward mobility. Presumably, the threat of this eventuality becomes apparent too late in one's life-cycle to make any difference in fertility behaviour. (Most people probably count on their families staying together rather than breaking up.)

For those in Class III also, large families are not without advantages, particularly if the family remains loyal to its patron in the patron-client relationship that prevails:

> Survival mediated through patronage is likely to provide employment for all employable members of the loyal household or none. The increased generation of labour-power within the household will be thus encouraged by the consideration of obtaining the maximum share of whatever patronage is available, if and when it is provided. No anti-natalist pressure would exist on these counts (Adnan, 1982: 105).

Adnan's thesis, then, is that in some societies the stratification system may be such that no strata stand to gain by fertility reduction; in such societies high fertility levels are likely to continue indefinitely. More interestingly, Adnan seems to espouse the idea that if the stratification system is such that there is a *de facto* dominant class which exercises tight control over the economic opportunities that are open to members of all other classes, then the dominant class will shape the fertility orientations of all classes.

Geoffrey McNicoll (1982) advances a similar idea by positing that a sufficient condition for a change in fertility behaviour is a significant shift in the institutional and cultural environment in which fertility decisions are made. By way of illustration, McNicoll refers to the sharecropper-landlord relationship in the Asian countries. Traditionally, he points out, the relationship was not a simple contractual one. The responsibilities and expectations of the parties involved covered a wider domain, often involving the labour of family members, a credit system, and an element of social security. In such a setting, McNicoll believes, there was probably no incentive to deliberately control family size. When, as a result of tenancy reforms, the relation between the tenant and landlord became more narrowly defined, large families became problematic and consequently, deliberate birth control may have become common. It is logical to look for variations in the institutional context within which reproduction takes place for explanations as to why fertility varies. The differences in the kinship system have long been hypothesised as responsible for fertility variations in pre-industrial societies. Unfortunately, the theory as well as the empirical evidence concerning the relationship between fertility variations and kinship system remains rudimentary at best (Freedman, 1961–62; Burch and Gendell, 1970; Namboodiri, 1972). The case involving other institutions is not any different.

When the focus is on the system-versus-environment relationship, as in the case of McNicoll's suggestion, it is important to note that a feedback relationship may exist between the two. Thus, fertility often alters and is affected by the environment in which reproductive decisions are made. Family living arrangements may be altered by, and they influence, parity progression. Such feedback relationships have not received adequate attention in the existing literature.

II

Contrasting Views on Theoretical Orientations

From the attempt to specify crucial dimensions of change responsible for fertility transitions, let me turn to contrasting views regarding theoretical orientations that are considered helpful in improving our understanding of fertility dynamics. There are some who advocate a micro orientation, while others who believe that population questions are all macro questions and that macro questions should be answered in terms of macro factors. Micro orientations usually treat the individual or the couple as the unit of analysis, while the macro approach is concerned about aggregates (populations or subpopulations). Theoretical frameworks which give central importance to motivation and decision-making belong to the former category. Those who adopt macro orientations ask questions such as: What population characteristics account for the fact that in country A the fertility level is higher than in country B? I shall briefly discuss the decision-making orientation and then proceed to show in what respects the normative approach, which is often regarded as a macro approach, resembles and in what respects it differs from the decision-making approach.

Rational Decision-making Models

A rational decision-making model may be thought of as one that analyses human behaviour in terms of costs and benefits. It is assumed that there is a decision maker (who may be an individual or a collectivity) faced with two or more possible courses of action, from among which one is to be chosen. Each action has associated with it a set of outcomes (state of the world resulting from the action) which the decision maker evaluates and ranks from the most to the least preferred. The evaluation attached to a particular outcome is called utility. Utilities have certain properties (as far as rational decision-making models are concerned), transitivity being one. If I prefer two children to three, and three children to four, then I prefer two children to four. Another property is that two outcomes may have the same utility, in which case the decision maker is said to be indifferent between the two outcomes. In many situations, the outcome of an action may not be predictable with certainty (for example, a birth control method may or may not work). A decision is said to be made under risk if the decision maker is able to attach to each possible outcome a probability of occurrence. The choice made is said to be rational, when the decision is under risk, if that course of action is chosen whose expected utility (the sum of products of utility and the corresponding probability of occurrence of outcomes) is greater than or equal to those of others. The term rationality describes this relationship between actions, outcomes, and preferences rather than the mental processes such as the avoidance of superstition.

In order to use this approach to model actual decision-making, we need to know the decision maker's perceptions of the various possible actions, the outcomes of each action, with the probability of the occurrence of each outcome, and the utility if it occurs.

It is quite possible that people do not always perceive all the behavioural options available. Their position in the class structure, their limited knowledge, pressures from their peers, and other such factors often limit the options a decision maker perceives. Thus, having no children or only one child is likely to be ruled out from consideration altogether in some societies by members of certain subpopulations.

It is also likely that the decision maker may not be aware of all possible outcomes. The concept of unintended consequences captures this idea. Also, a decision maker's estimates of the probabilities of each outcome of a particular course of action may be well off the mark. Thus, McClelland and Hackenberg (1978) found that 75 per cent of a sample of Filipino villagers thought that the probability of a couple's next child being a boy is dependent on the sex composition of their children already born.

In view of these features, it is a common practice to characterise people's predictions of the outcomes of their actions and their probabilities as functions of factors such as personality variables, and background characteristics. (People with a high degree of optimism may underestimate the cost of children, while pessimists may overestimate it; the better educated are likely to be more informed as to the consequences of a particular contraceptive device than are persons with less education; and so on.)

A major problem in applying the rational decision-making model to family building behaviour lies in the measurement of utility independently of the actual behaviour observed. It has been suggested that the techniques developed to measure perceptions of physical stimuli can be employed to measure reactions to action outcomes also (Meeker, 1981; Hamblin, 1971, 1974; Shinn, 1971; Terhune and Kaufman, 1973). The use of such techniques is yet to be demonstrated on a large scale.

In decisions such as those concerning marriage and contraception, more than one decision maker is involved. A number of decision rules are possible in such situations. Meeker (1981) distinguishes six: (*a*) rationality in the strict sense (each participant acting in such a way as to maximise his or her own expected utility); (*b*) altruism (each participant choosing that course of action which benefits the other participant the most); (*c*) competition (each aiming to maximise the absolute difference between the benefits accruing to self and those accruing to others; (*d*) group gain (each helping to maximise the total benefit of the group); (*e*) reciprocity (each helping to equalise the benefits of the participants); and (*f*) status consistency (choosing that course of action which benefits higher status participants more and lower status participants less).

A catalogue such as this helps to consider the decision rule adopted in a particular situation as a variable. It makes some sense to postulate that a decision maker's choice will be affected by the particular decision rule he or she adopts (see in this connection, Leibenstein, 1981, for an elaboration of the suggestion that the mode of decision-making be used as an explanatory variable when analysing human fertility). But it is one thing to suggest that this or that factor be included in an analytic framework and quite a different matter to carry out that suggestion in actual practice. The problems of identifying the actual decision rule adopted in particular instances remain difficult to tackle, especially in matters connected with contraceptive practice and parity progression.

It has long been suggested that fertility decisions are sequentially made: seldom is the case that a couple decide at the outset of their reproductive career how many children they want to have and when to have each child and stick to those decisions no matter what happens later (see Namboodiri, 1972, 1983 and the literature reviewed therein). It is true that most couples have an idea about the number of children they want and about the ideal child spacing, but these ideas change over time as experience accumulates and as factors beyond one's control set in. While the notion of a sequential model appeals to common sense, not much progress has been made in developing the model on a sound empirical or theoretical basis. Also, the question whether the complexity characteristic of the model and its data requirements is commensurate with the additional insight the model offers about fertility dynamics remains to be satisfactorily examined.

Decision-making models can take into account the normative pressures impinging on the decision maker. This can be seen by examining the approach of the new household economics to decision-making. In this approach it is assumed that the decision maker is a maximiser. The utility derived from the consumption of various items is the thing that is maximised: the consumer (decision maker) chooses that combination of quantities of consumption items which maximises utility. Each consumption item is produced by the consumer in accordance with a production process using his/her own time and market goods and services. The production process is influenced by the technology available, the organisational context involved, and a number of relevant characteristics of other persons and collectivities. Also, the amount of expenditure is constrained by the amount of resources available, be it time or money. In so far as it is recognised that the production of each item of consumption is influenced by selected characteristics of the 'relevant other,' it is legitimate to claim that the decision-making model thus conceptualised is capable of taking into account normative pressures. How many children an individual chooses to have, how she brings them up, when and whether she chooses to use contraception, and matters like that are influenced by what the individual perceives to be the standard behaviour in these respects. It is explicitly assumed that such influences are mediated through what the individual decision maker perceives to be the costs and benefits involved in adhering to the norms. In this respect, the decision-making approach parallels the normative approach.

The Normative Approach

There are some who approach fertility model building from the position that individuals are free agents, and there are others who approach it from the position that individuals are mere cogs in a machine. Some rely on an implicit or explicit model of the actor who makes everyday decisions based on his/her own interpretations of the environmental changes. Some, on the other hand, rely on the assumption that what individuals do is determined by their socialisation and training, and the social control by which they are subjugated. The decision-making approach outlined above uses the former position, while the normative approach discussed here espouses the latter.

In the normative approach, the usual starting point is the assumption that, 'Human society regulates the behavior of members, through norms . . . . The individual acquires norms through indoctrination' (Ryder, 1979: 198). Applying this general proposition to reproduction behaviour, it is held that:

> Each reproductive act may be thought of as an implicit contract. In making a decision to have a baby, the costs and benefits perceived depend on the terms of that contract, because it spells out the rights and responsibilities of (the parties involved, e.g., parents and children). The contract exists because the occurrence of a child is consequential to others beyond the immediate decision makers, and those others are prepared to administer rewards and punishments to insure that the individual decisions correspond with the collective will (Ryder, 1980: 198).

When put in this way, it becomes difficult to separate the decision-making approach from the normative approach because both are founded on the postulate that the choice behaviours of individuals are influenced by the perceived costs and benefits associated with the different options available—costs and benefits to the individuals themselves as well as to others. In what essential respect, then, is the normative approach different from the decision-making approach? It appears that the crucial distinction is in terms of the level of analysis. One remains at the macro level, while the other remains at the micro level.

> As a demographer, my research object is by definition aggregate in character. A population is defined as an aggregate . . . .
>
> The questions which are justification for the magnitude of support for fertility inquiry are in my opinion demographic questions. Why is fertility higher in one population than in another? Why does fertility decline more in one population than in another? . . . .
>
> If one begins with a macro question—why does one population have higher fertility than another, or why does a population have lower fertility than it once did—it seems obvious that one must look for a macro answer. The populations clearly differ or change in some respect, and our task is to determine the relevant respects. With this perspective, it is an interesting puzzle why the answer should be sought at the micro level.
>
> (One possible explanation is the stance of) the sociologists who are at root reductionist in persuasion. Possessed of the ineluctable truth that no explanation at the sociocultural level can be complete without a demonstration of an explicit instrumentality at the individual level, and suspicions that properties of the collectivity are myths invented by Durkheimians, and ideologically offensive as a threat to the concept of free will, they proceed directly to the finest level of detail attainable, unaware or uncaring that they will drown in the morass (1980: 199–201).

If one adopts the macro approach, then, as already mentioned, the model building strategy is guided by questions such as—Given two or more populations which differ in their fertility levels, what population characteristics account for their difference (in fertility levels)? Given a mixture of populations or sub-populations, how could one separate the mixture? The data needed to answer the second question usually come from individuals and couples.

> [The] respondent is an agent in the process of inquiry, contributing a description of her reproductive behavior to the construction of a macro distribution, and signifying in response to questions about ascribed status the groups to which she belongs, and thus the aggregates for which it is appropriate to construct such distributions (Ryder, 1980:199).

The appropriate aggregations are presumably those which differ among themselves in systems of socialisation and social control. So the research strategy involves identifying subpopulations which differ in respect to what their members are taught as appropriate demographic behaviour and the strictness with which adherence to appropriate behaviour is enforced.

The main problem with this approach is that it is not easy to separate subpopulations according to systems of socialisation and social control. The use of background information such as religious affiliation, ethnicity, rural-urban residence, and birth cohort permits only crude classifications. Do the Nayars of central Kerala form a meaningful subpopulation for the purpose? Should those of Cochin heritage be separated from those of Travancore extraction? Are the Irish Catholics living in South Side Chicago illustrative of the subpopulations the analysts should be looking for?

There is also the problem that a reproductive unit might come under the influence of different systems of socialisation and social control in a sequential manner. (Think of the migrants to South Side Chicago from another neighbourhood in Chicago or elsewhere). If this happens, do those who move carry with them their old habits? If they do, how soon after they move do they discard their old habits and adopt new ones? Or is it the case that what one learns as an adolescent dictates what one does throughout the rest of one's life? Are those born and brought up in a rural setting likely to behave in the same way in an urban setting as their counterparts who remain in their familiar rural environment? An associated question is—If we were to adopt a macro approach, how would we handle the implication of population turnover? (In-migrants are not necessarily similar to out-migrants in their past history.) These and similar issues remain to be examined.

There is also the issue as to why a particular system of socialisation and social control develops. One answer may be that, for every meaningful aggregate (for instance, the Catholics of Irish extraction living in South Side Chicago), there exists an optimum distribution of reproduction behaviour (for instance, an optimum distribution of the first conception interval), and that there are systems of socialisation and social control, any one of which can guarantee a given distribution of reproductive behaviour. From this viewpoint, the emergence of a particular system of socialisation and social control is a matter of natural selection. The problem is one of convincing oneself of the validity of the basic postulates that lie behind this viewpoint. Is there such a thing as an optimum distribution of behaviour that can be regarded as a unique characteristic of a given aggregate? It may very well be that all aggregates are tolerant of a wide variety of distributions of behaviour (as long as, say, for example, the tail ends of the distributions are narrow). Also, no given system of socialisation and social control can guarantee that extreme behaviours (such as very long first conception intervals) would be kept to a minimum.

It is also worth noting that, much as they dislike it, macro analysts are often drawn willy-nilly to using micro notions in the interpretation of their findings. As Alexander Rosenberg (1980: 34) says,

> (A)ny one who finds the explanation of one set of macrosocial facts in the citation of another set must eventually commit himself to an underlying

mechanism to connect them; and when the mechanism consists in moral constraints and the appropriateness of behavior to circumstances, as it does with Durkheim, the appeal to psychological states becomes almost unavoidable.

Thus, the macro approaches, particularly those which are built around the concept of social norms are not free of problems. In fact, such approaches seem to raise more questions than they answer.

III

Brief as this survey has been, several points have emerged: a large number of theories are available in the literature; it has not been the practice so far to refine cumulatively any single theory. Instead, investment of efforts has been in the direction of developing separate proposals or orientations; concerted attempts in evaluating the extant theories are yet to be undertaken. The protocol for such exercises remains to be agreed upon; and while the volume of written material on theories and theory building in demography has increased tremendously since Rupert Vance discussed *Theory for Demographers*, it cannot be said that much progress has been made during this period with respect to building a 'binder' for the accumulating data on population change.

REFERENCES

ADNAN. (1982). Conceptualising fertility trends in peripheral formations. *In* C. Hohn and R. Mackensen (Eds.), *Determinants of fertility trends: major theories and new directions for research*. Ordina edition for the International Union for the Scientific Study of Population.

BURCH, T.K. and M. GENDELL. (1970). Extended family structure and fertility: some conceptual and methodological issues. *Journal of Marriage and the Family*, 32: 227–36.

CALDWELL, J. (1976). Toward a restatement of demographic transition theory. *Population and Development Review*, 2: 321–66.

———. (1982). Wealth flow theory of fertility decline. *In* C. Hohn and R. Mackensen (Eds.), *Determinants of fertility trends: major theories and new directions for research*. Ordina edition for the International Union for the Scientific Study of Population.

COALE, A.J. and E.M. HOOVER. (1958). *Population growth and economic development in low-income countries: a case study of India's prospects*. Princeton, NJ: Princeton University Press.

COALE, A.J., B. ANDERSON, and E. HARM. (1979). *Human fertility in Russia since the nineteenth century*. Princeton, NJ: Princeton University Press.

DAVIS, K. (1963). The theory of change and response in modern demographic history. *Population Index*, 29: 345–65.

FREEDMAN, R. (1979). Theories of fertility decline: a reappraisal. *Social Forces*, 58: 1–17.

———. (1961–62). The sociology of human fertility: a trend report and bibliography. *Current Sociology*, 10/11: 35–121.

HAMBLIN, R.L. (1971). Ratio measurement for the social sciences. *Social Forces*, 50: 191–206.

LEIBENSTEIN, H. (1981). Economic decision theory and human fertility behaviour: a speculative essay. *Population and Development Review*, 7: 381–400.

McClelland, G.H. and B.H. Hackenberg. (1978). Subjective probabilities for sex of next child: US college students and Philippine villagers. *Journal of Populations*, 1: 132–47.

McNicoll, G. (1982). Institutional determinants of fertility change. *In* C. Hohn and R. Mackensen (Eds.), *Determinants of fertility trends: major theories and new directions for research*. Ordina edition for the International Union for the Scientific Study of Population.

Meeker, B.F. (1981). Fertility decision-making theories in interpersonal behaviour. *In* T.K. Burch (Ed.), *Demographic behaviour: interdisciplinary perspectives on decision making*. AAAS Selected Symposium 45. Boulder, Co: Westview Press.

Namboodiri, N.K. (1972). Some observations on the economic framework for fertility analysis. *Population Studies*, 26: 185–206.

————. (1983). Sequential decision making and the life course. *In* R. Lee (Ed.), *Determinants of fertility*. New York: Academic Press.

Rosenberg, A. (1980). *Sociobiology and the preemption of social science*. Baltimore: Johns Hopkins Press.

Ryder, N.B. (1980). Where do babies come from? *In* H.M. Blalock (Ed.), *Sociological theory and research: a critical appraisal*. New York: The Free Press.

Shinn, A.M. (1971). Measuring the utility of housing: demonstrating a methodological approach. *Social Science Quarterly*, 52: 88–102.

Terhune, K.W. and S. Kaufman. (1973). The family size utility function. *Demography*, 10: 599–618.

Vance, R.B. (1952). Is theory for demographers? *Social Forces*, 31: 9–13.

Willis, R. (1982). The direction of intergenerational transfers and demographic transition: the Caldwell hypothesis reexamined. *Population and Development Review* (Supplement), 8: 207–34.

4

Some Observations on the Regulation of Fertility

Albert I. Hermalin

This paper briefly treats three issues concerning the regulation of fertility: the levels and trends in developing countries, the issues in modeling fertility regulation, and the new approaches to measuring the costs of fertility regulation.

Levels and Trends in Fertility Regulation

One important by-product of international programmes for comparable high quality fertility or KAP-type surveys such as the World Fertility Survey or the Contraceptive Prevalence Survey is the availability of good estimates of contraceptive prevalence. If these are combined with the estimates from service statistics and individually sponsored surveys, one can obtain reasonably accurate and comparable estimates of prevalence for large numbers of developing and developed countries, together with the distributions by type of method, though there are cautions to be observed in assembling and interpreting such data. The tables developed by Larson (1981), Nortman and Hofstatter (1980), and the United Nations publications paint a fairly clear picture of the world's contraceptive prevalence scene as of the mid- to late seventies whose main contours may be sketched as follows (unless otherwise indicated, figures cited refer to percentage of women of reproductive age in a union currently using any contraceptive):

(*a*) There is a considerable range in prevalence among developing countries. There are nearly a dozen countries with prevalence rates over 50 per cent; these have gained parity with a number of European countries which are at the lower end of the developed country distribution. At the same time, prevalence rates below 10 per cent were still observed in nearly ten developing countries and that number may grow with surveys that are yet to be released.

(*b*) In many countries, a high proportion of users rely on modern methods.

Acknowledgement This paper is based on a talk presented at the Conference on Fertility Research and Population Policy in Developing Countries, National Academy of Sciences, Washington, DC, November 1982. It derives from a longer essay, 'Fertility regulation and its costs'. *In* R.A. Bulatao, R.D. Lee, P.E. Hollerbach and J. Bongaarts (Eds.), *Determinants of fertility in developing countries: a summary of knowledge*. New York: Academic Press, 1983. Chapter 19, pp. 518–60.

For example, a UN (1981) analysis of twenty World Fertility Survey countries shows that in fifteen, two-thirds or more of the users were using sterilisation or a modern reversible method.

(*c*) Despite this, there is little correlation in the developing countries between the overall prevalence rate and the proportion of users relying on traditional methods. Stated otherwise, countries which rely on traditional methods show a wide range of prevalence, as do countries with a high reliance on modern methods.

(*d*) An interesting sidelight is that in some respects, the *pattern* of contraceptive use seen in the developing countries is more modern than that seen in the more industrialised countries. That is, in a number of developed countries (Belgium, France, Czechoslovakia, Poland, Spain, Romania, Ireland, Yugoslavia), traditional methods account for a higher proportion of total use than in many of the developing countries for which we now have data. At the same time, an analysis by Leridon (1981) has shown a substantial shift toward modern methods among the developed countries in the 1965–75 period. Clearly, for these highly motivated couples, the safety, convenience, and effectiveness of the modern methods were attractive alternatives to the traditional methods they were using. These patterns also suggest that for many of the couples in the developing countries, these same features helped them overcome their reluctance to initiate fertility regulation. In other terms, the availability of modern methods, improvements in the IUD, the pill, and techniques of sterilisation, and their active promotion appear to be a significant contributor to the prevalence rates attained in many developing countries. (The relation of method-type and method-mix to prevalence is an intriguing question meriting more attention, and a further analysis of the differences between the developing countries and the industrialised countries in this regard may provide useful insights.)

(*e*) It is difficult to gauge *trends* in prevalence and distributions by method because comparable surveys at useful time intervals are available for relatively few countries. Nevertheless, the data I have been able to collect show that some fifteen countries have increased their prevalence rates by 20 points or more within a decade or less. Generally, increases in contraceptive prevalence have been accompanied by shifts to more modern techniques; the rapid rise in the use of female sterilisation in a number of countries is noteworthy.

Without being overly dramatic, it is probably fair to say that a modest revolution in the adoption of contraception occurred in a significant part of the developing world between the mid-sixties and mid-seventies. It is probable that as late as 1970, population experts would have underestimated the levels of prevalence that emerged by the mid- to late seventies. One should also keep in mind that the data we are examining are five to seven years old, and that a current survey may reveal another quantum jump. The momentum for continued increases over the last seven years is suggested by several factors:

(*a*) There is a considerable potential for diffusion across social groups. Very high levels of use among educated subgroups have been achieved in a number of countries. For example, in ten of seventeen World Fertility Survey countries analysed by the UN (1981), three-quarters or more of the exposed women with ten years or more of schooling who wanted no more children were currently using contraception, and in two other countries the proportion was between 60 and 75 per cent.

(*b*) The use of contraception for spacing has been steadily increasing. The same UN analysis also showed that in ten of seventeen countries, 25 per cent or more of *all* exposed women who wanted more children were currently using contraception—that is, they had adopted contraception for spacing.

(*c*) There seems to be a persistent vigour of family planning efforts in a large part of the developing world. The UN monitoring reports (1980) reveal that 85 of 116 developing countries surveyed supported access, directly or indirectly, to modern methods of birth control; and the Nortman and Hofstatter Factbook (1980) estimates that over 90 per cent of the developing world's population resides in countries which support family planning either to reduce population growth or to promote health. Also in recent years, the family planning delivery system has been undergoing rapid change with greater promotion of commercial retail sales, integration of family planning services with primary health care and development programmes, and with an increased emphasis on community-based distribution systems. Judgment on the effectiveness of these various approaches must await further research, but they point to the sustained interest in many quarters in promoting the availability of contraception.

Contraception is, of course, only one form of fertility regulation and it is worth remembering that other mechanisms restricting fertility are in effect in many developing countries. Accurate data on abortion are elusive, but we know it is a significant factor in a number of countries, adding substantially to the overall level of control. An important question meriting greater attention is the effect of increasing contraceptive availability on the incidence of abortion. David (1983) suggests that there are a number of possible patterns:

> The relative roles of contraception and abortion vary within and between countries. In some countries, abortion is the primary means of fertility control; in others, increased use of contraception lowers the incidence of abortion; and, in still others, abortion and contraception increase together. These patterns suggest a possible transitional path from high to lower abortion rates. With the passage of time, contraceptive practice improves and resort to abortion declines although it is never eliminated (pp. 691–92).

Breastfeeding in the developing world, according to last reports, is also an important regulating mechanism. Comparative data based on the World Fertility Survey by Ferry (1981) show that a very high proportion of children (above 80 per cent) are breast-fed in nearly all the countries, with the likelihood of breast-feeding and its duration generally higher in Asia and Africa than in Latin America and the Caribbean.

Issues in Modeling Fertility Regulation

One way to evaluate a field or subfield is to examine its theory or conceptual framework, the manner in which key concepts are measured, and the degree to which there is concordance between empirical analyses and the theoretical structure. Embedded in the framework guiding the National Academy of Sciences panel on the determinants of fertility is a rather simple conceptual model which states that the motivation to limit childbearing and the costs of regulation determine the degree of fertility regulation. This is a rather simple three variable problem which has received relatively little direct empirical attention. Before pursuing the possible reasons, one must keep in mind the larger framework of which this is a part. Motivation is determined by the balance of demand and supply; actually it would be more accurate to say that motivation is *defined* by supply and demand. Under this framework, it is assumed that once the actual or likely number of surviving children reaches or exceeds the desired number, a couple is motivated to some degree to control their fertility (Figure 1).

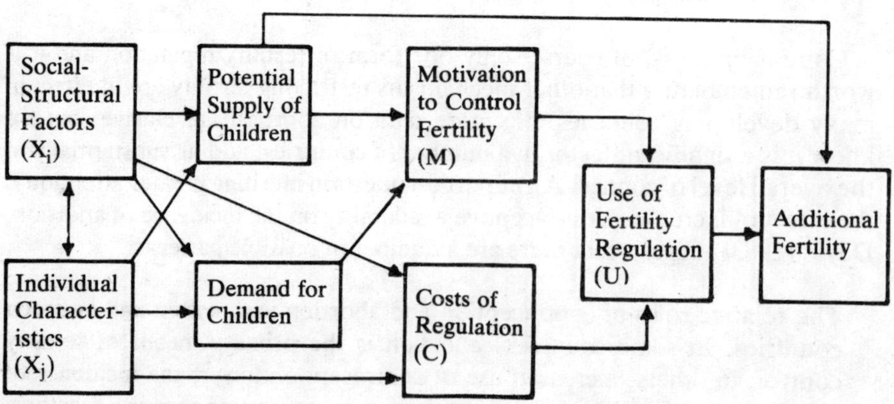

FIGURE 1 Basic Model of Factors Determining Additional Fertility

Supply and demand are determined by two *sets* of exogenous or predetermined variables: the first could be termed social structural, and includes such variables as levels of development, family planning inputs, community social organisation, and laws, norms, and customs governing abortion, breastfeeding, and marriage; the second set consists of individual characteristics covering socio-economic, biological, and demographic factors. Elements of these two sets not only determine supply and demand but they also have a direct effect on the costs of fertility regulation. It should be noted that since motivation flows from the balance of supply and demand, the model assumes that fertility regulating behaviour will be for the purpose of limiting births rather than spacing. Given the rise of contraceptive use for spacing in the developing countries as noted earlier, the need for modifications in our conceptual framework to accommodate this form of behaviour deserves high priority.

Although it is possible to estimate the entire structural model just described, many studies focus on selected aspects. Even though only a portion of the model may be estimated, attention to the whole model is useful in selecting and operationalising variables and in choosing appropriate techniques of estimation. The part of the model which concerns us is how motivation and costs combine to determine regulation and on the exogenous factors (individual and structural) which determine the level of motivation and costs. This sub-model is portrayed in Figure 2.

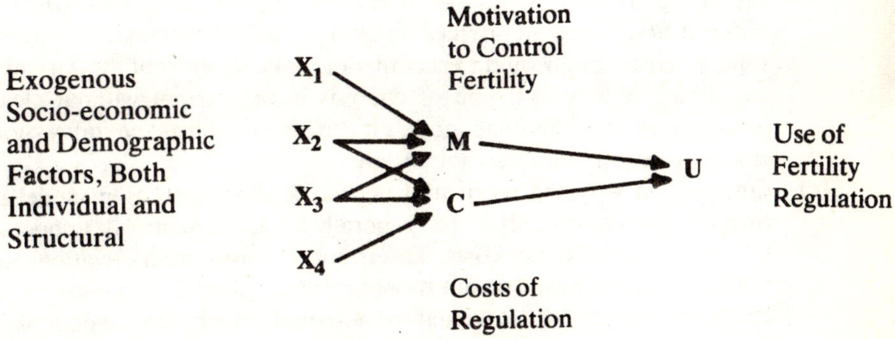

FIGURE 2 Basic Model of Factors Affecting Fertility Regulation

A comparison of the two figures (the general framework and the sub-model) reveals some important guidelines for empirical research:

(*a*) The exogenous factors set forth as determinants of motivation must also determine supply and demand, since supply and demand *define* motivation. In other words, if a variable is not considered a determinant of supply or demand, it should not be used as a determinant of motivation. Note that this is not true of costs; the variables that influence motivation need not be the same as those which influence costs.

One implication of this, for example, is that it makes little sense when studying motivation to include the number of living children as a determinant. This variable is already embedded in the measure of motivation, as when we calculate actual minus desired number, or when we ask women whether more children are wanted. The result is to create an artificially strong effect for this particular variable and to produce misleading inferences about the strength of other independent variables.

(*b*) Given adequate measurement of motivation and cost variables, socio-economic, demographic and other factors do not enter as direct effects on fertility control. These variables can affect both motivation and costs, but their relation to fertility control should be derivable from these effects and the effect of motivation and cost on control.

This calls into question the practice of employing a hierarchical decomposition of variance explained, in which socio-economic and demographic variables are entered first and then the per cent of variance in the use of contraception (or a similar measure) explained by motivation is calculated. Usually it is found that motivation explains relatively little compared to its zero order effect. This type of result is not surprising given the model we are examining. Insofar as the variables entered first in the hierarchical analysis largely determine motivation (and/or costs) one would expect them to take up most of the variance explained. When this type of analysis is carried out without clear attention to an underlying model, it can give the mistaken impression that motivation is not very important.

(*c*) One reason why the motivation, costs, fertility regulation model is rarely employed directly is that generally it has been difficult to operationalise the concept of costs. This is not the case with motivation. As noted, measures based on questioning women whether they want more children—comparing the actual (or potential) number of children with the desired number—are available and have been employed in analyses. Once such a measure of motivation is employed, it should be recognised that any other variable included as a determinant of regulation is serving as a proxy for cost, if one accepts the framework given in Figure 2. This suggests that models which combine a measure of motivation and a cluster of socio-economic and demographic variables may not be too meaningful unless these variables can be justified as measuring dimensions of cost.

(*d*) How we measure motivation will require increasing attention, particularly with greater prevalence and wide use of contraception for spacing. A simple example will reveal the problem. If motivation is measured as actual less desired number of children, the actual number may well be determined in part by the prior use of contraception. If we then use motivation as a determinant of current contraceptive use, there is an inherent circularity or simultaneity. This is why some analysts like Easterlin and Crimmins (1982) and Boulier and Mankiw (1980) in employing this type of model have estimated 'potential supply' rather than using the actual number of children. 'Potential supply' is not something that can be observed, couple by couple, but must be estimated by some means. Greater attention to the need for such an estimate should pay good dividends.

These examples highlight a more general problem. Collectively, we seem to be somewhat schizophrenic in our use of theory to study the process of fertility regulation. We understand these frameworks, give them wide lip service, articulate the need for a better theory and then frequently ignore the theory we have in the actual execution of research. This contributes to a state of disarray within the field and hinders systematic progress. As is well known, theories are arbitrary constructs designed to organise and explain a range of phenomena. They are neither true nor false but,rather, useful or not useful. Their ultimate utility can only be determined through systematic exploration, including careful measurement of the concepts and appropriate modes of analysis. In this way, we can identify the features of an existing theory that seem to work well, and pinpoint areas where new and revised formulations are needed.

New Approaches to Measuring Costs

It was mentioned earlier that one of the reasons for rarely employing the motivation, costs, fertility regulation model directly is the difficulty in operationalising the concept of costs. We have no difficulty in classifying costs into economic, social opinion, health, and psychic categories, but we have measured relatively few of the specific items successfully, nor incorporated them adequately into our analyses. This stems, I believe, from our tendency to individualise concepts that are more properly viewed as macro or global characteristics, which, in turn, derives from our over-reliance on surveys as the sole source of information. For a long time, knowledge of contraceptive methods was the prime measure of costs; more recently, interest has centred on the perceived availability which has led to the inclusion of questions on the knowledge of an outlet for various methods and time and/or distance to these outlets.

There is an inherent simultaneity of effects, however, between the elements of costs measured in this way and the contraceptive use which confounds analysis and limits programme or policy inference. This simultaneity arises because prior and/or current use affects reported knowledge and attitudes, although these reports are taken as affecting current use. Since previous knowledge and attitudes cannot be established from a cross-sectional survey, the inherent simultaneity cannot be unravelled. This means that analyses which incorporate current knowledge and attitudes as determinants of use are likely to produce biased and misleading results.

One solution to this problem is to carry out a more longitudinal analysis in which an individual's knowledge, attitudes, and other aspects of costs, as well as her contraceptive behaviour can be traced over time. This type of design is very expensive and not without problems in terms of maintaining the panel and avoiding contaminated responses, though variants of this design could be possible in certain settings.

I believe that a more satisfactory and feasible approach stems from recognising that many elements of cost are properly viewed as characteristics of communities or areas, and to incorporate this perception into our measurement and analysis. Consider the elements of market costs: these include the availability of information and services, the method mix, the prices, the access to outlets in terms of time or distance, and the services provided. An examination of this list indicates that these are essentially programme or macro-characteristics and the same value of each could be assigned to each community or small geographic cluster. The same may be said for many of the social and cultural factors involved in the cost structure. Social norms, opinions, and values are by their very definition characteristics of aggregates.

The measurement of these macro aspects of costs represents relatively few difficulties compared to those at the individual level. The actual distance from the centre of each community or small geographic cluster to each type of outlet can be measured directly, as has been done in Thailand in connection with the recent Contraceptive Prevalence Survey, and in some other countries. Information about other macro characteristics may be gleaned from the records of the family planning programme and private providers, or it may have to be obtained by making a special effort. The cost of doing this should not be great since the data are needed only for the communities that are sampled as part of the fertility survey. Limiting any special investigation to select communities and to salient items that are variable across communities, should make this type of undertaking relatively inexpensive and quick in comparison to the collection of individual data.

Once this type of data is in hand, it is possible to carry out a multilevel analysis which combines data on the characteristics of individuals with those of their setting. I think many of us would accept, as a general framework, that most demographic behaviour of interest is a function of the individual characteristics as well as the social and economic structure in which individuals exist. Multilevel analysis enables one to operationalise this framework and to examine the social structural effects while controlling for individual variation. Equally important, it allows one to examine the *interactions* between individual and aggregate effects. For example, it is reasonable to assume that where family planning services are readily available, it will be easier to convert a desire for no more children into contraceptive use. This suggests that the differentials in use between those who do and those who do not want more children should be greater in areas of high availability than in areas of low availability. Entwisle et al. (1982) recently completed such a multilevel analysis for the rural areas of Thailand and found exactly this pattern.

Multilevel analysis can also provide considerable operations research guidance to policy makers and programme administrators by identifying those aspects of the programme structure that have a greater and lesser impact on contraceptive use. Inferences derived from data on the *perceived* availability and similar survey questions are much more limited in this respect and, as noted, may even be misleading.

By advocating the case for the macro-nature of many elements of cost, I do not wish to imply that there are not important individual cost factors. Indeed there are, and how individuals perceive the aggregate structure is also of interest. Nor does measuring costs as a macro characteristic automatically remove all problems of simultaneity and proper inference. Yet when the entire situation is carefully analysed, I believe we will find that there are strong conceptual as well as pragmatic grounds for measuring these structural elements and considerable pay-off from employing these in multilevel analyses.

Conclusion

There is a connection, albeit somewhat tenuous, between the three issues addressed in this paper. The level and patterns of fertility regulation are changing rapidly in the developing world. By paying closer attention to our conceptual frameworks and systematically exploiting them, and by being open to orderly modification in our modes of analysis, I believe we will be in an excellent position to gain valuable insights into the theoretical and practical issues posed by these changes.

REFERENCES

Boulier, B.L. and N.G. Mankiw. (1980). An econometric model of the demand, supply, and regulation of fertility. Paper presented at the annual meeting of the Population Association of America, Denver.

David, Henry P. (1983). Abortion: its prevalence, correlates, and costs. Chapter 23. *In* R.A. Bulatao and R.D. Lee with P.E. Hollerbach and J. Bongaarts (Eds.). *Determinants of fertility in developing countries: a summary of knowledge*. New York: Academic Press. Pp. 679–719.

Easterlin, R.A. and E.M. Crimmins. (1982). An exploratory study of the 'synthesis framework' of fertility determination with WFS core questionnaire data. Paper prepared for the International Union for the Scientific Study of Population, International Population Conference, Manila.

Entwisle, B., A.I. Hermalin, P. Kamnuansilpa and A. Chamratrithirong. (1982). *A multilevel model of family planning availability and contraceptive use in rural Thailand*. Population Studies Center Research Report. Ann Arbor, Michigan: University of Michigan.

FERRY, B. (1981). *Breastfeeding*. WFS Comparative Studies No. 13. London: World Fertility Survey.

LARSON, A. (1981). *Patterns of contraceptive use around the world*. Washington, D.C.: Population Reference Bureau.

LERIDON, H. (1981). Fertility and contraception in 12 developed countries. *Family Planning Perspectives*, 13: 93–102.

NORTMAN, D. and E. HOFSTATTER. (1980). *Population and family planning programmes*. 10th ed. New York: Population Council.

UNITED NATIONS. (1980). *World population trends and policies 1979 monitoring report*. Volume II, ST/ESA/SER.A/70/Addl. New York: United Nations.

———. (1981). *Variations in the incidence of knowledge and use of contraception: a comparative analysis of world fertility survey results for twenty developing countries*. New York: United Nations.

5

The Value and Cost of Children: Converging Theory and Research

JAMES T. FAWCETT

The value and cost of children is a new research area, or at least an area that has been newly labelled and defined. According to Berelson (1976), psychologically-oriented studies of the value of children represent one of five promising new lines of inquiry in the general area of determinants of fertility. Microeconomic studies, which also deal with the value of children, are listed by Berelson as another of the five new research developments. In this paper I will discuss both the psychological and microeconomic perspectives, but these will be examined within a more general framework. My approach will be purposely non-technical, with the aim of providing an overview of what this research area is, what has been accomplished, and what remains to be done.

Conceptual and Theoretical Approaches

As suggested above, the value and cost of children is a topic that is usually associated with psychological and economic research and theory, within the general area of determinants of fertility. This particular terminology—value and cost of children—has in fact been used most often by psychologists and economists, but the topic itself is hardly new or limited to these academic disciplines. What is new, however, is a more systematic and more theoretical approach to the micro level determinants of fertility. First, I will discuss some central features of the micro level approach, and then proceed to elaborate this approach within the broader context of population studies. By micro level approach, I mean a theoretical or conceptual framework that focuses on the individual and stresses motivation and choice in childbearing behaviour. This seems to me a common core in the various disciplinary perspectives, but beyond that common core there remains much diversity.

One indication of this diversity is the variety of terminology that is used in connection with the value and cost of children. For example, consider the expressions indicating the positive and negative orientations toward children and childbearing:

POSITIVE ORIENTATIONS

Value of children	Tastes for children
Satisfactions of children	Preferences for children

Advantages of children	Demand for children
Rewards of children	Motivation for parenthood
Functions of children	Reproductive motivations
Utility of children	Childbearing motivations
Need for children	Fertility motivations
Desire for children	

NEGATIVE ORIENTATIONS

Costs of children	Disvalues
Dissatisfactions	Disutilities
Disadvantages	Negative motivations

The expressions in each group cannot always be used interchangeably, but in most cases it is difficult to make clear distinctions among them. Terminology in this area of research is loose and non-uniform, and suffers from a lack of systematic definition, either theoretical or operational.

A basic assumption underlying most micro level research in this area is that the positive and negative orientations toward children are important elements in a cognitive and behavioural process that is related to the number and timing of births. The process is indicated by several kinds of terms:

Decision	Purposive
Choice	Planful
Rational	Volitional

As part of the process, children are compared to other possible sources of satisfaction.

Alternatives to children
Substitutes for children
Functional equivalents of children
Child surrogates
Tastes for alternative goods or services
Preferences competing with children

A simplified version of the system just discussed is presented in Figure 1.

In different theoretical approaches, this basic model is elaborated in various ways. For example, a consideration of the alternatives may be regarded as interacting with values and costs, or may be entered separately as a factor affecting choice. In psychological theories, particular attention is given to the ways in which children or alternatives satisfy basic needs or more general values. In economic theories, the emphasis is upon income or resource constraints, in relation to the cost of children. In all micro level approaches there are also important intervening elements not shown in this simplified model, such as the 'quality' of children desired and the availability and psychological cost of birth control. And, of course, the larger social, cultural and situational context is taken into account, with feedback and interaction effects shown among the various components of the model. As noted earlier, the common core is a framework that focuses on the individual and stresses motivation and choice in childbearing behaviour. Some additional assumptions of the micro level framework are:

(*a*) That people act in anticipation of future rewards and costs.
(*b*) That children provide important satisfactions in life, but not without cost.
(*c*) That people differ with respect to the types of satisfactions and costs that are important to them.
(*d*) That these differences are related to both internal psychological factors and external social and economic factors, which also affect the desirability and availability of alternatives to children.
(*e*) That the balance of satisfactions and costs of children changes over the life-cycle.
(*f*) That choices about children versus alternatives are made by most people at some point in the life-cycle, although not necessarily for every birth.

I will now turn from this discussion of micro level concepts and assumptions toward a broader theoretical view, pointing out the differences and the similarities among the four approaches within the field of population studies that deal with the value and cost of children at different levels of conceptualisation. I

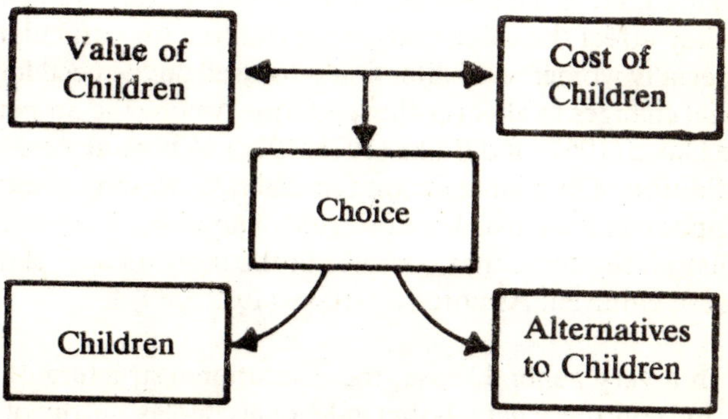

FIGURE 1 Basic Model of the Value and Cost of Children

have labelled these approaches—the sociodemographic, the social-structural, the microeconomic, and the social-psychological. I will provide a brief characterisation of each and will cite some relevant sources, but will not attempt to do justice to the richness of thought and the diversity of research within any of the four approaches.

In the *sociodemographic* approach, emphasis is given to long-term changes in the social and economic circumstances that produce corresponding changes in the value and cost of children (as well as an awareness of the alternatives), and thus a demographic transition. This approach is heavily historical and inferential, although the ideas are also applied to contemporary cross-sectional data and in some historical studies attempts are made to assess explicitly the value and cost of children (for example, Banks, 1954). I have described the essence of this approach elsewhere (Fawcett, 1972: 5–7):

> In sociology and demography, the satisfactions and costs of children have been discussed as a partial explanation for the demographic transition. The shift from a rural, agricultural society to an urban, industrialized one implies a reduction in economic benefits from children and an increase in costs. Other changes, such as higher levels of education, tend to increase awareness and desires related to alternatives to children. Changing family structures, along with alterations in economic activities of the family, tend to make large numbers of children dysfunctional.

From the sociodemographic approach, then, we obtain the view that the value and cost of children is a major factor intervening between socio-economic change and demographic change. However, the value and cost of children is conceptualised in very broad terms and is usually inferred, not measured.

The *social-structural* approach tends to be more contemporary and action-oriented. Here the emphasis is on the norms, values and opportunity structures as they affect the value and cost of children for particular segments of society, especially women, and attention is focused on the need for selective, purposeful social changes to alter fertility patterns. Among the proponents of this approach are Davis (1967) and Blake (1971; 1972). A central focus of this approach is the definition of familial and non-familial roles within a society and the availability of opportunities to assume one role or another. The social structure is seen as a constraining force that operates at the individual level through the value and cost of children. As noted by Blake (1972: 280):

> In a very general sense, the institutional structure of every society defines and controls what it is that individual couples 'get out of' having children—the rewards or utilities for having a family, and how much couples must sacrifice to have them—the costs. Thus, it is fair to say that all of the specific variables that are said to 'affect' family-size preferences—urbanization, educational level, possession of modern consumer goods, female labour force participation, or whatever, actually may be said to affect the utility of children and their costs.

There is obviously substantial similarity between the sociodemographic and social-structural approaches. However, I perceive important differences in that the social-structural approach gives greater emphasis to mediating factors (roles, norms, the utilities and costs of children) and is oriented more towards the individual level (reproductive motivations) and the question of choice—more specifically, the lack of choice owing to social-structural constraints.

The *microeconomic* approach, by contrast, assumes that choices among competing alternatives are available and are made within a framework of allocation of limited resources. This distinction between the microeconomic and social-structural approaches led to the now famous remark by Duesenberry (1960: 233), 'Economics is all about how people make choices. Sociology is all about why they don't have any choices to make'. The reconciliation of these two viewpoints has yet to take place, but it seems to me that some convergence is discernible in the area of fertility studies. For example, an economist (Turchi, 1975) has noted in a recent critique of microeconomic theories that 'although fertility decisions are allocative decisions, they take place in a social and psychological context that can be ignored only at great peril.'

Although the particular derivations of microeconomic theory tend to be complex, the underlying framework is straightforward. A summary has been provided by Easterlin (1975: 54):

> The conventional theory of consumer behavior views the individual as trying to maximize satisfaction, given a range of goods, their prices, and his own tastes and income. In the application of the theory to fertility analysis, children are viewed as a special kind of good, and fertility is seen as a response to the consumer's demand for children relative to other goods. In the last few years, a special variant of this approach has emerged, deriving chiefly from a 1965 article by Becker and distinguished by use of the concept of a 'household production function'.

In microeconomic theory, social and psychological dimensions are subsumed under such concepts as satisfactions, utility, tastes and preferences. As with psychological theories, there is an assumption that children provide certain kinds of satisfaction and that similar satisfaction can be gained from other goods, services or activities. Turchi (1975: 108–109) states:

> Thus, it is argued, if parents choose to have children they must have weighed the psychic rewards from bearing and rearing children against the rewards from other activities which might have been undertaken instead. This, then, is the fundamental rationale for the adaptation of consumer theory to fertility analysis . . . . The most important assumption is that the decision-maker behaves rationally.

As suggested above, preferences and tastes indicate, in the fertility context, the combination of goods and children that is preferred. This has predictive meaning, however, only when the possibility of 'rational' choice is present. For this reason, Easterlin (1975) has made the distinction between situations of 'social control' and 'individual control,' indicating that household decision theories of fertility would apply mainly in the latter situation, where fertility is not fully controlled by biological and social forces. (Similar distinctions have been made by others, for example, Bourgeois-Pichat, 1967; Fawcett, Albores and Arnold, 1972.) Easterlin also notes the need for sociological research in connection with economic theories, especially on the issue of 'taste formation'. However, I think it is fair to say that neither he nor the other economists have faced squarely the question of whether tastes can be more formally incorporated in microeconomic theories, with different types and patterns of tastes being directly measured and having specified interactions with the other elements in the theories. This has now become a salient question, in the light of recent research from a social-psychological perspective.

The *social-psychological* approach emphasises the needs of individuals that are fulfilled by having children, the alternative ways of meeting those needs, and the interactions among the psychological, social and economic values and costs of children. The social-psychological approach tends to rely on self-reports of motivations—that is, the values, costs and alternatives that are consciously perceived and articulated—which is both a strength and a limitation. The strength lies in the possibility of providing empirical support for certain common assumptions, such as that rural parents have a more utilitarian view of children, as well as in providing opportunities for entirely new analyses based on the patterns and strength of motivations. The limitation of this approach is primarily methodological. One must have serious reservations about the reliability, validity and completeness of self-report data on this topic, for reasons that are well-known and need not be elaborated here. These methodological problems set boundaries on what can be achieved and dictate caution in the interpretation of results, but they do not argue against the ultimate usefulness of the approach. In fact, many of the problems can be overcome in the long run through methodological controls, the use of multiple methodologies, and the replication of studies.

The social-psychological approach is, in a sense, the broadest of the four approaches discussed here, since it attempts to incorporate all dimensions of the value and cost of children that are relevant to childbearing decisions. However, these dimensions are usually assessed only from the perspective of the actor—perceptions of changes in the economic value of children across generations, expectations of the economic burden of a particular family size, and awareness of social pressures and constraints. It remains an open question whether this approach can produce knowledge that is of equal or greater value, scientifically and practically, than the other approaches mentioned.

The theoretical rationale for the social-psychological approach has been provided by Hoffman and Hoffman (1973). Other general perspectives can be found in Arnold et al. (1975), Berelson (1973), and Fawcett (1972). There is also a burgeoning empirical literature, although much of the work in this area is on-going and thus reported only in a preliminary fashion. The most ambitious effort is a cross-national study known as the Value of Children project. Comparative results from an initial round of exploratory studies in six countries are reported in Arnold et al. (1975); the results from a second round of studies based on national samples in eight countries are not yet available, although preliminary findings from selected countries are beginning to appear (for example, Hoffman, 1975, 1976; Fawcett, 1976).

The four approaches to the value and cost of children discussed above—sociodemographic, social-structural, microeconomic, social-psychological—have distinctive features, but they also have much in common. I think that one can realistically be hopeful in that each will gain from the other and ultimately these efforts will contribute new depths to our understanding of why people desire children and how they reconcile conflicting goals in the process of family formation.

Current Knowledge about the Value and Cost of Children

Although systematic research on the value and cost of children is a recent development, the quantity of research generated is impressive. In this section, I will review some of the important findings. I will draw heavily upon the social-psychological approach since that is the area of my own interest and experience, but I will also selectively cite findings based on the other approaches. First, I will examine the different ways of categorising the value and cost of children, then I will discuss the way those dimensions differ across groups and how they relate to other variables.

In order to study the value and cost of children, a scheme for classifying different types of values and costs is needed. Although something is lost in reducing the complexity of childbearing motivations to a limited set of conceptual categories, much is also gained. The discussion of the topic becomes clearer, guidelines are provided for the development of research instruments and designs, comparisons across studies become more feasible, and prospects are advanced for the formulation and testing of a comprehensive set of hypotheses. For all these reasons, a number of classification schemes have been developed in the past few years. I will cite here the scheme developed from our Value of Children study, as an example of the specific topics that are studied within this area of research.

POSITIVE GENERAL VALUES

(1) *Emotional benefits*: Happiness, love, companionship, fun; also viewed in reverse as relief from strain and avoidance of boredom or loneliness.

(2) *Economic benefits and security*: Benefits from children's help in the house, business, or farm, from care of siblings, and from sharing of income; old age security for parents, including economic support, physical care, and psychological security.

(3) *Self-enrichment and development*: Learning from the experience of childbearing; becoming more responsible and mature; incentives and

goals in life; being viewed as an adult, a grown woman or man; self-fulfilment; feeling of competence as a parent; being needed and useful.

(4) *Identification with children*: Pleasure from watching growth and development of children; pride in children's accomplishments; reflection of self in children.

(5) *Family cohesiveness and continuity*: Children as a bond between husband and wife; fulfilment of marriage; completeness of family life; continuity of family name and traditions; producing heirs; having future grandchildren

NEGATIVE GENERAL VALUES

(1) *Emotional costs*: General emotional strain; concern about discipline and moral behaviour of children; worry over health; noise and disorder in household; children as nuisance.

(2) *Economic costs*: Expenses of child rearing; educational costs.

(3) *Restrictions or opportunity costs*: Lack of flexibility and freedom; restrictions on social life, recreation, travel; lack of privacy; restrictions on career or occupational mobility; no time for personal needs and desires.

(4) *Physical demands*: Extra housework, caring for children; loss of sleep; general weariness.

(5) *Family costs*: Less time with spouse; disagreements over rearing of children; loss of spouse's affection.

LARGE-FAMILY VALUES

(1) *Sibling relationships*: Desire for another child to provide companionship for existing children; enriching the lives of children; avoiding an only child.

(2) *Sex preferences*: Specific desire for a son or daughter; desire for a certain combination of sexes among children.

(3) *Child survival*: Concern that existing children may die; need for more children to have enough survive to adulthood.

SMALL-FAMILY VALUES

(1) *Maternal health*: Concern that too many pregnancies or pregnancy when the mother is beyond a certain age, is bad for the mother's health.

(2) *Societal costs*: Concern about overpopulation, belief that another child would be a burden to society.

These dimensions cover many, but not all, aspects of the value and cost of children. Different schemes are useful for differing theoretical approaches, and the application of a particular conceptual scheme is likely to vary according to the availability of data or the methodology used in collecting data. At the present stage of research these schemes should be regarded as heuristic devices, to be modified and improved as new knowledge is accumulated. (For examples of other schemes, see Berelson, 1973; Blake, 1972; Hoffman and Hoffman, 1973; Kirchner, 1976; Townes, 1975.)

The usefulness of such conceptual schemes in generating new knowledge has been demonstrated in recent research. As a critical first step, it has been shown that the value and cost of children can be meaningfully measured using direct assessment techniques. That is, people are able to articulate their positive and negative orientations toward children in structured and unstructured interview situations, in response to various projective testing instruments, and through self-administered scales and questionnaires. By such means a substantial amount of data on subjective orientations toward children has been collected in recent years. Of particular interest are the findings on differentials in the value and cost of children across major social groups and the relationships with fertility and other relevant variables. It is these differentials and relationships that indicate the potential usefulness of research on the value and cost of children, since this kind of knowledge leads toward the development of causal models on family formation. A better understanding of causal processes will, in turn, lead to improved prospects for effective intervention policies to regulate fertility trends.

In this review I will be selective and will not discuss the methodological problems or other shortcomings of the data. The review is illustrative, intended mainly to show the direction in which this new research area is moving. This review will be discussed under the following headings: class and culture; sex of parent and child; and stage of family formation.

Class and Culture

Different patterns of the value and cost of children have been shown for persons belonging to different social classes. The most striking contrasts appear in comparisons of urban and rural groups. With respect to positive values, urban parents (especially the more educated) emphasise the psychological and emotional benefits that derive from interacting with children and by observing and guiding their growth and development. The feedback effects on the parents' own sense of competence and maturity are also important. Rural parents, by contrast, emphasise the economic and practical benefits to be derived from children, including the long-term benefit of old age security. On the negative side, urban parents are more concerned about the opportunity costs and restrictions on freedom, along with the economic costs, while rural parents tend to give maximum weight to economic factors and the physical burden of child care. This general pattern of class differentials has been shown across a number of countries. The contrast is greatest between the urban middle class and the rural respondents, with the urban lower class falling between these two groups on most dimensions. However, the urban lower class is distinctive in showing the highest level of concern about the economic costs of childbearing, reflecting the unfortunate combination of low incomes, high fertility, and the high cost of living in urban areas.

These attitudinal findings obtained from the Value of Children project, based on surveys conducted in Asia and the United States, are extended by the findings obtained from studies in other areas using different methodologies. In Africa, for instance, Caldwell (1967) has shown that the perceived economic value of children in rural areas is related to the prevalence of subsistence farming and to school attendance rates. In India, econometric techniques have been used by Rosenzweig and Evanson (1975) to show that fertility is directly related to factors affecting the economic returns to child labour and to child labour force participation, while being inversely related to school attendance rates. In Indonesia and Nepal, village studies by anthropologists have documented the actual economic contribution of children in rural households, for both household maintenance and directly productive work, leading to a tentative conclusion that households with more children are economically more efficient in certain types of peasant societies (Nag et al., 1976). In Taiwan, Mueller (1972) has shown that the ideal family size and the practice of family planning are meaningfully related to the perceived economic cost and utility of children, and she has interpreted these findings in relation to the effects of economic development on the perceptions of the value and cost of children.

These findings highlight a general pattern of relationships between social, economic and environmental factors and the motivations for childbearing. Expectations about the benefits to be derived from children and the costs entailed are shown to differ according to the background and circumstances of the parents. The micro level research thus confirms the inferences about motivations that have been derived from the macro level perspectives of demography, sociology and economics. It is hardly surprising that differences in objective conditions are paralleled by differences in subjective preferences, but the pattern emerging from the research findings enhances confidence in the validity and usefulness of the micro level approach. Possibilities, thus, emerge for going beyond broad class differentials to an analysis of the more specific factors influencing both group and individual behaviour. Examples of such factors will be discussed later, but first another broad classification variable—culture—deserves mention.

In the cross-national study of the value of children, the similarities between comparable social classes across countries were more impressive than the differences attributable to nationality or culture. However, two dimensions that seem to reflect cultural differences are worth noting—religion and son preference. In the Philippines, which was the only Catholic country studied, positive orientations toward children and large families were pervasively strong, tending to override class differences. In the Philippines and in Buddhist Thailand, disapproval of abortion was extreme. In Korea and Taiwan, where sons enjoy a special cultural significance partly because of religious beliefs and rituals, the preference for sons was strongest. From the evidence gathered, it is apparent that general value systems can also exert an important influence on the perceived value of children and in some circumstances can carry greater weight than the more immediate economic and social factors affecting the family. These general value systems can be partly understood through the measurement of the value orientations of individuals—and that should be an important element in research on the value and cost of children—but the more holistic perspective provided by the cultural or societal levels of analysis should also be taken into account.

Sex of the Parent and Child

One would assume that women and men should have different orientations toward children because of the importance of childbearing and child care factors in the traditional differentiation of sex roles. This expectation is to some extent borne out by the data, but we were surprised in our research at the relative paucity of sex differences. Within social classes, men and women seem to have a generally similar orientation toward children. Where differences did occur, however, they could in most cases be readily ascribed to the effects of socially defined sex roles.

Women, who spend more time with children and tend to have a narrower range of alternative social relationships, generally give greater emphasis to the emotional relationship between parent and child and to the value of companionship from children. Not surprisingly, women are also more concerned than men about the emotional and physical burdens of childbearing. On the economic side, women are more likely to stress expectations of economic help from children, perhaps reflecting their general economic dependence on others. On the other hand, men are more concerned than women about the economic costs of raising children, probably because they have the major responsibility for the economic welfare of the household.

Preferences for sons and daughters also are related to the definition of sex roles and to the sex of the parent. In general, sons are desired especially to serve long-term economic, social and religious functions—to help in the house, business or farm, to continue the family name, and to provide security to the parents in their old age. Daughters, on the other hand, are desired especially for the practical help they provide in the household while they are still children and remain with the family. There is also a tendency for fathers to want sons and mothers to want daughters for purposes of companionship. Sons and daughters are also desired for reasons related to the stereotypic ideas about the personality characteristics of boys and girls, and partly because of these ideas there is usually a desire to have children of both sexes.

These preferences for daughters and sons as revealed by the interview data in the Value of Children project, are affected by cultural and religious values, as discussed earlier, and are also strongest in rural, traditional societies where role differentiation is more pronounced. Anthropological studies have documented the activities carried out by boys and girls of different ages in the rural areas (Nag, Peet and White, 1976). Studies undertaken in Javanese and Nepalese villages have shown that girls aged nine years and above spend a greater proportion of their time working than do boys, and between fifteen and nineteen years of age their total work contribution nearly equals that of an adult male. The relatively greater contribution of girls as compared to boys, is due mainly to the fact that girls spend a greater amount of time in household maintenance work (such as food preparation and child care); they also spend somewhat more time on directly productive work than boys, partly because boys are more likely to be in school and are less available for work. Although relatively little systematic research of this kind has been conducted, the available evidence and informal observations lend support to the connection between sex role differentiation in the household and sex preference as a motivational force in family formation. An important question for the future is the extent to which equalisation of sex roles will reduce the strength of this motivation, especially if beliefs about sex-related personality differences persist.

I will now turn to an important issue for those concerned about population—number of children—which I will discuss in relation to the stages of family formation.

Family Formation

Although it has often been noted that the motivational context for child-bearing changes with the birth of each successive child (for example, Hass, 1974; Namboodiri, 1972), there have been relatively few efforts to assess motivational change directly. Research on the value and cost of children provides a framework for carrying out such analyses. Though the real need is for longitudinal data, but even cross-sectional studies can provide some important insights. I recently undertook an analysis of motivations for the first and second child (Fawcett, 1976), using the limited data available.

A general consideration that seems significant to me is the heavily symbolic meaning of the first child. The first birth signifies for the parents such powerful factors as adulthood, parenthood, virility or femininity, fulfilment of the marriage, and the establishment of a family within a community of families. The anticipation of this change in status is revealed in the findings of a recent study of the value of children in the US (Hoffman, 1976). When childless women were asked 'How is a woman's life changed by having children?', the most common responses pertained to adult status, becoming a woman, assuming responsibility, and fulfilment in life. These motivations, which are largely a reflection at the individual level of socially prescribed roles, exert a strong pull toward the critically important first birth, which sets in motion other influences toward increasing the number of children. The social motivations are reinforced by more personal ones, not the least of which is a kind of curiosity motive—to experience childbirth and parenting, to create a new life and reproduce the self, to witness the growth and development of one's own child. At the same time, negative motivations are relatively weak at this stage. The most important appears to be loss of freedom or opportunity costs, but for most people this affects the timing rather than the occurrence of the first birth.

The effects of the first birth on motivations for a second child are of particular interest. There is some evidence that the peak of childbearing motivations occurs between the first and second birth (Campbell *et al.*, 1976). The major elements of motivation for a second child appear to be:

(*a*) the perceived need for a sibling for the first child, based heavily on negative stereotypes about the only child;

(*b*) the increased salience of sex preference, usually in favour of having a child whose sex is the opposite of the first child and particularly relevant for the parent of the opposite sex;
(*c*) a heightening of the importance of emotional benefits, apparently based on the feelings experienced in connection with the first child; and
(*d*) a *decrease* in the salience of opportunity costs, because the major burden of this cost has already been accepted with the first birth.

The first birth thus activates several strong motivations for a second child and decreases at least one important constraint, thereby impelling people toward an increase in family size. In addition, the change in social and familial roles that results from the first birth facilitates further childbearing. The wife is less likely to be working after the birth of the first child and, according to Hoffman's (1976) research, women with one child have higher scores on a scale of traditional sex-role ideology than do women with no children, they have less decision power in the family, they receive less help from their husbands in housework, and they have more contact with relatives. With these powerful forces at work, it is hardly surprising that there are only a few one child families in societies where economic or other constraints are not highly salient.

The process of family formation after the second child is, at once, less clear psychologically and more important demographically. From a psychological standpoint, most needs would appear to have been met by having two children. Yet many people go on to have three or more children, and it is precisely these children that represent the growth in population. It is here that the challenge lies in relating the value and cost of children to the variety of situational and personal factors that have a potential influence on the decision to stop having more children. Some progress has been made in this direction.

It seems clear, for instance, that those who expect long-term economic benefits from children are not usually deterred by short-term cost considerations in the choice between two or three children. Rather, the expectation of long-term gains tends to push family size towards the limits of current economic capacity and even beyond, to the point where the standard of living of the family declines substantially with each additional birth. In many societies, the association of sons and not daughters with economic security plays an important part in this desire for large families. For those who have more than three or four children, the factors that eventually impel a decision to stop having children appear to be heavily economic, but that is probably too simple an explanation for a decision that is multiply determined.

Parents who choose to have a truly small family—one or two children—generally do not expect economic benefits from their children, are more oriented toward a psychological appreciation of childbearing, and appear to be influenced heavily by quality and convenience considerations. Both quality and conveniences stand out as powerful mediating variables—quality in the sense of wanting to provide better care and greater opportunities for children, convenience in the sense that a small family is easier to manage and interferes less with the parents' non-familial activities. Both quality and convenience have economic components—the costs of educating children, the possibility of the wife's income forgone—but they are cast more in a framework of choices among alternatives that each offer psychological rewards, rather than choices among different levels of sacrifice at a low standard of living.

These appear to be some of the major considerations which influence decisions regarding the number of children, but, of course, the process is extremely complex and the elements differ across individuals and groups. To disentangle the variety of positive and negative motivations at work in this process will require more careful research than has been done so far, with particular attention to the issue of defining psychologically meaningful subgroups that would be expected to value children differently and have different levels of access to alternatives. Here, too, progress is being made, in looking at factors such as the level of aspirations for children, expected or permanent income, the ages of children in relation to the parents' ages, the mother's work history and expectations, the characteristics of the community of residence, and so on. Many of the tools for an improved understanding are now at hand, but the appropriate use of those tools in research design and analysis is only beginning.

At this point a cautionary word is in order. The generalisations made above on the basis of the research findings on the value and cost of children are merely generalisations: they omit many qualifications and details, they are drawn mainly from one research project, and they are cited here only to illustrate the nature of this research. A variety of books and articles may be consulted for further detail and for other disciplinary perspectives (Arnold et al., 1975; Ben-Porath 1974; Bulatao, 1975; Easterlin, 1975; Fawcett, 1972; Mueller, 1972; Schultz, 1973, 1974; Turchi, 1975).

Needs for Additional Knowledge

Future research on the value and cost of children should not only build upon and expand current efforts; it should reach out in entirely new directions, integrate the various disciplinary approaches, and include evaluative research to test the usefulness of work in this area. I will discuss future needs within the framework of cultural extensions, analytic approaches, and applications.

Cultural Extensions

Although much of the research on the value and cost of children has been conducted in a cross-cultural context, it is still culturally limited. The most glaring omission is research in socialist societies, where one would expect quite different findings stemming from factors such as the relative equality of sex roles, the provision of old age security and other social benefits, an ideology of individual actions based on the concept of common welfare, and the assumption by the state of many childbearing costs. Studies of societies with a strong socialist ideology, such as China, would represent extreme cases in assessing the effects of social-structural variables as against psychological need variables. Comparisons of countries with different social-political systems would yield important new insights into the interaction between society and the individual, with respect to the value and cost of children.

There is also a need for studies in a wider range of cultures and across a broader spectrum of development. Major research projects that include measures of the value and cost of children have been carried out in Belgium, the Netherlands, Germany, the United States, Turkey, and in nine countries of East and Southeast Asia. More limited projects have been conducted in Chile and Nigeria. To my knowledge, systematic research on the value and cost of children has not yet been done in any Arab country (although a study is planned for Tunisia) and relatively little information is available from Latin America, Africa, and the South Asian subcontinent. To obtain a clearer picture of the effects of cultural influences and development patterns, research on the value and cost of children should be extended to these areas.

Analytic Approaches

Most of the research on the value and cost of children has been based on only two methodological approaches—the collection of interview data through surveys, or the analysis of existing survey and census data using econometric techniques. In addition, a few important village-level studies have been conducted or are under way; in these studies observation, surveys, and other methods of data collection are combined. Two such studies have been completed in Indonesia (Hull, 1975; White, 1976), at least two or more are in progress in Bangladesh, and several others which deal partly with the value and cost of children are being carried out in Asia and Africa under the auspices of the Institute of Development Studies, Sussex, and the Department of Demography, Australian National University.

In my view, there is a pressing need for additional community-level studies, in which efforts are made to relate the subjective evaluations of children, data on the actual costs and benefits of children, and the various aspects of the local social structure and economy. It is in these studies that data collection and analysis can be most closely matched to theory, and through such studies a realistic picture of the interplay of different types of influence can emerge. Of course, village-level studies also have their limitations, but they are now needed to complement the large-scale surveys that are under way.

Another important need is the design of original studies to collect data tailored to the concepts and hypotheses of microeconomic theory. This call has been made often, but I am aware of only a few serious efforts in response. Turchi (1975: 122) put the case succinctly:

> . . . currently existing data sets are incapable of providing the basis for adequate assessments of the explanatory and predictive power of the economic models. The implications are clear: economists and others interested in the actual application of the microeconomic theories to fertility will have to generate their own data. Until appropriate data become available, the microeconomic models of fertility are likely to be noted more for their potential than for their payoff.

I must also repeat here the comment made earlier, that longitudinal studies are essential to achieve an understanding of the life-cycle changes in childbearing motivations. This again is a call often made, and here, too, some slight progress is evident. I am aware of several studies in the US and one in Malaysia that deal with motivations in a longitudinal context. But this is only a beginning, compared to the need.

Beyond the methodological issues, there is great scope for conceptual and theoretical advancement. I will list just a few of the issues that need intensive attention.

The quality of children. This is a central concept in theory and in policy, but it is used and measured in different ways. We need to define more clearly what is meant by 'child quality' and how it relates to other concepts.

Interaction of economic costs and benefits. For example, how do expectations of long-term gains affect the perceptions of short-term costs? Do opportunity costs have effects similar to those of direct economic costs?

General value orientations and the value and cost of children. The measurement of subjective expected utilities is appropriate in this area, but probably works only for sophisticated respondents. Other measures are needed to assess how children and alternatives satisfy important general values.

Substitutes for children. Knowledge about the activities, services or goods that compete with children is limited, yet this is a major issue for population policy. This overlaps with the question of general values, but can also be treated as a separate topic.

The transition point. Ultimately, research should provide a better understanding of the point at which costs are perceived to outweigh benefits, thus moving the individual toward a decision to limit childbearing. Knowledge about the factors leading to the transition point will suggest policy levers to shift the point earlier in the family building cycle.

APPLICATIONS

Knowledge about the value and cost of children to parents is useful for its own sake, representing as it does a fundamental aspect of human social behaviour that is little understood. But such knowledge is also useful for more immediate practical purposes in connection with the concern about population growth.

Knowledge about motivations for childbearing can be useful in at least three ways: to improve the effectiveness of population education, information and persuasion campaigns; to suggest various kinds of incentives and disincentives that may influence motivations directly; and to understand the effects of development programmes that may influence motivations indirectly. Knowledge about the value and cost of children has a more general policy usefulness, which I will discuss later.

In the area of education-information-persuasion, knowledge about the perceived value and cost of children can be regarded as audience research, suggesting guidelines for the content of messages that will be salient and effective for particular groups. For example, research on the value and cost of children has been used to develop a training manual for social workers engaged in family planning counselling; research findings have been incorporated in a package of materials for training educators in the field of population education; and recommendations have been made in research reports for the development of different persuasive appeals directed toward groups with differing motivational profiles (see, for example, Arnold et al., 1975). In a very straightforward way, this research can be used to bring about greater congruence between the content of communications and the motivations of the target audience.

Likewise, knowledge about the value and cost of children can be used to design or evaluate incentives and disincentives that may affect the perceived balance of the values and costs of children. For example, the Taiwan experiment with educational bonds that are provided as an incentive to limit family size might be expected to be effective in any country where perceptions of child quality are linked to education, where people have high aspirations for education of their children, and where the costs of education are shown to be salient in family size decisions. Research can also indicate in a general way the extent to which financial disincentives (such as tax provisions) are likely to be effective, by an analysis of the perceived financial burden of children and the perceived trade-offs between short-term costs and long-term benefits (for example, Mueller, 1972).

Perhaps, most important is the understanding of the effects of development programmes on fertility motivations. This can be achieved to some extent by a careful analysis of data obtained from on-going studies around the world. As these studies proceed, insights will be gained into the motivational effects of different opportunity structures and employment patterns, agricultural activities and production practices, educational and literacy levels, and so on. As Repetto (1976) has pointed out, there are also opportunities for using natural experiments to test the impact of development policies on the value and cost of children. For example, land reform projects could be studied to assess their effects on the 'pension motive', that is, the extent to which land-ownership is a substitute for expectations of old age security from children. In a similar vein, model education projects could be studied to assess their effects on quantity versus quality trade-offs in fertility motivations. In a variety of ways, the interactions between development and the value and cost of children could be analysed in a concrete and practical way through imaginative field studies.

These few examples indicate the potential usefulness of research on the value and cost of children. Yet, it must be noted, many of the programme and policy actions that are indicated may also be adopted on a common sense basis, without the detailed empirical knowledge that is emerging from research. Perhaps with this in mind, Berelson (1976) has stated that the policy relevance of research on the value of children has yet to be demonstrated. Is this area of research really worthwhile, in a strictly pragmatic sense?

My answer would be in the affirmative, but not because I believe that research on the value and cost of children will produce policy prescriptions or programme innovations that would not otherwise emerge. Rather, I view the importance of this research in the light of its contribution to the general dialogue about population problems and population policies. By its focus on the individual and on the issues of motivation and choice, this research has helped to turn the policy dialogue toward the question of how social and economic policies may guide or constrain individual choice. By substituting the knowledge about motivations for speculation, it has clarified and made more concrete a central issue in the continuing debate about population policy. By demonstrating the multiple ways in which children provide important life satisfactions, and by highlighting the absence of alternative sources of satisfaction for a large majority of the world's people, it has given emphasis and meaning to the human welfare dimension of population policy. The policy dialogue is beginning to shift away from supply toward demand, away from the provision of family planning services and toward an appreciation of childbearing motivations and other motivations, in the context of improved family and child welfare. Research on the value and cost of children is a significant element in this movement toward people-centred population policies, and this, I believe, is its important and lasting contribution.

REFERENCES

ARNOLD, FRED, et al. (1975). *The value of children: a cross-national study*. Vol. 1, Introduction and Comparative Analysis. Honolulu: East-West Population Institute.

BANKS, J.A. (1954). *Prosperity and parenthood: a study of family planning among the Victorian middle classes*. London: Routledge & Kegan Paul.

BEN-PORATH, YORAM. (1974). Notes on the micro-economics of fertility. *International Social Science Journal*, 26 (2): 302–14.

BERELSON, BERNARD. (1973). The value of children: a taxonomical essay. *In The Population Council Annual Report 1972*. New York: Population Council. 17–27.

————. (1976). Social science research on population: A review. *Population and Development Review*, 2 (2): 219–66.

BLAKE, JUDITH. (1971). Reproductive motivation and population policy. *BioScience*, 21 (5): 215–20.

————. (1972). Fertility control and the problem of voluntarism. University of California International Population and Urban Research and Institute of International Studies Reprint No. 442, Berkeley. Reprinted from *Scientists and World Affairs*, Proceedings of the Twenty-second Pugwash Conference on Science and World Affairs. 7–12 September 1972, London, n.d.: 279–88.

BOURGEOIS-PICHAT, JEAN. (1967). Social and biological determinants of human fertility in non-industrial societies. *Proceedings of the American Philosophical Society*, 3(3): 160–63. *In* Richard Easterlin, An economic framework for fertility analysis. *Studies in Family Planning*, 6 (3): 54–63.

BULATAO, RODOLFO A. (1975). *The value of children: a cross-national study*. Vol. 2. Philippines, Honolulu: East-West Population Institute.

CALDWELL, JOHN. (1967). Fertility attitudes in three economically contrasting rural regions of Ghana. *Economic Development and Cultural Change*, 15 (2) Part I: 217–38.

CAMPBELL, F.L., et al. (1976). Shifts in values associated with fertility decisions following childbearing. Paper presented at the Workshop on Psychosocial Aspects of Fertility and Fertility Regulation. Montreal: 27–29 April 1976.

DAVIS, KINGSLEY. (1967). Population policy: will current programs succeed? *Science*, 158: 730–39.

DUESENBERRY, JAMES. (1960). Comment on Gary S. Becker's 'An economic analysis of fertility'. *In Demographic and economic change in developed countries*. Universities-National Bureau of Economic Research, Conference Series 11. Princeton, N.J: Princeton University Press.

EASTERLIN, RICHARD A. (1975). An economic framework for fertility analysis. *Studies in Family Planning*, 6 (3): 54–63.

FAWCETT, JAMES T. (Ed.). (1972). *The satisfactions and costs of children: theories, concepts, methods*. Honolulu: East-West Population Institute.

————. (1976). The value and cost of the first child. Paper presented at the Conference on the First Child. Asilomar, California: 23–26 March 1976.

FAWCETT, JAMES T., et al. (1972). The value of children among ethnic groups in Hawaii: exploratory measurements. *In* James T. Fawcett (Ed.), *The satisfactions and costs of children: theories, concepts, methods*. Honolulu: East-West Population Institute. 234–59.

HASS, PAULA H. (1974). Wanted and unwanted pregnancies: a fertility decision-making model. *Journal of Social Issues*, 30 (4): 125–65.

HOFFMAN, LOIS W. (1975). The value of children to parents and the decrease in family size. *Proceedings of the American Philosophical Society*, 119 (6): 430–38.

————. (1976). Effects of the first child on the woman's role. Paper presented at the Conference on the First Child. Asilomar, California: 23–26 March 1976.

HOFFMAN, LOIS W. and MARTIN L. HOFFMAN. (1973). The value of children to parents. *In* James T. Fawcett (Ed.), *Psychological perspectives on population*. New York: Basic Books. 19–76.

HULL, TERENCE H. (1975). Each child brings its own fortune: an inquiry into the value of children in a Javanese village. Unpublished doctoral dissertation. Canberra: Australian National University.

KIRCHNER, E.P., et al. (1976). Anticipated parenthood satisfaction and dissatisfactions associated with children at four development stages. Paper presented at the Workshop on Psychosocial Aspects of Fertility and Fertility Regulation. Montreal: April 1976.

MUELLER, EVA. (1972). Economic motives for family limitation: a study conducted in Taiwan. *Population Studies*, 26 (3): 383–403.

NAG, MONI, et al. (1976). Economic value of children among Javanese and Nepalese peasants: an anthropological inquiry. Unpublished paper, Columbia University.

NAMBOODIRI, N. KRISHNAN. (1972). Some observations on the economic framework for fertility analysis. *Population Studies*, 26 (2): 185–206.

REPETTO, ROBERT. (1976). A survey of policy prospects and research approaches in the relationship between fertility and the direct economic costs and value of children in the less developed countries. *In* Ronald Ridker (Ed.), *Population and development: the search for selective interventions*. Baltimore, Maryland: Johns Hopkins Press for Resources for the Future, (in press).

ROSENZWEIG, MARK R. and ROBERT EVANSON. (1975). Fertility, schooling and the economic contribution of children in rural India. Paper presented at the Annual Meeting of the Population Association of America. Montreal: October 1975.

SCHULTZ, THEODORE W. (Ed.). (1973). New economic approaches to fertility. *Journal of Political Economy*, March/April, 81 (2): Part II.

————. (Ed.). (1974). Marriage, family human capital and fertility. *Journal of Political Economy*, March/April, 82 (2): Part II.

TOWNES, BRENDA D. (1975). Measurement of values associated with birth planning decisions. Paper presented at the Annual Meeting of the American Meeting of the American Psychological Association. Chicago: September 1975.

Turchi, Boone A. (1975). Microeconomic theories of fertility: a critique. *Social Forces*, 54 (1): 107–25.

White, Benjamin. (1976). Production and reproduction in a Javanese village. Doctoral dissertation, Columbia University.

II

Culture and Fertility

6

Cultural Emphasis in the Mysore Population Study

C. CHANDRASEKARAN

Introduction

The Mysore Population Study was jointly undertaken by the Government of India and the United Nations nearly three decades ago.[1] The study was essentially an experiment in the use of a sample survey of households to measure the trends and characteristics of the population and to investigate their interrelations with the process of economic and social change in an area undergoing economic development.

The Mysore Study was unique in its concept, scope and operational procedures and continues to be the most comprehensive field survey ever undertaken to study the interrelations of population and economic and social factors. The need for developing similar studies in other parts of India as well as in other countries was emphasised by the United Nations Population Commission when it recommended that the field study in the Mysore State should be regarded as an initial step in a programme of studies on the interrelationships of demographic, economic and social factors.[2]

Over the years, the Mysore Population Study has been read widely and cited quite often. Several of the procedures developed in the study have been applied in other investigations. The Indian National Sample Survey made use of some of the techniques developed in the Mysore Population Study to improve the completeness with which data on births and deaths were obtained. The procedures used in the Mysore Study for the recording of data on fertility and family planning practices helped in the development of the IUSSP 'model questionnaires' and the Core Questionnaire of the World Fertility Survey.[3]

Yet, it is felt, that some of the cultural issues involved in designing and developing the Mysore Study and the substantive and technical questions they

[1] *The Mysore population study: a cooperative project of the United Nations and the Government of India*. Population Studies No. 34 (New York, United Nations, 1961).

[2] *Reports of the population commission*, Documents E/1313 para 30 and E/1711 para 17 (New York, United Nations).

[3] The World Fertility Survey is an international research programme undertaken by the International Statistical Institute in collaboration with the United Nations and in cooperation with the International Union for the Scientific Study of Population. Further information about the publication can be obtained from Publications Office, International Statistical Institute, 428 Prinses Beatrixlann, P.O. Box 950, 2270 AX Voorburg, Netherlands.

raised do warrant an emphasis in an effort to assist in the proper designing of 'interrelationships' studies in the field of population and it is with this objective that the following discussion is presented.

Choice of Population 'Contrast' Groups

A fundamental issue in the study of 'interrelationships' is the extent to which social and economic development affects the dynamics of population growth, especially its fertility and mortality profiles. The importance of this issue is accentuated by the fact that in a number of developing countries basic data on births and deaths are not routinely obtained, and even if obtained, are not accurate enough to be of much avail in understanding such relationships. The assessment of fertility and mortality levels and their variations with economic and social factors affected by the overall development, therefore, are of primary importance.

The findings gleaned from studies of fertility and mortality differentials associated with individual social and economic factors are frequently found in demographic literature and often get top priority in discussions on 'interrelationships'. The considerations underlying the population groups selected for study and the nature of development whose influences are investigated are given secondary importance, if these are at all taken into account. Witness for instance, the suggestion made by Professor P.C. Mahalanobis that the funds available for the Mysore Population Study could be merged with those of the Indian National Sample Survey spread all over India and the data obtained used for the study of 'interrelationships'.[4] The practice of using the data obtained from a national sample to study differentials, as was also the case with the World Fertility Survey, is of relevance in this context.

Cultural factors impinge on fertility and mortality levels—more so in the case of fertility—and a 'culturally homogeneous' population provides the proper base for studying the effect of contrasting developments on demographic characteristics. The concept of 'culturally homogeneous' population would, perhaps, be considered as a mere theoretical abstraction by some, in that differences in cultural practices can be found even among population groups living close together, such as, in a village or in even smaller areas such as a 'hamlet'. Nevertheless, it is also recognised that marked cultural differences exist between population groups living far apart. In India, for example, the existence of marked differences in the average age at marriage of women living in different parts of India has been well documented. Under Indian conditions, 'caste' introduces some cultural homogeneity, especially if it is considered along with language. Geographical proximity along with the use of the same language can be expected to bring in a certain amount of cultural homogeneity. Such considerations led to the decision to locate the UN-Government of India Interrelationships Study within a limited geographical and linguistic area.

Another important reason for selecting the erstwhile Mysore State as the

[4] This suggestion was made by Professor P.C. Mahalanobis to me.

area for this study was the belief that this princely state had made tremendous progress in its health programmes and in its efforts at agricultural and industrial development. This impression was, in fact, justified by the findings of the 1951 Population Census of India which revealed that during the decade 1941 to 1951, the average population of the Mysore State increased by 23.7 per cent as compared to 13.3 per cent for the whole of India during the same period.

The nature of development in the Mysore State also lent itself to stratification by geographical areas of the rural part of the Mysore State. For purposes of the Mysore Population Study, six rural zones were identified: Zone I—Rural hill area—villages with anti-malarial operations; Zone II—Rural hill area—Villages without large-scale malarial operations; Zone III—Rural plains (tank irrigated areas); Zone IV—Rural plains—canal irrigated areas with large-scale malarial operations; Zone V—Rural plains—canal irrigated areas without large-scale anti-malarial operations; and Zone VI—Rural plains—dry-farming areas.

Within each zone, administrative units were stratified according to the size of the population. All communities with less than 5,000 population were classified as villages. Those with populations of 50,000 and more were classified as 'cities'. Communities having a population between 5,000 to 50,000 were classified as 'towns'. Towns were further subdivided into three groups according to size namely, 5,000 to below 10,000; 10,000 to below 25,000; and 25,000 to below 50,000.

Due to the limitation of funds available, the Survey was confined to villages selected from Zones I, II and III, fifteen 'towns' of intermediate size (10,000 to 25,000) located in these zones, and Bangalore City, the largest city in the Mysore State.

The principle of 'cultural homogeneity' used in selecting the Mysore State for the study was also extended to grouping the municipal divisions of Bangalore City into different strata primarily on the basis of religion and caste composition—with over 35 per cent Muslims, with over 35 per cent Christians, with over 35 per cent Scheduled Castes, with predominantly other Hindus and with a male literacy rate of over 60 per cent, and with other Hindus and with a male literacy rate of below 60 per cent. In the traditional Indian setting, religious affiliation influences behaviour patterns especially in respect of the determinants of fertility, and it was considered important to use religion as a factor for stratification. Chief among these influences is the religious sanction to have several wives among the Muslims as opposed to strict monogamy among the Christians. Although till recently there was no bar on Hindus to have more than one wife, such a practice was not looked upon with favour by the community. The social dichotomy that exists between the Scheduled Castes and other Hindus together with its attendent effects stressed the importance of effecting further stratification of Hindus using these characteristics.

The three rural zones, the towns and the five strata into which Bangalore City was divided provided various contrasts of development within a relatively homogeneous cultural set-up to study 'interrelationships'. This procedure of using areas of contrasting development as units for studying the interrelationships of population and development is at variance with the procedure often used—of classifying different areas on the basis of a number of indices of development and using a regression analysis to study the relationships. Using areas as units as was done in the Mysore Population Study permitted, at least conceptually, the taking into account of the totality of effect generated by developmental efforts. The broad nature of development associated with an area was often well recognised and was the reason for its inclusion in the 'interrelationships' study; however, there are possibilities of obtaining additional information about the development of an area through fresh efforts. The characteristics of development in the rural zones were obtained through a comprehensive village schedule which elicited information about each village selected for the survey. In addition, the household schedule which elicited information about the characteristics of each household surveyed as well as of the members of the household, also provided information relating to the social and economic characteristics of the various zones and strata surveyed.

In addition to contrasting the demographic characteristics of the three rural zones, of the rural areas with those of the 'towns' and Bangalore City, and of the various strata within Bangalore City, the Mysore Study could also work out the differentials in mortality and fertility rates as related to the social and economic characteristics within each zone or stratum. The characteristics used, for example, education and literacy status, mother tongue, religion and caste were provided by the household schedule.

Influences of Culturally-oriented Factors on Fertility

It is well recognised that fertility is culturally oriented and that a deep study of fertility cannot afford to overlook related cultural factors. The Mysore Study provides numerous examples highlighting the importance of including culturally oriented factors. The following discussion relates to two such sets of factors—cultural factors affecting the length of married life; and cultural factors affecting the interval between births.

Cultural Factors Affecting the Length of Married Life

Age at marriage The age at marriage assumes a special interest in India where marriage is almost universal and where it has been customary for girls to be married at a very early age. In the Mysore State, as in India generally, different religious communities and the various Hindu castes have different traditions in this matter. Contrary to popular belief, anthropologists have shown that some of the castes in the Mysore State had a tradition of late marriage for women but socially important groups had followed the custom of child marriage.[5]

[5] A detailed discussion of this subject can be found in L.K. Ananthakrishna Iyer and H.V. Nanjundayya, *The Mysore tribes and castes*, Vols. 1–4 (Mysore, Mysore University, 1928–35). See also M.N. Srinivas, *Marriage and family in Mysore* (Bombay, New Book, Co., 1942); E. Thurston, *Castes and tribes of Southern India* (Madras, Government Press, 1909).

The progressive forces in India have long attempted to put an end to the practice of child marriage. In the Mysore State, the Infant Marriage Prevention Regulation was passed in 1894, prohibiting the marriage of girls below eight years of age and of boys below fourteen years. After the Sarda Act was passed in British India in 1930, an effort was made in the Mysore State to enact legislation fixing the minimum age of marriage for a girl at fourteen years and for a boy at eighteen years. This effort failed, but in 1955 (after the Mysore Population Study had been undertaken) a revision of the Hindu marriage law, which had validity throughout India, established the minimum age of marriage at fifteen years for girls and eighteen years for boys. Recently, the minimum age at which girls can marry has been legally fixed as eighteen years; for boys it is twenty-one years. Under the pressure of legal measures and the slowly changing force of public sentiment, the proportion of marriages contracted at very tender ages has been decreasing in the Mysore State, as in the whole of India. The data obtained in the Mysore Study has made it possible to trace this trend over a period of half a century among various categories of the population, and to analyse its relationships with other factors in the changing economic and social situation. Information regarding opinions as to the age at which a woman should marry was also obtained in order to permit a comparison of the patterns of thought and practice in this regard.

The median age at marriage of the birth cohorts of younger women as compared with those of older women showed that the age at marriage of women had been gradually increasing over a long period of time in all the zones. The increase has been more rapid in the urban than in the rural zones. In Bangalore City, the median age at marriage for women born during the fifteen years period prior to 1908, that is, between 1893–1907 was 14.3 years or about one-and-a-half years younger than the median for the cohort of women born during 1928–1932.[6] Approximately the same increase is indicated by the medians for the towns, but the increase was only about one half as large in the three rural zones; that is, an increase of less than one year. The median age at marriage of women in Zones I and II was higher than that in Zone III and the difference was as great among the cohorts born during 1893–1902 as among those born during 1928–1932. It appeared therefore, that the higher median age for women in the rural hills as compared with those in the rural plains was of a long-standing nature.

The five strata of Bangalore City showed differences in the median age at marriage for the most recent birth cohorts of women, pointing to the practice of late marriage among Christians, Muslims and the more literate and well-to-do groups than in the rest of the urban population. Stratum 2 (over 35 per cent Christians) had uniformly the highest median value, ranging from about 15.6 to 17 years for women of the various cohorts. Stratum 1 (over 35 per cent Muslims) also had median values above the city average for all cohorts.

[6] The earliest birth cohort of women with sizeable numbers were those aged between 50 to 59 years at the time of the Survey (undertaken for the most part in 1952) that is, those born during the period 1893 to 1902. Those born after 1902 were divided into five-year age groups according to the estimated year of birth.

Stratum 4 (other Hindus, male literacy above 60 per cent) also had a relatively high median age at marriage for the youngest birth cohort of women but not for the older cohorts. It is this stratum which showed the largest increase in the median age over the period considered and it is noteworthy that the greatest part of the increase was found in the difference between the median ages at marriage for cohorts born in 1923–1927 and 1928–1932—an increase of 1.3 years. In Strata 3 and 5 (over 35 per cent Scheduled Caste, and other Hindus, male literacy below 60 per cent respectively), the median age at marriage for the most recent birth cohort was relatively low, though the figures indicate a gradual increase in the course of time for these strata as well. It should be noted that Stratum 3 had the largest percentage of illiterates, the poorest housing, and the largest percentage of persons in the lowest economic status categories among the various strata of Bangalore City.

The Mysore Study also highlighted the fact that in addition to the differences in the age at marriage of women in the different religious groups—Muslims have a higher median age at marriage than the Hindus and the Christians marry later than the Muslims—the several caste groups among the Hindus also showed some differences. The *Adi-Karnatakas* and the *Adi-Dravidas* recorded higher median values than the *Brahmins*.[7] During the course of the twentieth century, however, the pressures of tradition have relaxed and there is less variation in the median ages at marriage of women in the different groups, although practically all of them now marry later than they formerly did.

In the case of males, the data obtained from the survey indicates that the age at marriage in the Mysore State tended to be slightly higher than in India as a whole, though lower than what is typical in most of the industrialised Western countries. It was exceptional for males in the Mysore State to marry before they were eighteen years of age; but more significantly the median age at marriage of males, even in the rural zones was around twenty-four years for all cohorts born between 1893–1902 and 1923–1927. The analysis also highlighted the relative consistency of the median age at marriage of males during the last half century, and the absence of any important differences in the ages at which men married in the different areas and categories of the population. The relatively high age at marriage of males as compared with the age at marriage of females is culturally interesting but its significance is also of demographic importance.

Ban on widow remarriage The ban on widow remarriage imposed by Hindu tradition is fairly well-known, but the extent to which such a ban affects the fertility of Indian women was not well understood. The Mysore Study not only yielded some quantitative data in this regard but it also brought out clearly that as development tends to reduce the mortality rates, there is a built-in structural effect to increase the fertility of Indian women.

This finding resulted partly because data on the marital status (and on the

[7] The *Adi-Karnatakas* and the *Adi-Dravidas* belong to the Hindu 'scheduled' castes. While the *Adi-Karnatakas* are considered indigenous, the *Adi-Dravidas* are believed to have migrated from the adjoining areas. Traditionally, persons from both the castes were engaged in menial work.

number of times married) of women was obtained through the household schedule and partly because of the judicious innovations at the analytical stage. Not only was fertility as determined by the number of children born during the reproductive life of ever-married women taken up for study, but ever-married women were also divided into four groups to bring out the effect of widowhood and multiple marriages on fertility: (*a*) currently married plus widowed or separated after reaching the age of forty-five (once married only); (*b*) currently married but married more than once; (*c*) widowed or separated (once married only) before reaching the age of forty-five; and (*d*) widowed or separated, married more than once.

The average number of children ever born to women above the age of forty-five was found to vary from 4.7 to 5.6 in the various zones and strata. The weighted average for Bangalore City was 5.3 children, for the towns 5.6, and for the rural zones 4.8 children. This finding—urban fertility being higher than rural fertility—was against accepted concepts when the Mysore Population Study was undertaken and would have had the unfortunate effect of casting doubts on the validity of the basic data obtained during the survey, had it not been for the supporting evidence provided by the innovative analysis referred to earlier.

The fertility of Group (*a*) women who were over forty-five years at the time of the survey was approximately six children in Bangalore City and about the same in the three rural zones together. The fertility level of such women was slightly higher in the towns. As such there was hardly any difference in the fertility level of urban and rural ever-married women, if the marriage remained unbroken till they were forty-five years of age, which is generally assumed to be the end of a woman's reproductive life. The relative frequency of women in Groups (*a*) and (*c*) in Bangalore City and the rural zones provided clear evidence that widowhood was more frequent and numerous in the rural zones as compared with Bangalore City, and was the chief cause for the higher fertility of all ever-married women in the urban areas as noted earlier.[8]

It could be observed from this discussion that the Mysore Population Study inquired into the incidence of multiple marriages, though not on widow remarriage. The data obtained did indicate that not only among the Muslim and Christian women but also among the scheduled and backward classes, second or higher order marriages were sizeable in number. With modernisation, this factor could become important even among the non-backward castes of the Hindus.

Cultural Factors Affecting the Interval Between Births

The Fertility Schedule used in the Mysore Population Study was confined to interviewing women in Bangalore City and Zone III (rural plains, tank irrigated areas) and focused on the various cultural factors which could affect the

[8] C. Chandrasekaran, 'Fertility trends in India.' Proceedings of the World Population Conference (Rome, 1954) 1: 827–40.

interval between births. One of these factors is the practice of women going to their parents' home for delivery and continuing to stay there till the child grows old. An explicit question was asked if the woman interviewed had left her husband's home for delivery and if so, the duration of separation from the husband's house was ascertained. The findings showed that the practice was more common among the Hindus as compared with the Muslims, but even among the Hindus it was fairly frequent only for the first delivery. In Bangalore City 54 per cent of the women and in the rural plains 61 per cent of them had gone to their parents' home for the first delivery. For the second and subsequent deliveries, the percentage was lower, averaging about 12 per cent. The Fertility Schedule also examined two other cultural factors which since the Mysore Study was undertaken, have begun to draw considerable attention namely, nursing practices and the length of lactation amenorrhoea. The latter question was asked of the last two terminations only and its utility was reinforced by the experience gained in the 'Pilot Study on the Rhythm Method of Family Planning' which was undertaken in a rural area near Bangalore around the same time as the Mysore Population Study. As part of the pilot study, the taboo on sex practices was also studied.[9] A major finding of this study was that in the rural area surveyed, coitus was initiated only quite a few days after the woman had had her purification bath, which was usually taken on the fourth day after the start of the menstrual period. The interval between the onset of menstruation and the beginning of coitus was almost ten days. This finding was of great significance for the propagation of the rhythm method which calls for abstinence at about the very time during the intermenstrual period, when couples were used to having coitus. It has also some relevance for studying the effect of abstinence as traditionally practised in Indian society on inter-birth intervals as well as on overall fertility. The Mysore Study made no attempt to investigate the pattern of coitus, and procedures other than the survey technique used in the Mysore Study are, in fact, necessary to throw more light on the influence of abstinence on fertility.

Values of Family Planning

Prior to the onset of the Mysore Population Study, very few investigations on family planning had been undertaken covering wide cross-sections of the population in any part of the world. The few well-known surveys carried out in the West were the Indianapolis Study, another study sponsored by the Royal Commission on Population, England, and the one referred to by Raymond Pearl in his *The Natural History of Population*, and these were limited to sections of the population who could be expected to definitely cooperate in the survey.[10] The latter two studies were confined to women attending clinics while

[9] C. Chandrasekaran, 'Cultural patterns in relation to family planning in India.' Report of the Third International Conference on Planned Parenthood (Bombay, 1952).

[10] P.K. Whelpton and Clyde V. Kiser (eds.), *Social and psychological factors affecting fertility*, Vol. 5, Concluding Reports and Summary of Chief Findings from the Indianapolis Study. Reprinted

the Indianapolis Study was limited to the white population in a city. The Mysore Population Study was historic in that it was the first cross-sectional field survey undertaken in any developing country, and it included family planning as a basic issue of investigation. Indeed, this attempt should be considered a bold one in as much as at that time open reference to family planning was a taboo in India as in most other developing countries.

The confidence with which questions on family planning were introduced in the Fertility Schedule should be attributed to the successful experience of the author of this article, who directed the Mysore Population Study, in obtaining data on family planning practices from different sections of the Calcutta population and in surveying the health status of several rural populations while working at the All-India Institute of Hygiene and Public Health, Calcutta. The part of the Fertility Schedule related to family planning included questions on the knowledge and practice of methods of family limitation by women and their attitudes and motivations regarding family size. A separate schedule was also used to study the attitudes and motivations regarding family size of the husbands of women who were administered the Fertility Schedule.

One of the major findings of the Mysore Population Study was the desire, both in the rural areas and in Bangalore City, for both men and women not to have too many children. 59 per cent of the women in the 'rural plains' and 72 per cent of the women in Bangalore City did not desire to have more children if they had at least three living children; the corresponding percentages for men were 48 and 56 respectively. The significance of these findings, at a time when the Government of India was deliberating on the intensification of the family planning programmes,is self-evident. They also disproved effectively the views held by several influential leaders that family limitation would not be acceptable to the rural people, emphasising at the same time the need for undertaking careful studies in order to understand the prevailing cultural norms.

The motives for desiring or not desiring more children were investigated in the Mysore Population Study by presenting a number of propositions to the respondents and asking them whether they 'accepted' each proposition and in case it was accepted whether it was a motivation or not for their wanting or not wanting more children, as the case may be. The propositions relating to wanting more children were presented in terms of eight possible advantages. Those pertaining to 'not wanting more children' were presented in terms of ten propositions related to possible disadvantages in having more children—five were related to economic reasons and five to non-economic reasons.

Large sections of men and women in both Bangalore City and the rural plains were motivated by the 'inability to provide even the bare necessities for their families' and 'the expectation that the family's level of living would fall,' for not wanting more children. 'To make the home happier,' 'to ensure family survival'

---

trom the *Milbank Memorial Fund Quarterly*, Vol. 32 (1954), Vol. 33 (1955), Vol. 34 (1956) and Vol. 36 (1958). E. Lewis-Fanning, *Report of an enquiry into family limitation and its influence on human fertility during the past five years*. Papers of the Royal Commission on Population, Vol. 1 (London, 1949). Raymond Pearl, *The natural history of population* (Oxford University Press, 1939).

and 'to be taken care of in old age' were often given as the motivations for wanting more children. It is doubtful if more sophisticated approaches for studying the motivations affecting family size would have proved more revealing than the procedures used in the Mysore Population Study, as later experience has shown. It is also significant that the list of motivations included in the Mysore Study covered as broad a spectrum of possible motivations as that found in the literature subsequent to the undertaking of this study.

An attitude related to family limitation revealed by the Mysore Population Study was that after a certain age, women and men should not have any more children. Both men and women also expressed views on the age of a woman at the birth of the first child. In Bangalore City the better educated and economically better off class of the population favoured a shorter period of family building, with a later beginning and an earlier ending than was considered appropriate by persons of lower educational and economic status. All the same, there was evidence that a sizeable proportion of the persons interviewed did not have any definite opinion on the age after which no more children should be born. However, in contrast to several other cultures, including those in the West, it is only in India that both men and women express the attitude that procreation should cease after a person advances in age. The significance of this attitude has not yet been well understood. It can, however, be hypothesised that when childbearing starts at an early age, women do not like to become pregnant when their daughters are old enough to do so.

The majority of the respondents even in Bangalore City and a large section of those in the 'rural plains' zone indicated that they had no knowledge of the possibility of avoiding or delaying pregnancy by any method other than abstinence. Some men and women mentioned 'pills' and 'injections' as methods they had known and considering that the Mysore Population Study was undertaken before hormonal contraceptives were discovered, the reference to 'pills' and 'injections' was possibly to some abortifacients or to methods heard of in vague conversations. That the exposure to hospitals and clinics had apparently played a part in spreading the knowledge of contraceptives became evident as 'sterilization of wife' was the method most well-known both among women and men in the 'rural plains' and Bangalore City.

The Mysore Population Study showed that although there was a widespread desire for limiting the number of children born, the motivation to achieve this end was weak, and family planning was being attempted by cultural practices which indirectly influenced fertility. Abstinence, which had a cultural sanction, was practically the only method used to avoid pregnancy.

Use of Cultural Traits in Improving the Accuracy of Survey Data

The survey procedures to be used and, in particular, the form of the questionnaire to be developed and its mode of administration have to take into account the cultural setting in which the survey is to be undertaken. In the Indian context, people are not used to remembering dates of past occurrences according to the Western Calendar, much less to the recording of these dates in that Calendar and various devices have to be found to help them recapitulate past events and the times when they occurred in order to study them according to modern demographic practices. The Mysore Population Study laid great emphasis on the improvement of the accuracy of the data collected by assisting the interviewee to recapitulate past events through using culturally-oriented questions and through the use of ancillary questions, and by assisting the interviewer to make use of these responses in a systematic manner. Attempts were also made to assess objectively the accuracy of the data obtained in the study.

The use of culturally-oriented questioning can be illustrated by examples taken from the household schedule of using well-known religious festivals as bench-marks for determining the timing of the occurrence of vital events, by avoiding direct questions on the number of children born to a woman or the number of children living at the time of survey, and by obtaining information on deaths among members of the household through indirect questioning.

Religious Festivals as Bench-marks

The household survey was carried out between December 1951 and August 1952 in the different areas selected for the survey. In order to increase the accuracy with which the timing of vital events could be recorded, the festival of Ganesh Puja was used as the bench-mark and those events were recorded which occurred after this Puja was celebrated on 15 September 1950. The events that occurred during the twelve months between 15 September 1950 and 14 September 1951 were taken into account for the calculation of yearly rates. The dates of occurrence were sometimes obtained in terms of the Indian Calendar using information on the *nakshatra* (star) for the birth of the baby and the *thithi* (phase of the moon) for a death—events which are often well remembered—and then converted into dates in the Western Calendar.

Number of Children Born and Living

In the Indian setting, it is recognised that there is a resistance to answering truthfully the question on the number of children living especially if the number is large, and this tendency is also likely to affect the response to the question on the number of children born. Procedures which avoid direct questioning are likely to yield more accurate responses. The Mysore Survey adopted two procedures which although were statistical artifices designed to improve and check the accuracy of the responses, were also helpful in avoiding direct questioning regarding the number of children born and living.

Information about the number of children born was obtained by asking three separate questions—the number of children living with the mother listed in the household; the number of her children living elsewhere; and the number of her children born alive but who died later. The total of the responses to these three questions gave the number of children born alive to the woman and this figure was presented to the respondent for verification. Further, the sum of the responses to the first two questions indicated the number of living children. This method of obtaining the required information by asking several questions necessarily diverted the attention of the respondent from the main purpose of the inquiries, namely, to ascertain the number of children born and the number of children living and very likely increased the accuracy of the information obtained.

It was also possible to check if the entry made against the number of children living with the mother tallied with the information given by the list of members of the household and their relationships recorded in an earlier section of the household schedule for this purpose.

Recording of Deaths

Recalling a person's death is by no means a pleasant experience and this is more so when the death relates to the spouse or to one's own child; however, valuable information can often be elicited through indirect questioning. With this objective in view, a number of ancillary questions were included in the household schedule. To cite an example: if the entry against 'marital status' showed that a person was a widow or a widower, an ancillary question was asked to ascertain if the person became a widow or a widower after the Ganesh Puja of 15 September 1950, the bench-mark date. In case of a positive response, a death which was needed to be recorded in the household schedule would become known. A separate box was provided in the household schedule to check if this death had been recorded when deaths in the household were ascertained, and if not, the entries had to be corrected or an explanation as to why the death was not recorded had to be entered. Another illustration of using ancillary questions was to ask if a baby born in the household was still alive; if not, the death of an infant or child which should have been recorded in the household schedule would come to light. As in the case of the death of a spouse, a box was provided in the household schedule to ensure that the information so obtained would be taken into account in the final recording of the deaths of the members of the household.

The Mysore Study verified the accuracy and reliability of the data obtained through the survey by using several procedures. The cross-check method and the application of 'Inter-penetrating Samples' technique were among the methods used for the purpose.[11] Some of the results of the investigations have been cited here to bring out the extent to which difficulties in obtaining accurate data, including those arising from sub-cultural resistances, were overcome through judicious planning.

[11] C. Chandrasekaran and W. Edwards Deming, 'On a method of estimating birth and death rates and the extent of registration,' *Journal of the American Statistical Association*, Vol. 44, No. 245, pp. 101–15; P.C. Mahalanobis, 'Recent experiments in statistical sampling in the Indian Statistical Institute,' *Journal of the Royal Statistical Society*, 1946, pp. 325–70.

The cross-checking of the data on births obtained through the household survey with those on the Registration Lists showed that the survey had recorded 91 to 94 per cent of the events in the three rural zones, the average being 93 per cent. For deaths, the corresponding percentages ranged from 83 per cent in Zone II to 94 per cent in Zone III and the average was 90 per cent. No significant relation was found between the completeness of recording and the population size of the village, sex, religion or (in the case of births) age of the mother. As regards the age of decedents, the cross-check showed that the recording of deaths in the age group 15–44 years was less complete than in older and younger age groups.

An important finding gleaned from the analysis of the data obtained from the use of 'inter-penetrating' samples was that the interviewers did not differ significantly in their performance. Their training and other factors,which had a bearing on the procedures used in the survey, helped them to overcome many of the cultural barriers to obtaining reliable data through field inquiries.

Thus, the paper has focused attention on the importance of taking into account cultural settings and cultural factors in studies of the interrelationships of economic, social and population factors, with the assistance of suitable illustrations from the Mysore Population Study. In Indian settings, geographic proximity and linguistic homogeneity could provide a suitable base population in which 'contrast' groups could be identified. The importance of taking into account cultural factors especially in the study of the dynamics of fertility has been highlighted by reference to their influence on factors affecting the length of married life, the interval between births and the knowledge and practice of family planning methods. The need for orienting survey procedures and questionnaire design to suit the cultural background of the population surveyed has been emphasised.

7

Determinants of Religious and Caste Differentials in Fertility: A Village Study

K. MAHADEVAN

The major population of South Central India belonged to several cultural groups namely, the Caste Hindus, the Harijans, the tribals and the Muslims. Census, Sample Registration Scheme and certain other studies (UN, 1961; Rele and Kanitkar, 1979; Mahadevan, 1979) have confirmed the existence of differential fertility among these communities. According to the Census of India during 1961 to 1971 period, the Muslims experienced an increase of 30.8 per cent as against the decennial growth rate of 28.7 per cent for the Hindus (Census of India, 1971). Another cross-cultural study showed that the Harijans have the highest fertility (CBR: 37) as against the lowest fertility for the Gounders (CBR: 24.5) and the other numerically smaller groups together had an intermediate level of fertility (CBR: 32.7; Mahadevan, 1979). Similarly, several other studies have confirmed differential fertility across different cultural groups in India (Davis, 1951; UN,1961; Driver, 1963). Nevertheless, negligible efforts have been made subsequently to identify various determinants of fertility behaviour. No doubt certain differential socio-economic status factors were partially attributed to existing fertility differences, but several deep-rooted cultural and social factors relevant in the Indian context have not been explored. Therefore, through this micro level study an effort is made to identify major dimensions of the value of children, sex roles, economic activity, social structure, and their influence on fertility behaviour. For realising this goal, an in-depth study was carried out in a representative village in the Andhra Pradesh State (South India) during 1982–83. However, a sub-sample alone was studied for the research questions dealing with the value of children and time-budget data.

Village Setting

Ramapuram is a village located near Tirupati in Chittoor District of Andhra Pradesh. The name Ramapuram is very common to several villages in the state. This village and the adjacent Harijan hamlet were selected for the study because of the representative character, economic development, caste and religious composition and above all the anticipated fertility differences among

Acknowledgements I am grateful to Professor Ronald Freedman for the constructive suggestions offered for the improvement of this paper. I also acknowledge the valuable services rendered by Ms. V. Gowri, Audinarayana and D.A. Naidu for data collection and analysis.

the Caste Hindus, the Harijans and the Muslims. As in the other villages, the Harijans do not live in the village proper, but far away from the boundary of the village, because they were traditionally the untouchables. But the Caste Hindus and the Muslims live in close proximity in segregated localities. Geographically, Ramapuram lies very close to the highway that connects Tirupati and Chittoor. It is sixteen kilometres away from Tirupati town. Its eastern, western and southern directions are flagged by hills with open fields on the north. Located at a height of 1,010 feet from the mean sea-level, the climate of the village is generally temperate throughout the year and very hot (40°C) during summer. The annual rainfall has decreased to ten centimetres during the last five years.

Historically, Chandragiri fort situated near this village was founded by the Yadavaraya chief who lived in 1000 A.D. However, the Muslim population could have reached this village probably during the period between 1646 to 1782 A.D. During the earlier period, the Sultan of Golkonda took over Chandragiri and in the 1780s Hyder Ali of Mysore captured the whole of Chandragiri area. This village was in existence during the period of the Vijayanagar empire dating back to 1592 when Chandragiri became the capital of this dynasty. In other words, the existence of this village in all probability dates back to as early as 1000 A.D. (Rao, 1976).

The population of the village according to the 1981 Census is 2,500; the Caste Hindus constitute the largest single group with 1,820 persons followed by 400 Muslims and 180 Harijans. The average household sizes are: 5.65, 5.80 and 5.90 for the Caste Hindus, the Harijans and the Muslims respectively, and 5.80 for the village as a whole. During the last two decades fifteen Muslim families and two Hindu families have migrated to Madras and Tirupati in search of employment. Obviously, the Muslims are more migratory than the Hindus. The population is relatively young since 45 per cent is below the age of fifteen years.

The habitat of the three communities is characteristic of the same feature seen elsewhere in the country. The Caste Hindus reside around the two Hindu temples; their houses are built extremely close together on both sides of a kutcha street. The Caste Hindu families are exclusively seen in their own localities. Adjacent to their locality but separated by a road, the Muslims occupy another cluster of houses on the eastern side of the village adjacent to a mosque. Most of the houses are thatched though there are a few modern terraced and tiled houses. However, all but a few educated Harijans live far away from the main village in thatched but neat huts; a few modern houses have been added recently by the social welfare department.

Social Structure

The social structure is conspicuously stratified among the Hindus and Muslims along traditional lines. The Caste Hindus are represented by the Yadava (herding community), Besta (fishermen), Reddy (farmers), Gandla (oil mongers) and Vaisya (businessmen), and a numerically small group of village officials (Pillais) and the service caste. Although the Yadavas are numerically greater, economically most of these castes have a more or less similar status. There is no social distance and intermarriage between these communities but they exchange food items and dine together during feasts and festivals. The Muslims are represented by the Shaikes and the Sayeds. The Shaikes are the converts and the Sayeds are the Muslims with a long-term tradition. They rarely intermarry and when marriage does take place between them, the children born out of the wedlock belong to the sect of the father. The Shaikes are economically prosperous and have a better social status. The Harijans belong to a single caste of Mala subcaste.

Interactions between the Hindus and the Muslims are very frequent and relationships between them are very cordial. The equality of social relationship exists among them. Economically, these communities are more or less equal. They own similar acres of land but some Muslims have entered into business. Entrepreneurship is relatively more among them. Fraternal relationship is visible between these two communities. Young Muslim boys address elderly Hindus as 'Anna' meaning elder brother and vice-versa. This term of address reflects the cordial and informal relationships that exist between these two communities. However, their children address a senior Harijan by name and the latter use respectable terms of address for the Muslims and the Hindus. The Muslims and the Hindus dine together and exchange food during their festivals and religious and marriage ceremonies, but do not dine with the Harijans. However, their relationship with the Harijans has undergone major changes. Almost fifteen years ago most of the Harijans were bonded agricultural labourers of the Caste Hindus and the Muslims. When the nearby forest was almost denuded and firewood became a rare commodity, the Harijans travelled to far-off places, brought and sold firewood to other people and started earning more money independently. As a result they slowly became professionally free and economically independent. The employment opportunities available for unskilled labourers at the Tirupati temple complex, further improved their position and made them economically independent. Today, several Harijans are engaged as peons, clerks, and other class four employees in the T.T.D. and S.V. University at Tirupati. This has been possible because of the existing reservation in education and employment opportunities at the intervention of their Panchayat President who became the first Harijan from this village to reach the position of a Divisional Revenue Officer.

Quick communication within the village is effected mostly through the women. When they go out to draw water from the common open well or tube-well they have to wait for sometime before their turn. While these women wait to draw water, they exchange day-to-day news. The men exchange news when they go to the road side shops for shopping, and spend their leisure time in common meeting places and on the farms. People returning from the nearby towns of Chandragiri and Tirupati disseminate current news to all those on the homeward journey. Modern communication facilities are very poor. There is only one telephone connection at the local post office which is seldom used. Since the literacy rate is very low only one English and four Telugu dailies are brought out regularly. One-fifth of the families own either radio or transistor sets along with the community radio set. Cinema is very popular among the Caste Hindus and the Harijans and to a lesser extent among the Muslims. Since it is the most extensively used mass media by a majority of the people, young and old, men and women, there is tremendous scope for introducing educational newsreels for their social development through this powerful media.

Even though leadership is perceived to be important within each community, there is no common leader for the village as a whole. Traditional leadership in terms of seniority of age or occupation and economic status, among the Caste Hindus and among the Harijans for limited social purposes such as arranging marriages, religious and ritual functions, in decision-making is observed. Of course, they have a role in deciding on the location of common social welfare schemes and also for settling intra-village disputes between families and communities. However, the newly emerging leadership among the Harijans is relatively more functional. Nagaiah is the first Harijan to have reached the position of a Revenue Divisional Officer (RDO). Though he studied only up to the secondary school, his hard work and his ambition, together with his determination to compete with the advanced communities and the provision of reservations, enabled him to attain the high position that he holds now. He is respected by the other Harijans and they are proud of his achievements. Most of his children received higher and technical education. He is referrent to most Harijans. On his initiative several Harijans were able to get jobs in Tirupati town in several offices. He is an influential Harijan leader. His achievements have inspired the other Harijans to pursue education and seek non-traditional employment. He wields a lot of influence in the entire Harijan community.

The Muslim priest who functions almost like a leader exercises social control over most of his fellowmen in this village. He conducts prayers five times a day at the mosque. The prayer conducted on Friday noon is specially important because, besides reciting the prayer, he speaks on a common theme such as how to attain heaven and avoid hell, different types of sin, and the ideal Muslim way of life, quoting anecdotes from the *Quran* or other texts or even from day-to-day life. On certain occasions he talks against family planning but is unable to substantiate his arguments by quoting any passage from the *Quran*. However, he conveniently interprets the Muslim religion against family planning partially because his formal schooling is very little except the knowledge that he has of Urdu language and literature dealing with Islam. He argued that 'Allah gives life and food for all creatures. He gives children not only a mouth to eat but also two hands to work. By adopting family planning we go against what Allah creates and supports.' Although he preaches against family planning, he does not object to abstinence and safe-period. In his personal conversation with certain individuals who had adopted family planning he gave the same interpretation and expressed the view that they should not have adopted sterilisation. They told him that if he had advised them earlier they would not have accepted it. All this data confirms the importance of the Muslim priest as a leader who exercises great influence in this community in decision-making on several aspects of life including the acceptance of family planning. Therefore, special programmes may have to be evolved to educate first and foremost the Muslim priests to change their attitude and to minimise their interference in health and family planning matters.

FERTILITY BEHAVIOUR

The total fertility rate of the population as a whole is 3.9, but the Muslims have the highest fertility (4.5) and the Harijans the lowest (3.4), while the Caste Hindus occupy an intermediate position (3.9). However, fertility differences between the Caste Hindus and the Harijans are significantly lower compared to the Muslims. The same trend is observed with regard to the mean number of living children. Though loss of infants is 0.5, 0.4 and 1.0 respectively among the three cultural groups, it is almost 1.0 baby each for the Muslims and 0.5 baby each for the other two groups. The highest fertility rate of the Muslims is expected but the lowest fertility rate of the Harijans is rather surprising because almost all the earlier studies have revealed very high fertility rates for the Harijans all over India. In this state also, fertility is very high for both the Harijans and Muslims. However, their ideal family size ranges from two to three children for all the groups and the ideal number of sons ranges from 1.5 to 1.8 only. Since this population, in general, has two or more children and the possibility of early childhood mortality is reduced drastically, their fertility will stabilise more or less at the present level.

TABLE 1 Fertility Patterns among the Caste Hindus, Harijans and Muslims

Variables	*Cultural Groups and Mean Values*			
	Caste Hindus	*Harijans*	*Muslims*	*Total*
Number of live births (TFR)	3.9	3.4	4.5	3.9
Number of living children	3.4	3.0	3.5	3.3
Ideal number of sons	1.5	1.5	1.8	1.6
Ideal family size	2.6	2.9	3.2	2.9
Total (N values)	(20)	(20)	(15)	(55)

THE ECONOMY

The village economy is entirely based on agriculture, which is not adequately backed by modern technology in many aspects. Therefore, the prosperity of the village is not substantial. However, the proportion of poor is very negligible; land is owned by over 70 per cent of both the Caste Hindus and the Muslims as against 70 per cent of landless Harijans. Land owners, on an average, possess three acres of land which is either partially or fully irrigated. The cost of an acre of irrigated land ranges from Rs 15,000 to Rs 30,000 depending upon the fertility of the soil and the irrigation facility (1983). Most of the land-owning people have installed electric pumps for irrigation. Today, all the farmers have stopped using the conventional method of drawing water from an open well with the help of animals. An overwhelming proportion of them have their own wells while others share the common well. Landless people with the exception of the Harijans cultivate land that is leased out to them. The Harijans, on the other hand, work as casual labourers in the farm and non-farm sectors since they do not own any land. As rainfall is seasonal and is less now-a-days, agricultural operations are fully dependent upon underground water. Animals which form the backbone of the economy are used for ploughing, transport and obtaining manure. Modern ploughs, fertilizers, and pesticides are extensively used in agriculture.

Agricultural pursuit is remunerative to most farmers, though the productivity of land is comparatively low by modern standards. This may be because the holdings are small, only manual efforts are made by the owners themselves besides using locally available resources, and manure. Further, throughout the year, one crop after another is grown in their fields. The land is very seldom allowed to remain fallow. The crops grown are largely sugarcane, paddy, groundnut and vegetables. Paddy is raised twice annually and a vegetable or groundnut crop is also raised in between. These crops are mostly high yielding varieties. An acre of land produces up to thirty bags of paddy (2,250 kgs) which may fetch Rs 3,000 at the present market price (1983). Expenses per acre including electricity charges, fertilizers and labour together may account for Rs 1,500. Thus a balance of Rs 1,500 remains as profit for every one acre of land when paddy is grown. For sugarcane, per acre yield for one year goes up to Rs 7,000; this is slightly more profitable than paddy. Vegetables are produced in large quantities and sold at very cheap rates to the middle-men who sell them at higher prices in the nearby town. Since the productivity of these two major crops is very little, there is considerable scope for improvement in agricultural economy, if recent innovations in agricultural technology are appropriately adopted.

The development of animal husbandry is a parallel and extensive agricultural activity for most of the Caste Hindus and the Muslims but not for the Harijans; 50 per cent of the Caste Hindus, 20 per cent of the Harijans and about 70 per cent of the Muslims own one or more cattle—they own indigenous buffaloes and cows. Dairying, thus, is a major activity of the Caste Hindus and the Muslims but a very minor one as far as the Harijans are concerned. This profession has become popular in the village over several years probably because the single largest group of population belongs to the Yadava caste who are traditional herders. Though only rarely do they own high yielding animals they have extensively started adopting modern practices of animal husbandry like artificial insemination. The cost of the present local varieties of animals ranges from Rs 500 to Rs 5,000 but most of them cost less. The cost of maintaining animals is very marginal as most families are able to obtain straw, groundnut cake, and grass from their own farms. Animals are often looked after by children and partly by housewives and no wage labour is involved in this activity. An average buffalo or cow gives four litres of milk per day and each litre costs Rs 2.50 at the local milk cooperative society. They sell three-fourths of the milk daily and earn additional income to supplement the income from agriculture. Occasionally, some of the dairy animals are also used on the farm.

Thus, animal husbandry forms a major supplementary and beneficial occupation for most of them. However, there is much scope to double or even triple the production of milk in this village from the present daily sale of a hundred litres, if better animals, fodder and more financial and technical support are provided to them.

EDUCATION

An interesting observation is that the educational status of the Harijans has surpassed that of the two economically advanced communities. It is contrary to their educational attainment at the state and national level. Of the five post-graduates in the village, four are Harijans; the only M.Phil. student and the doctor are Harijans, a doctor and an engineer have married two daughters of the Harijan leader Nagaiah but such cases are not seen among the Muslims and the Caste Hindus. With regard to the literacy level also Harijans are slightly better off than the Caste Hindus and the Muslims in general. However, women in general are largely illiterate; the Muslim women are more illiterate than the rest of the population. What could have been the major factor that contributed to the rapid educational achievement of the Harijans? Nagaiah's occupational advancement and the attainment of professional and technical education by most of his sons and daughters together decisively inspired several other Harijans to pursue their studies with greater enthusiasm and seriousness. Of course several other factors have also contributed to their awakening in education. They are illustrated in the case study of Ramaiah, a Harijan youth, in the following lines.

After completing his post-graduate studies, Ramaiah is currently pursuing the M.Phil programme. His father and maternal uncle are engaged in non-traditional professions, namely, village officer's assistant and cashier in a temple trust (T.T. Devasthanams) respectively. Their new professions have made them relatively more modern and better informed about higher education. Their interest and financial support were the crucial factors which enabled him to pursue higher education. Ramaiah's maternal uncle wanted him to marry his first daughter. Ramaiah on his part was interested in education for the concern he had developed for his community compared to the better-off people in the other two communities, the inspiration he got from the Harijan leader Nagaiah, the increased opportunities for the Harijans to get a job under the reservations scheme and the scholarships, etc., offered by the government. His interest in education was also enhanced by the example of his elder cousin who had secured a Master's degree. Though at first he was only interested in obtaining a degree, later on he desired to go in for higher education and work for both his Master's degree and an M.Phil. His ambition, thus, gradually rose and now he is interested in becoming a lecturer. Though he received a government fellowship worth Rs 40 throughout school below class ten, Rs 400 and Rs 600 annually till intermediate and graduation, and a monthly fellowship of Rs 200 and Rs 400 for post-graduation and M.Phil. respectively, this was not a major factor for fostering his ambition to pursue higher education. Over and above they have no landed property and a great demand exists for child labour to work on the farms of the Caste Hindus and the Muslims. Similar factors also influenced several other Harijans in this hamlet.

For higher education, a high school is located only four kilometres away and a university sixteen kilometres away from the village. Neglect of female education is serious which is on account of the traditional belief that it is undesirable to send girls to school when they are nearing puberty. Widespread illiteracy among women is a major pronatalist force among all the communities and is exceptionally greater among the Muslims. Muslims hardly send their girls to school till class V even though Urdu medium classes are organised for them. Another factor that discourages girls' education is the early age at marriage (soon after puberty) that widely prevails in the village. While the Muslims get their daughters married soon after puberty, the Caste Hindus and the Harijans get them married at the age of 16 years and 15 years respectively. The demand for child labour adds to the neglect and drop-outs in education. A Muslim respondent remarked that if schools were located in their locality and had privacy and were run by lady teachers, there is scope for educating Muslim girls and adults. There is an urgent need and scope for adult education, separately for the three communities.

Wealth, Modernisation and Fertility

Wealth Most of the Caste Hindus and the Muslims have wealth in the form of landed property, own house, and animals as against the Harijans who own such wealth to a very limited extent. The data in Table 2 is consistent with the hypothesis that wealth in general has a somewhat negative association with fertility behaviour for the communities whose cultural influence on fertility is relatively less as in the case of the Caste Hindus and the Harijans but not for the ethnocentric Muslims. For the population as a whole also wealth has a negative association. The trend is remarkably consistent even though the sample size is small. But wealth has an inconsistent or a somewhat positive association with fertility when the cultural influence on fertility is great as in the case of the Muslims. For the Muslims wealth cannot influence fertility independent of culture. Influence of culture on the Muslims' pronatalist behaviour has been more or less confirmed in the preceding section. For the Caste Hindus and the Harijans a negative relationship is observed between fertility behaviour and property, ownership of animals, and family income. But in the case of the Muslims, the influence of landed property and family income confirm only an inconsistent relationship; whereas with the number of animals owned fertility shows a definite direct relationship. However, when the factor of landed property is not applicable to the Muslims, which is the main source of income here, people of this community exceptionally limit their family to a smaller size. Thus, wealth has different types of relationship with fertility behaviour depending upon the cultural background of the population (see Table 2).

Modernisation Modernisation is unusually greater among the Harijans as compared to their counterparts in the rest of India and also as compared to the

TABLE 2 Mean Distribution of Live Births by Selected Socio-economic Characteristics of the Respondents

Variables	*Three Cultural Groups and Mean Live Births*			
	Caste Hindus	*Harijans*	*Muslims*	*Total*
Income				
≤3000	5.0	4.1	4.0	4.3
	(3)	(7)	(2)	(12)
3001–5000	4.4	4.0	3.8	4.0
	(5)	(3)	(4)	(12)
5001–7000	2.3	3.6	5.8	3.9
	(4)	(5)	(4)	(13)
7001+	3.6	3.2	4.0	3.6
	(8)	(5)	(5)	(18)
Land-owning				
No land	4.8	3.9	2.8	3.8
	(6)	(15)	(4)	(25)
≤1 acre	2.0	3.5	5.8	4.5
	(1)	(4)	(5)	(10)
1–2 acres	3.7	—	4.5	4.0
	(7)		(4)	(11)
2+ acres	3.0	3.0	4.5	3.3
	(6)	(1)	(2)	(9)
Animals.				
No animal	3.7	4.0	4.0	3.9
	(10)	(16)	(5)	(31)
1–2 animals	3.7	2.8	4.4	3.8
	(6)	(4)	(7)	(17)
3+ animals	4.0	—	5.3	4.6
	(4)		(3)	(7)
Education				
Both illiterate	5.0	4.8	4.6	4.8
	(2)	(6)	(5)	(13)
Husband literate+	4.1	3.3	4.0	3.7
wife illiterate	(10)	(12)	(4)	(26)
Both literate	3.0	3.0	4.5	3.6
	(8)	(2)	(6)	(16)

Caste Hindus and the Muslims. Since a greater proportion of Harijan men are educated (60 per cent) as compared to the Caste Hindus (50 per cent) and the Muslims (27 per cent), this factor makes a qualitative change in the relatively greater levels of modernisation among them. This has also paved the way for more Harijan men to take up non-traditional occupations in a limited way with the existing concessions and reservations acting in their favour. Educational achievement is very low among women in general. In short, advancement in education together with the changes in occupation have led to a higher degree of modernisation among the Harijans which, in turn, would have depressed their fertility very significantly. This was, of course, accelerated by their community leader as cited earlier. But more than occupation, education acts as a decisive anti-natalist influence for the Harijans. Though the Caste Hindus are closer to the Harijans in terms of the level of education of men, this factor did not depress their fertility to any great extent because of several other factors, namely the need for child-labour, and the cultural influence of maintaining lineage and owning property. Such economic and cultural factors have very little relevance for the Harijans. Since only a limited number of Harijans own land, the concept of lineage is not important for them. For Muslims education did not make much headway in depressing their fertility, since the educated Muslims are also very traditional in their attitude. Thus, it may be probable that landless or poor people with less ethnocentrism on culture like the Harijans,change their fertility rapidly toa very low level with the advancement of education and other modernisation as compared to a land-owning or rich and culturally complex peasant community like the Muslims and the Caste Hindus.

Muslim Culture and High Fertility

The Muslim men, women and children in Ramapuram are generally traditional and fatalistic in their behaviour. They conventionally pray five times a day as prescribed for the Muslims. Of course, persons who are preoccupied with work and women in general pray only in the morning, noon and evening. Various cultural factors, prominently seen among them, greatly influence their pronatalist behaviour which is also supported by Dudley Kirk (1966). These factors are: (*a*) the concept of hell and heaven; (*b*) belief in sin and fatalism; (*c*) desire for extended kinship; (*d*) greater value attached to sons; and (*e*) overall influence of religion and the priest in different aspects of life.

The concept of hell and heaven is developed among the Muslim population right from early childhood; initially parents and later the priest inculcate this concept in the minds of people. Friday prayer at noon is very often the occasion when the priest educates the Muslims about what action would lead a person to heaven or hell. There are seven types of heaven and hell each for the Muslims. Depending upon the seriousness and nature of the actions that the Muslims perform they may go to lower or higher levels of heaven or hell. According to this priest, people who go to the higher levels of heaven can enjoy life very well after death if they do everything according to Muslim conventions. One respondent remarked that 'in the highest level of heaven they can see beautiful ladies in seminaked form even after wearing 70 layers of sarees and enjoy them.' On the other hand, the priest and other elderly Muslims preach that going to other religious centres, non-observance of 'Namaz' (prayer), and adoption of family planning will only lead people to hell. This concept of hell and heaven is widely prevalent among the rich and poor, men and women but the intensity of belief slightly decreases when people are educated or poverty stricken or have other priorities which make them overlook the traditional concept of hell and heaven. This cultural factor, as mentioned earlier, acts as a strong pronatalist force among the Muslims in general.

Belief in fatalism and sin are extensively noticed among the Muslims irrespective of educational status. A post-graduate in commerce who had obtained his degree in 1976 is still unemployed. To a question why he is still unemployed he replied that 'it is my fate and what else'. Fatalism is attributed to what they are destined by God. Sin is a frightening concept because all the sinners are expected to go to hell after death. Those who believe in the concept of hell and heaven and attach importance to this concept would naturally pay greater attention to avoid committing sin. One eligible couple remarked 'whatever number of children are destined for you that will be born and how can you prevent it? If you resort to interrupting the God-blessed conception, you will be doing a sin and may go only to hell.' Such a belief is deep-rooted among several Muslims with the exception of a negligible number of educated individuals, or those who have widely travelled and are modernised and have received proper education on sex and conception from credible persons such as Muslim doctors or other family doctors.

The desire for extending kinship ties and getting social and economic opportunities and benefits are seen among the Muslims and others alike. Naturally, through child birth and marriage kinship ties are extended. Since the traditional norm of social and economic security is weakening its function, parents are facing a risk in obtaining old age security. In the past such a situation did not arise because land was available in plenty, the cost of living was not high and the responsibility to look after parents in their old age was not neglected by the children. The changes in social life compel the parents to have at least two sons. A respondent said that if he had two sons at least one would take care of them (the parents) in their old age. Another observation is that, very often, daughters-in-law become a problem for the parents. A common feature noticed among the Muslims here is that after marriage the man tends to attach more importance to his wife's views than to his parents. The existing unfriendly atmosphere between parents and daughters-in-law inhibits the parents to develop close ties and make frequent visits to the house of their son's in-laws. On the other hand, their sons-in-law normally have cordial relationships with them because of their daughters, who have always enjoyed good relationships with the parents. According to several respondents, sons-in-law are more affectionate than their own sons. Hence parents use the extended kinship ties through their daughters for socialising, seeking affection and emotional support and having mourners at the time of death, which is valued by majority of the Muslims. Therefore, Muslim parents are not prepared to limit their family size to one or two sons but would like to have at least one daughter in order to fulfil their social goals. Daughters are valued as much as sons by some Muslims (chiefly by land-owning or rich class) as they are useful during times of crises. Of course, parents do not like to have many daughters because of the expenses involved in getting them married and giving them a dowry.

The salient features of the Muslim life style provide for the greater control of the priest and religion on their already traditional pattern of life. Their secluded habitation, purdah system, frequent interaction with Muslims themselves, discouragement of higher education for girls and several other factors inherent in the Muslim culture together act as a pronatalist force, because these factors do not facilitate the development of a modern and scientific approach to life.

Family Size Norm and Values

The concept of small family norm is yet to develop among the majority of the Caste Hindus and the Muslims with the exception of the Harijans. In fact, Muslims believe that this norm itself is an anachronism to Islam. There is no social sanction for it now, but it is gradually developing. The older generation had, on an average, six children and never at any point of time thought of limiting the family size in all the cultural groups. However, the present generation expressed the desire to limit the family size to two sons and a daughter and would have a fourth child only if all the three children happened to be boys or vice versa. Therefore, the present family size norm ranges between three to four children depending upon the value of children. However, this generation hesitates to produce a fifth child. But they may limit their family size to three children if all the three children happened to be daughters. In short, they do not have a definite norm for a small family but the preference is conditioned by the value of children. No doubt they attach relatively greater value to sons as against daughters. The mean number of sons preferred is 1.5 for the Caste Hindus and the Harijans and 1.8 for the Muslims. Further study revealed that most of them preferred, however, a high level of ideal family size—2.6, 2.9 and 3.2 children among the Caste Hindus, the Harijans and the Muslims respectively, which indicates that all of them would like to have at least one daughter (see Table 1).

The greater value attached to sons and the relatively lesser value attached to daughters is uniformly mentioned by all the three cultural groups. But the intensity of value for more than one son and one daughter is manifested by the Muslims alone. There are several value dimensions which account for the existing differential value of children. They give top priority to old age security irrespective of their background. Most of the people (75 per cent) who own a house and/own landed property (mostly the Caste Hindus and the Muslims) give equal importance to son(s) to continue the lineage and inherit the ancestral property. They desire not only one son but would like to have at least two sons. Similarly, people who are skeptical of getting old age support from one son based on the past experiences of their family also prefer two sons to make sure that they would get old age support from one or the other son. This is because some daughters-in-law create problems for the old parents and tend to neglect them particularly when marriages are arranged outside the circle of permissible relatives. Daughters-in-law, it is reported, create fewer problems for the old people when marriages are arranged between cousins or uncles and so on. But more data is needed to confirm this hypothesis. On the basis of such experiences one respondent commented that 'one son is not a son and one eye is not an eye,' which clearly means that they want at least two sons. Another Muslim respondent remarked that 'if daughter-in-law is good he is even prepared to stop with one son and a daughter.' Land owners, economically well-off and business people desire even more than two sons primarily to cultivate the land or to look after the family business. For instance, a Muslim respondent stated that she was not repenting for having given birth to five sons and a daughter. To quote her, 'I can use one fellow for looking after land, another to take care of animals, a third to educate and get a good job, a fourth to take up priesthood and fifth one to assist the parents and so on.' This is so because they do not consider any job to be inferior and unworthy. They have confidence that their children can somehow make a reasonably good living. Their aspirations are reasonable and the desire for comfort is limited. All these values and aspirations together manifest as greater son preference among the Muslims followed by the Caste Hindus and the Harijans. Hence the value of children and particularly son(s) produces high fertility among the Muslims and the Caste Hindus but not among the Harijans. Other dimensions of the value of children are the economic value of children by getting dowry, and the cultural role they play in performing death rituals.

Value of Children and Fertility

Though there are several dimensions to the value of children as discussed earlier, old age security is the prominent and universal one for the population as a whole. It is discussed at three levels: (*a*) sources of financial support; (*b*) degree of dependence on children; and (*c*) number of sources of old age security other than children. Together they account for an in-depth and extensive coverage to influence the fertility behaviour.

The value of children in terms of dependence on children and particularly on son(s) for old age support, promotes fertility among the three cultural groups (see Table 3). When old age security through wealth, business, pension, life insurance, gold, etc., are available, the Caste Hindus and the Harijans decisively reduce their fertility whereas the Muslims do not. Even among the Muslims, when the dependence on sons for security is higher, they manifest higher levels of fertility (see Table 4, item 2). The increase in the number of sources of old age security has a negative association with fertility behaviour. The relationship between the existing number of sources of old age security and fertility behaviour provides sufficient policy issues to introduce or expand different feasible old age security measures for rapidly controlling fertility in developing countries (see Table 3) like India.

A Case Study

In a Yadava family the parents have only one married daughter. They own two acres of wet land and cattle. They are fairly well-off with a terraced modern building. When their daughter had given birth to two sons she wanted to adopt sterilisation. She is educated and so is her husband. But she could not take the decision alone as her husband and parents were very keen that she has a daughter. Both had different reasons for this. The parents felt that old age security and support can also be provided by a daughter, since they themselves received security from their only daughter as they did not have a son, whereas the husband felt that there was nothing wrong in having another child as he and his wife owned enough landed property. Having another child would not affect their economic well-being at all. In this case, the immediate factor that influenced them was kinship but old age security, physical support of parents and possession of landed property operate through kinship. Though this is an exceptional case, the situation is similar with regard to the desire for son(s).

TABLE 3 Mean Distribution of Live Births by Old Age Security to the Respondents

Old Age Support (N)	*Caste Hindus (20)*	*Harijans (20)*	*Muslims (15)*	*Total (55)*
1 *Source of Financial Support Needed During Old Age*				
(*a*) Help from children alone	4.6 (9)	4.1 (8)	4.3 (3)	4.4 (20)
(*b*) Help from children and/or farm, business/savings/pension/ PF/LIC/gold, etc.	3.4 (11)	3.5 (12)	4.4 (12)	3.8 (35)
2 *Degree of Dependence on Children for Financial Support*				
(*a*) Rely a good deal	4.4 (8)	3.7 (14)	4.4 (11)	4.1 (33)
(*b*) Depends/uncertain	4.5 (4)	–	–	4.5 (4)
(*c*) Rely only a little	3.1 (8)	3.8 (6)	3.6 (4)	3.4 (18)
3 *Number of Sources Other than Children for Old Age Security*				
Exclusively children	4.6 (9)	4.1 (8)	4.3 (3)	4.4 (20)
2 sources	3.3 (8)	3.55 (9)	4.6 (11)	4.4 (28)
3+ sources	3.6* (3)	3.3 (3)	2.0 (1)	3.2 (7)

* In this category a person has exceptionally high fertility.

Demand for Labour

The demand for labour in Ramapuram is very great and underemployment is very marginal. Labour requirement may be broadly classified into (*a*) labour required for agriculture; (*b*) for animal husbandry; (*c*) for collection of firewood and roofing materials for household maintenance; (*d*) for domestic work; and (*e*) for other miscellaneous jobs like house construction/maintenance, workshop, petty shop and several other traditional jobs. All these labour requirements are met indigenously. Labour required for agriculture, dairying, collection of firewood and roofing materials, construction and manual work is mostly productive. Household maintenance includes human effort which is partially productive and partially indirectly productive as it supports other production activities. Division of labour by sex, age, caste and religion are conspicuously seen.

Grownup men from the Caste Hindu and Muslim families are mostly engaged in the cultivation of sugarcane, rice and groundnut on their farms, the Harijans are either engaged as wage-labour or work in offices as assistants or are involved in the collection of firewood. Men are chiefly engaged in ploughing, levelling, digging, watering and supervision. These activities are largely carried out by the owner-cultivators while ploughing, weeding, manuring, harvesting are performed by wage labourers namely, the Harijans and small and marginal farmers. Of course time spent by way of idling and taking rest is not for a very long period considering the same situation prevailing elsewhere in India. However, leisure time is more in the case of the Caste Hindus and Muslims compared to the Harijans. Since a major share of the agricultural activity is carried out by the farmers themselves, agriculture is remunerative to them.

Women from the Caste Hindu and Muslim families spend maximum time in doing domestic work probably till the age of 45; these tasks include processing food, carrying the meal to the fields for their husbands and children, looking after children, fetching water, washing utensils, cleaning the house, feeding animals, etc.; they seldom work on the farm with the exception of those who are landless. Muslim women adhere to this routine very strictly till the end of their reproductive span. Harijan women, however, work on farms belonging to others and are engaged in weeding, transplanting, harvesting and other manual work. Cooking is done in a traditional way using firewood and local vessels, which is highly time consuming. Commonly used local vessels are earthenware pots, brass and aluminium vessels. However during the recent past, dehusking of paddy and flouring of cereals have been done through machines which has minimised the drudgery of women. However, grinding and pounding are still done manually. Fetching water for cooking, bathing, drinking, and for animals is a tedious task for women. Of course, children do assist them partially in these tasks. Women also have to look after the younger children in the midst of their tight schedule of household work but surprisingly they spend very little time looking after the younger children since the older children share this responsibility with their mothers, while the Caste Hindus and the Muslims spend more time in child care, the Harijan women do not. No doubt, wherever girls are available they share most of these activities of adult women. Though there are no wages for these tasks and direct profit and productivity is not visible, all other productive activities of men and children would suffer if the household work is affected in any manner (see Table 4).

Child Labour

Child labour is a norm to one and all families. The concept of child does not apply to physically grown persons and after the age of ten wages are given almost equally to children (10–14) on par with adult members. Of course, the involvement of child labour in household activities begins at the age of six and their involvement in productive activities outside the home begins at the age of

TABLE 4 Distribution of Respondents and their Husbands by their Mean Work Input per Day

Activities	*Caste Hindus*		*Harijans*		*Muslims*		*Total*	
	Husband	*Wife*	*Husband*	*Wife*	*Husband*	*Wife*	*Husband*	*Wife*
Domestic (sweeping, fetching, water, washing vessels, clothes, cooking, etc.)	—	6.27	0.07	7.15	0.2	8.18	0.09	7.20
Rearing children (preparing children to go to school, educating children, fondling, etc.)	—	0.65	0.37	1.25	—	0.6	0.12	0.83
Directly productive (cattle rearing, agri-culture, building construction (paid), office work,etc.)	9.6	2.9	8.2	2.29	8.2	2.2	8.67	2.46
Religious purposes (prayer, visiting temples, etc.)	0.2	0.1	0.21	0.2	0.85	0.44	0.42	0.25
Entertainment (playing/cinema/radio listening, etc.)	0.4	0.6	0.31	0.58	0.4	0.51	0.37	0.56
Idle time/rest	5.5	4.7	1.9	3.25	2.4	2.99	3.27	3.65
Sleep	7.3	8.5	9.67	9.03	8.33	8.56	8.43	8.70
Building constructior	—	0.2	0.21	0.16	—	—	0.07	0.12

eight. In most of the families, at least one child, preferably the first child, has dropped out of school in order to work.

Girls universally perform domestic tasks such as cleaning, cooking, fetching water, looking after other children, cattle rearing and thereby assisting their mothers. Invariably, they work in their own families and also do wage labour. They perform several activities differentially on the basis of age and wage. While very young girls take care of other children and assist in processing food, the older girls are largely engaged in domestic work such as cleaning, grinding, fetching water and cooking at home, besides weeding, transplanting, harvesting and looking after animals at home.

Since awareness about education is steadily increasing, both young and older children spend maximum time in their studies, which is a recent development. The older boys otherwise spend most of their time grazing animals and run errands such as carrying food, shopping and so on. They also work along with their fathers in most agricultural activities. Activities such as fetching grass, fuel and to some extent water are mostly carried out by grown up children. They are able to get rest for an hour-and-a-half at midday and after six o'clock in the evening. Very often children spend their late evenings in recreation activities like cinema. Though the younger boys (aged nine years) perform certain agricultural and domestic activities, they usually seek entertainment or remain idle (see Table 5). Young children spend more time in entertainment and idling compared to the time they spend on education which is a major unhealthy practice. This is probably because of the absence of proper schooling and child care.

Thus children from all communities are involved in different and common activities over a longer period of time, throughout the day and season. Seasonal variations in activities are very little because agriculture is a continuous activity in this village. Since widespread and extensive agricultural phenomenon is seen among all the cultural groups tabulation with fertility has not been attempted. The great demand for child labour in this rural society is highly productive in farm and dairy development, besides minimising the cost of obtaining fuel and rendering other help in the family. Hence, child labour acts as a pronatalist force.

TABLE 5 Distribution of Sons and Daughters of the Respondents of Three Cultural Groups by Children's Age and Average (Input) Number of Hours Spent in Various Activities Per Day

Activities		*Caste Hindus*		*Harijans*		*Muslims*	
		5–9 years	*10–14 years*	*5–9 years*	*10–14 years*	*5–9 years*	*10–14 years*
Sons	N=	7	12	9	8	8	8
Domestic work		0.5	0.58	0.04	0.18	0.47	0.3
Looking after younger siblings		1.9	0.4	0.25	0.08	0.28	0.06
Cattle rearing (own)		—	0.5	0.1	0.94	1.56	0.6
Agriculture (own)		—	2.3*	0.1	0.38	0.08	1.1
Education		3.9	6.75	4.75	6.25	4.7	5.55
Entertainment		3.5	1.2	5.7	3.5	5.7	4.6
Idle time/rest		2.4	2.1	0.27	0.37	0.3	1.2
Religious functions		—	—	—	0.14	0.2	0.37
Sleep		10.0	10.1	10.9	7.2	10.4	9.2
Daughters	N=	8	13	9	4	10	3
Domestic work		0.8	3.5	1.3	5.4	1.0	3.8
Looking after younger siblings		0.4	0.37	0.7	—	1.71	2.3
Cattle rearing (own)		—	0.3	—	—	0.13	1.7
Agricultural/paid work		—	—	—	1.06	0.07	0.13
Education		6.4	6.33	4.01	2.1	1.92	1.1
Entertainment		4.06	1.4	5.36	0.81	5.77	3.7
Idle time/Rest		1.6	2.2	0.53	4.3	0.06	0.9
Religious functions		0.06	0.2	—	—	0.36	0.9
Sleep		10.56	9.9	10.72	8.46	11.3	9.3

Case Study of Three Families

The time allocation and activities of the husband, wife, son and daughter of the Caste Hindu, Harijan and Muslim families provide a realistic understanding of the time allocation and various activities performed by parents and their children at the micro level. The husbands in the three families spend maximum time in productive activities. While the Caste Hindu and the Muslim men spend most of their time in agricultural pursuits, the Harijan men spend most of their time in collecting firewood and selling it to make a living. As seen in Figure 1 there are many features in common among the families. However, for the women in the three families, there are more differences in their major activities. While the Caste Hindu women spend more productive time in cattle rearing, the Harijan and the Muslim women perform manual labour and agricultural activities chiefly in the morning and in the afternoon. But there are many common elements in their domestic work. Daughters usually help their mothers in most of the domestic work, while the Harijan girls also bring firewood from the forest. Sons, on the other hand, usually assist their fathers mostly in agricultural activities and among the Caste Hindu and the Muslim families they also assist in rearing cattle. Sons in the Harijan family spend more time in collecting firewood in order to earn money, when they are not attending school (see Figure 1).

Role of Family Planning and Other Factors

Among the hundred Muslim families of this village only in two families six persons had opted for male sterilisation, even though the clients happened to be illiterate. In our sample of Muslim families only one person opted for sterilisation. They believe that sterilisation is a sin and is not desirable according to their faith. Several other factors have nullified this traditional belief regarding sin and fear of losing heaven. In one family, a relative who is a doctor working as regional health officer, motivated the clients to accept sterilisation overlooking the religious beliefs. In another family, a Muslim doctor working in the primary health centre area motivated two persons from the same family to adopt contraception. Now several years after sterilisation, they are satisfied with their decision regarding sterilisation and health status. To a question whether the adoption of family planning was in any way against Islam, both the respondents who had adopted family planning said no emphatically. They said, 'when our living children suffer for want of care, attention, food, etc., how can we bother about sin and losing the chance of going to heaven , etc.' These respondents have, so far, not discussed their family planning status with their Muslim neighbours but have done so with their close relatives who were, in fact, motivated by them to adopt contraception. Surprisingly, one of the adopters motivated three of his Hindu neighbours to opt for sterilisation. These Muslim respondents do not pray five times a day as specified by their religion but pray two to three times a day only. They do not fully accept the views of their priest on family planning. This change in their belief system was consequent to the education and motivation given by the Muslim doctors. They believe that Muslim doctors and other categories of family planning workers can more effectively motivate Muslims to adopt contraception than a non-Muslim pro-

FIGURE 1 Time Allocation and Activities of Members of Three Families by Religion and Caste

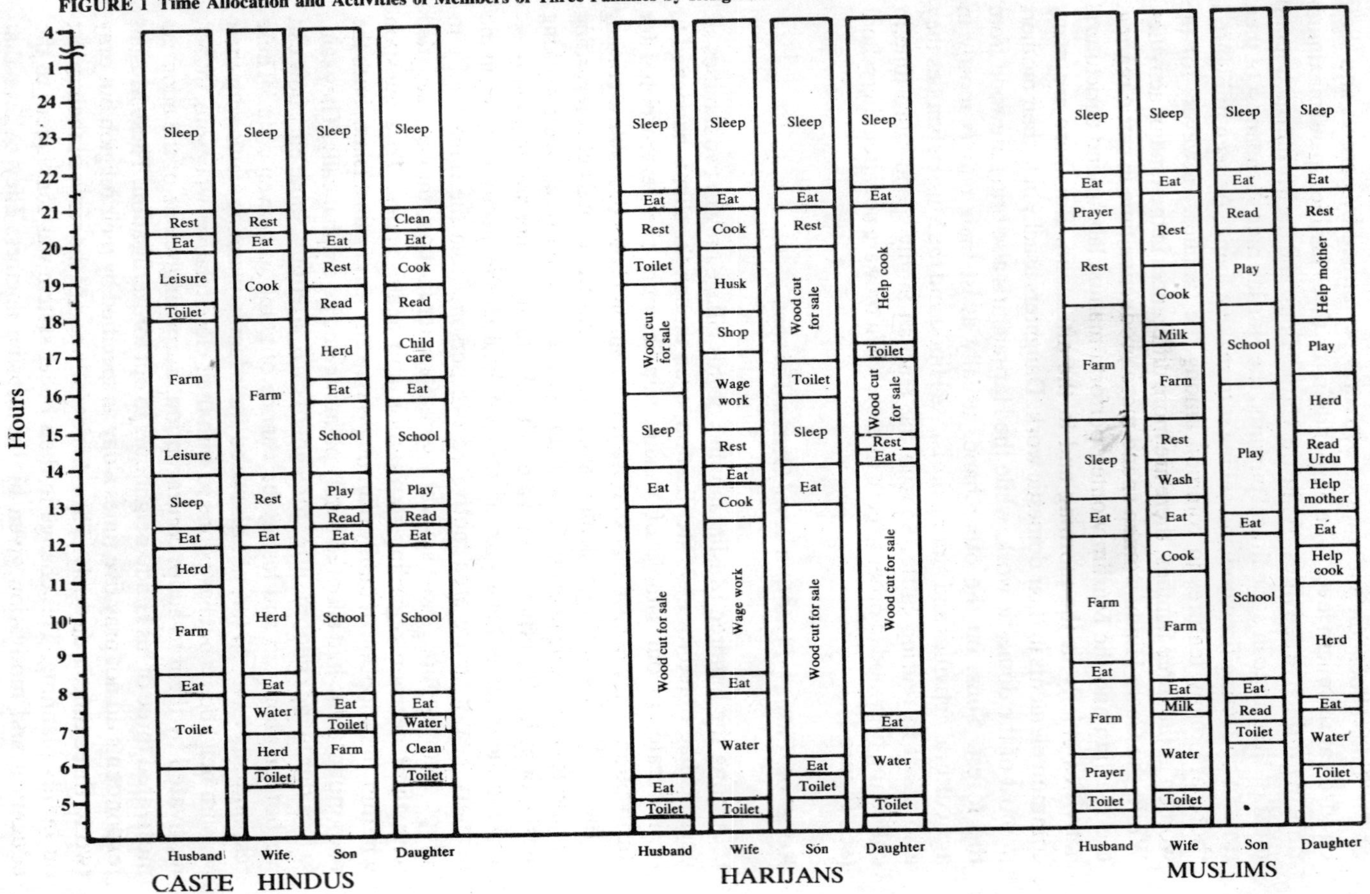

moter of family planning. This is an important policy matter for the consideration of administrators in a multi-religious country like India.

Family planning adoption varies significantly among the population of the three communities; there are eight adopters among the Muslims, seventeen among the Caste Hindus and nine among the Harijans in this village. The rate of adoption of family planning in this village as a whole, and by cultural grouping is far below the state and national average of adoption of family planning per year and also cumulatively. The present trend of adoption of family planning will increase and the Caste Hindus and the Harijans will stabilise (NRR) their population in a decade or two but not the Muslims if special efforts are not made to motivate them. However, adoption of family planning is of fairly recent origin and has spread over less than a decade and, therefore, it cannot explain much of the existing differential fertility among the three cultural groups.

The influence of the age at marriage is more or less same; the median age at marriage is 14 years for the Muslims and the Caste Hindus and 12 years for the Harijans. However, the duration of breast-feeding varies significantly among these communities; the mean duration of breast-feeding is 27.7 months for the Harijans followed by 19.1 months for the Caste Hindus and 16.7 for the Muslims, and it is 21.4 months for the population as a whole. Hence, the longer duration of breast-feeding among the Harijans is a major factor that depresses their fertility.

Thus, the differential influence of leadership, educational attainment, conformation or otherwise to cultural beliefs, roles of women and children, aspirations to be at par with the relatively developed community, child labour and the value of children, and occupational mobility that were noticed among the three cultural groups differentially contribute to their fertility differences. The Muslims' conformity to their cultural patterns, that is, an implicit faith in the advice of their priest, beliefs regarding family planning as a sin and attributing it for not attaining heaven in the future, poor educational background and demand for child labour, a desire for extending kinship ties together maintain their fertility at a very high level. The unexpected low level of fertility of the traditionally high fertility community, the Harijans, was largely due to their educational advancement, liberation from the traditional influence of upper castes, long duration of breast-feeding, occupational mobility and the benevolent influence and help rendered by their leader for the advancement of the Harijans in general. The low fertility of the Harijans, however, cannot be attributed to the effectiveness of family planning as their adoption level is much lower than that of the Caste Hindus. The Caste Hindus maintain a medium level of fertility falling between the Muslims and the Harijans. But their fertility level is closer to the Harijans' fertility level because relatively more of them have accepted family planning. Though they are traditional in their outlook and child labour is universal among them, they are likely to control their fertility level in the near future primarily because of scarcity of land, and the rapidly increasing educational level. Therefore, in order to promote fertility control in a community like this, changes in the attitude of religious leaders towards family planning, child labour, concept of belief, norms and values related to family formation, female education and an increase in the age at marriage, are important. Along with these social changes, strengthening of proximate variables like breast-feeding, and long birth interval may be considered besides vigorously implementing family planning programmes to reduce the overall fertility of this community.

APPENDIX I

Methodology

An in-depth study of Ramapuram village, which is also known as Aythepalli lies south of Tirupati town in Andhra Pradesh, India, was carried out for a period of one year during 1982–83. The village is situated about sixteen kilometres away from the famous Hindu pilgrim centre at Tirupati. Tirupati town, the abode of the most famous Hindu god, attracts 25,000 pilgrims, on an average, per day from various parts of the country. For serving the pilgrims, the temple has set up several organisations and industries which employ people from far and near Tirupati town. This village is also benefited by the different types of occupations which are open to the cross-section of population employed as sweepers, scavengers, peons, clerks, technicians, teachers and other administrative staff. The proximity of Tirupati pilgrim centre, thus, provides employment opportunities to men in various capacities, they receive medical facilities for their families, educational facilities for their children, loan facilities to construct a house and buy vehicles. Further, it is endowed with several other influences of the Tirupati urban complex, namely, frequent transport and communication facilities for shopping, exposure to mass media, particularly newspapers, radio and cinema. The utilisation of services like health centres and veterinary hospitals are a common feature as compared to several other villages in this part of the country. Thus, this village is relatively developed in the present context.

The universe of this study, Ramapuram village, has a population of about 2,500. The population is characterised by the Hindu and the Muslim religious groups and a group of forward and backward castes* revealing a distinct socio-economic dichotomy among them. Since several factors are common among the forward castes (Caste Hindus) they are grouped together for the purpose of study. Of the total population, all the eligible couples numbering fifty-five from the three cultural groups—the Caste Hindus (20), the Harijans (20) and the Muslims (15) with two or more living children and at least one child in the age group of 5–14 years, were included as respondents. However, five Muslims did not co-operate for the survey and dropped out after three rounds

* The forward castes include Reddy, Naidu and Balija communities. The backward castes consist of Ediga and Gandla caste groups.

of interview. This procedure was followed to study the various roles performed by the parents as well as by children in the family. Thus, Ramapuram village was purposively selected and all the eligible couples with the above mentioned characteristics belonging to the different religious and caste groups were considered.

The survey was carried out repeatedly in every alternative month throughout the year, thus covering six rounds of interview during the reference period. Information was primarily obtained from the wife and partly from the husband and children. However, observation was followed to document the various activities of all the members of the household. Though the original plan was to interview every household at least once in a month, it was difficult to continue because of the resistance expressed by a few respondents. Nevertheless, interview during the alternative months was made possible by using several rapport building techniques, particularly, involvement of a doctor to attend to the morbidity and health needs of the mother and children in all the families, besides seeking the help of influential leaders of the respective communities. A schedule was used for data collection and the interview was conducted by two trained investigators. However, most of the qualitative information was gathered by the author himself.

APPENDIX II

Per Cent Distribution of Respondents in the Three Cultural Groups by Selected Socio-Economic Variables

	Caste Hindus	*Harijans*	*Muslims*	*Total*
Education				
Both illiterate	10.0	30.0	33.3	23.6
	(2)	(6)	(5)	(13)
Husband literate +	50.0	60.0	26.7	47.3
wife illiterate	(10)	(12)	(4)	(26)
Both literate	40.0	10.0	40.0	29.1
	(8)	(2)	(6)	(16)
Occupation				
Traditional	80.0	40.0	66.67	61.8
	(16)	(8)	(10)	(34)
Non-traditional	5.0	50.0	13.3	23.6
Class IV	(1)	(10)	(2)	(13)
Non-traditional	15.0	10.0	20.0	14.5
Class III	(3)	(2)	(3)	(8)
Income				
≤3000	15.0	35.0	13.3	21.8
	(3)	(7)	(2)	(12)
3001–5000	25.0	15.0	26.7	21.8
	(5)	(3)	(4)	(12)

5001–7000	20.0 (4)	25.0 (5)	26.7 (4)	23.6 (13)
7001+	40.0 (8)	25.0 (5)	33.3 (5)	32.7 (18)
Land-owning				
No land	30.0 (6)	75.0 (15)	26.7 (4)	45.5 (25)
≤1 acre	5.0 (1)	20.0 (4)	33.3 (5)	18.1 (10)
1–2 acres	35.0 (7)	—	26.7 (4)	20.0 (11)
2+ acres	30.0 (6)	5.0 (1)	13.3 (2)	16.4 (9)
Animals owning				
No animal	50.0 (10)	80.0 (16)	33.3 (5)	56.3 (31)
1–2 animals	30.0 (6)	20.0 (4)	46.7 (7)	30.9 (17)
3+ animals	20.0 (4)	—	20.0 (3)	12.8 (7)
N =	20	20	15	55

REFERENCES

DAVIS, KINGSLEY. (1951). *The population of India and Pakistan*. Princeton: Princeton University Press.

DRIVER, E.D. (1963). *Differential Fertility in Central India*. Princeton: Princeton University Press.

CENSUS OF INDIA. (1971). Monograph No. 10. New Delhi: Registrar General, India.

MAHADEVAN, K. (1979). *Sociology of fertility: determinants of fertility differential in South India*. New Delhi: Sterling Publishers.

RAO, V. KAMESWARA. (1976). Chandragiri fort temples in Ithihas. *Journal of Andhra Pradesh Archives*. Hyderabad: Government of Andhra Pradesh, pp. 89–100.

RELE, J.R. and TARA KANITKAR. (1979). *Fertility and family planning in Greater Bombay*. Bombay: Popular Prakashan.

UNITED NATIONS. (1961). *The Mysore Population Study*. Population Studies No. 34. New York: United Nations.

8

Interrelation Between Fertility and Non-Economic Characteristics

P. KRISHNAN

Introduction

Human fertility is a complex phenomenon. The birth performance of a female is influenced not only by physiological factors, but also by a series of social, economic, cultural and psychological determinants which interact with each other as well as with the physiological factors. This paper attempts to study the interrelation between fertility and selected non-economic characteristics in Gujarat, India. Some of the characteristics studied are religious affiliation, age at marriage, present age, level of schooling and residence.

Since Davis' (1951) *The Population of India and Pakistan*, where the sociology of fertility in India has been brought out rather clearly, many a study on fertility has been carried out, resulting in routine reports. Theoretically based meaningful research, however, is rare. Recently, Mahadevan (1979) has looked at fertility in Tamil Nadu from a sociological perspective. An earlier piece of research on fertility differentials in Central India by Driver (1963) as well as the United Nations' (1961) classic on the population of Mysore may be pointed out here. Detailed fertility studies based on census data have been rather meagre. This is because a major non-economic but demographically relevant variable, namely, religion, is not cross-classified with others (for example, age, education). There have been, of course, special studies (for example, on Parsees by Karkal, 1977) done at the time of the 1971 Census. But a major source of information on fertility, collected at the time of the census and presented in the F-series of special tables (India, 1978), does not seem to have been, as far as this author is aware of, fully tapped. These tables present the distribution of 'currently married women and the related births' during the year before the census, by education, duration of marriage, age at marriage, present

Acknowledgements This is a revised version of a paper contributed to the conference on India's Population held at Oxford University in December 1982. The research was supported by a grant from the Social Sciences and Humanities Research Committee and the Department of Sociology, University of Alberta. Part of this research was done when the author was at the Centre de Estudios Demograficos y Desarollo Urbano (CEDDU), El Colegio de Mexico, while on sabbatical leave. The facilities provided by the CEDDU, El Colegio de Mexico are gratefully acknowledged. The author is grateful to Miss Lizy Kurian, Department of Sociology, University of Alberta, for research assistance and to Dr. Carlos Brambila of El Colegio for help with one of the log-linear models.

age, religion and residence. We analyse such data in order to assess the fertility dynamics in India.

Theoretical Model

THE FRAMEWORK

As noted earlier, fertility is influenced by several factors both individually and collectively. There are a number of sociologically relevant frameworks developed by Davis and Blake (1956); Freedman (1961–62); Hill, Stycos and Back (1959) and Tien (1968). Among the economic models are the ones developed by Becker (1960); Easterlin (1969) and Namboodiri (1972). Since non-economic characteristics are considered, the theoretical framework adopted here is from among the sociological models. As socio-psychological and demographic factors (for example, infant mortality) are not available for the analysis, the frameworks of Tien and Hill et al., cannot be put to use here. Thus, we are left with the Davis-Blake and the Freedman models, and as Freedman's framework is centred around norms, it is relevant here. Religious affiliation and social environment lead one to internalise certain norms and values regarding children. At times, as in India, economic necessity is a part and parcel of these norms (for example, the need for begetting a male issue as old age security and to perform funeral rites). It is easy to see how these norms determine the use of some, or all of the intermediate variables suggested by Davis and Blake. We will, therefore, consider only the Davis-Blake framework to develop our theoretical bases. A diagrammatic presentation of the model used here is given in Figure 1.

We have 'religious affiliation' (a poor indicator of religiosity or how deeply committed a person is), physiological factor 'age,' and rural/urban residence (environment) as independent determinants. The environment may interact with religion. It is seen that in the urban areas the variable religion is less significant than in the rural areas. This erosion of faith arises as a result of exposure to modernisation and interaction between people drawn from diverse backgrounds. Similarly, age and religion interact; the young tend to be less religious.

Between the set of Davis-Blake intermediate variables and 'religion,' 'age,' and 'environment,' we have 'norms' and 'education'. Norms are shaped by 'religion' and affected by 'age' and 'environment'. 'Education' is also governed by 'religion' and 'environment'. Its relationship with 'age' is direct, as primary level education, at least, is emphasised in India now. Similar reasoning can be applied to other variables as well. One can control one or more variables at a time. It is also possible to throw in various terms of interaction to make the model comprehensive.

PREVIOUS RESEARCH AND FINDINGS

A review of earlier research helps to formulate a proper relationship between

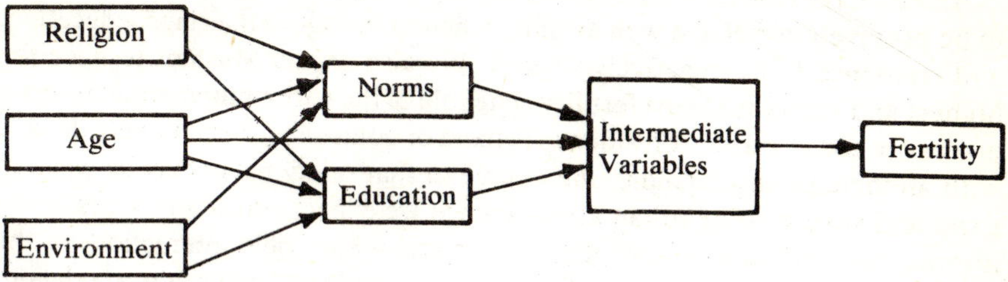

FIGURE 1 Theoretical Model-Flow Chart

fertility and some of the non-economic characteristics. An analysis of the 1951 Census data on fertility reveals that for Travancore-Cochin urban fertility (6.4 children on an average) is lower than rural fertility (6.6 children).

There is more information on fertility by religious groupings. Parsees have been known to have the lowest fertility in India, followed by Jains (see Davis, 1951). Majumdar's (1960) Kanpur survey showed that Moslem women, on an average, gave birth to 8.0 children as compared to 7.0 by Hindu women. Driver (1963) found that in Central India, the average number of children by religious groupings was 4.6 (Moslems) and 4.5 (Hindus). A similar pattern of Hindu-Muslim difference in fertility is oberved in the United Nations' Mysore Population Study. The UN study also pointed out differentials by level of schooling. It was seen that in Bangalore City, women with less than seven years of schooling (middle school) had, on an average, 5.3 to 5.5 children, while those who had ten or more years of schooling (high school, or more) had only 3.9 children.

There is more evidence of differential fertility by the age at the time of marriage in India. The Mysore survey showed that rural females who married between 14 and 17 years of age had 5.9 children, while those who married between 18 and 21 had only 4.7 children. The Kanpur survey revealed that the average number of children born to women who were below 15 years of age at the time of marriage was 6.9, and 6.0 for those who married after 19 years of age. Mukherjee's (1961) Calcutta survey, Balakrishna's study on Madras and Agarwala's survey of the villages near Delhi and Lucknow reveal (see Agarwala, 1967, pp. 38–39) that females marrying after the age of 19 have their fertility reduced by about 0.5 to 1 child on an average. Jain (1975) has noted that in rural Punjab females marrying below 18 had 5.7 children on an average, as against 5.2 for those married between 18–22 years and 4.4 for those married after 23 years of age.

Fertility differentials by caste have been dealt with in depth by Davis (1951) and in the Khanna study (Wyon and Gordon, 1971). This aspect is not included in our analysis. An overview of the major results in Indian differential fertility is available in Jain (1975).

Some Hypotheses

Based on previous research on fertility in India and elsewhere, we expect age to be positively correlated with fertility; whereas the age at marriage, education, and residence to be negatively related. We also expect Muslims to have the highest and Jains the lowest fertility. The role of the various interaction terms is not easily conjectured. If residence is fixed or controlled, fertility may decline with an increase in schooling; the degree of that relationship may be weak in rural and strong in the urban areas. Hence we expect education × residence interaction moderate in strength. Similar arguments yield interactions of the type: age × education, age × religion, age × residence, education × religion, residence × religion, religion × age at marriage to be moderate or weak. As mentioned earlier it is not easy to conjecture the nature of higher order interactions and, therefore, we have omitted them in the statistical analysis.

Data

The data cited here are for the state of Gujarat, from the special tables on fertility, collected in the 1971 Census. The results developed for Gujarat, it is hoped, would be reflective of the general nature of such relationships for the country as a whole. In the non-statistical section of the paper, various types of rates have been calculated and analysed. For the statistical model, births in 1970 have been cross-classified by a manifold contingency table and analysed by the log-linear technique.

In the demographic analysis section, Sikhs and Buddhists have been omitted, as no reliable measures of fertility could be developed with a small number of cases per cell.

Demographic Analysis

We present a simple demographic analysis here in order to elicit differentials in fertility. Since the data are categorised into several classes on the bases of several characteristics simultaneously, mostly standardised measures are shown.

Rural-Urban and Religious Differentials by Present Age

In Table 1, total fertility rates (TFR) based on the age span 13–47 years are shown for Gujarati women by religious affiliation and residence. The rural-urban differential is clear for all the religious groups and is in the expected direction. The largest differential is noticed for Hindus and the smallest for Christians. As regards religious groupings, Muslims have the highest fertility, and Jains the lowest. The fertility of Hindus and Christians is intermediate to those of Muslims and Jains. The rank order differs slightly, if residence is controlled. In urban Gujarat, fertility of the Hindus is the second lowest, followed by the Christians. In the rural parts of the state, Hindus have the second highest fertility.

TABLE 1 Total Fertility by Religion and Residence, Gujarat, 1970

Residence	*Religion*				
	Hindu	*Muslim*	*Christian*	*Jain*	*All Religions*
Rural	5,988	6,674	5,476	4,862	6,007
(Index Nos)	(99.7)	(111.1)	(91.2)	(80.9)	(100)
Urban	4,899	5,780	5,389	4,295	5,007
(Index Nos)	(97.8)	(115.4)	(107.6)	(85.8)	(100)

SOURCE: Computed from Census data

Using the TFR of all the religious groups together, index numbers are calculated for the four major religious groups considered here (Table 1). A close examination of the index numbers or the TFRs themselves, reveals that the differentials fall in a slightly narrow range in urban than in the rural areas. The Jain-Muslim differential is 1,812 per 1,000 women in rural and 1,485 per 1,000 in urban Gujarat. On the other hand, the Hindu-Jain differentials are 1,126 (rural) and 604 (urban) and interestingly enough, the Christian-Jain differentials are 614 (rural) and 1,090 (urban). While for the Hindus and Muslims, the shift from rural to urban decreases fertility in relation to the Jains, for the Christians an increase is noted.

Rural-Urban and Religious Differentials by Marriage Duration

Table 2 presents the TFR values obtained by standardising for marriage duration (30 + years of marriage duration is taken as 30–40 years).[1] We again see that urban fertility is lower for all the religious groups combined and for each of the major groups under consideration here. The largest rural-urban differential is noted for the Christians and the lowest for the Muslims.

Considering the religious groups again, we see that the lowest fertility is for the Jains and the highest for the Muslims. In rural Gujarat, the fertility differences among the Hindus, Muslims and Christians are, indeed, small. The Jains stand out as a separate group in this regard. In urban Gujarat, the differences are rather marked (Hindu-Muslim 633, in favour of Muslims; Hindu-Christian 662, in favour of Hindus; Hindu-Jain 1,280, in favour of Hindus; Muslim-Christian 1,285, in favour of Muslims; Muslim-Jain 1,913, in favour of Muslims; Christian-Jain 618, in favour of Christians). We may note,

[1] It is presumed that the data refer to currently married women and that the fertility referred to here is marital fertility. For Indian society, such an assumption seems reasonable.

then, that in the urban environment, fertility differentials are quite conspicuous among the religious groups.

TABLE 2 Marriage Duration Standardised Fertility Rate by Religion and Residence, Gujarat, 1970

Residence	*Religion*				
	Hindu	*Muslim*	*Christian*	*Jain*	*All Religions*
Urban	5,204	5,837	4,542	3,924	5,255
(Index Nos)	(99.0)	(111.1)	(86.4)	(74.7)	(100)
Rural	6,153	6,503	6,026	5,249	6,161
(Index Nos)	(99.9)	(105.6)	(97.8)	(85.2)	(100)

SOURCE: Computed from Census data

RURAL-URBAN AND RELIGIOUS DIFFERENTIALS BY AGE AT MARRIAGE

Total fertility like measures[2] are computed by age at marriage (28 + group is taken as 10 years in length) for all the major groups and Gujarat as a whole, and the index numbers for such measures are shown in Table 3. Here, too, the rural-urban differential is in the expected direction. As far as the differences in fertility by religious affiliation are concerned, the lowest and the highest levels are noticed for the Jains and the Muslims respectively. The urban Christians have lower fertility than the urban Hindus (the differential is rather negligible), while the rural Hindus have lower fertility than their Christian counterparts (differential not insignificant).

We may conclude that the rural to urban shift produces a change in the direction of religious differentials as far as the Hindus and Christians are concerned.

TABLE 3 Index Numbers of Fertility Rates Standardised for Age at Marriage by Religion and Residence, Gujarat, 1970

Religion	*Rural*	*Urban*
Hindu	99.3	98.6
Muslim	118.5	114.3
Christian	104.8	97.2
Jain	93.6	87.4
All Religions	100	100

SOURCE: Computed from Census data

[2] These TFR like quantities are computed exactly like the TFRs. Specific rates by age at marriage are multiplied by the appropriate lengths of age intervals and summed. These TFR like values will not equal the conventional TFR values obtained from age specific fertility rates.

Fertility Differentials by Religion, Residence and Education

The TFRs for rural and urban Gujarat by religious grouping and level of schooling are shown in Table 4. As one moves from education Group 1 to Group 4, the level of schooling increases.

In general, for all the religious groups, urban and rural, it is seen that fertility and schooling are negatively related. For the urban Christians and Jains, a slight curvilinearity is observed, when the fertility of educational groups is viewed along a continuum. The general pattern with respect to religious differentials, when schooling and residence are controlled, is as follows: in the urban areas, for those with higher levels of schooling (high school and above), the Hindus have the lowest fertility, followed by the Jains, Muslims and Christians in that order; for urbanites with some schooling (less than high school), the Hindus still have the lowest fertility, followed by the Jains and Christians, with the Muslims having the highest fertility. Among the illiterate urbanities of Gujarat, the Christians seem to have the lowest fertility, followed by the Jains, Hindus and Muslims.

TABLE 4 Total Fertility by Religion, Residence, and Education, Gujarat, 1970

Residence	Education	*Religion*			
		Hindu	*Muslim*	*Christian*	*Jain*
Urban					
	Group 1	5,427	6,003	3,192	4,403
	Group 2	4,364	5,348	4,773	4,452
	Group 3	3,864	4,214	4,649	3,882
	Group 4	3,490	NA	NA	3,865
Rural					
	Group 1	6,149	6,680	5,533	5,127
	Group 2	4,955	6,560	5,344	4,938
	Group 3	4,558	NA	NA	NA
	Group 4	3,135	NA	NA	NA

SOURCE: Computed from Census data

Legend: Group 1—Illiterate; Group 2—Literate but less than Matriculate; Group 3—Matriculate and above but less than Graduate; Group 4—Graduate and above

In rural Gujarat, the pattern regardless of schooling is: the Jains exhibit the lowest fertility and the Muslims, the highest birth performance. Among the illiterates, the Christians have lower fertility than the Hindus. But for those with a little schooling (less than high school), the Hindu women have lesser number of children as compared to the Christians.

It may be concluded, therefore, that education affects fertility in such a way as to change the general pattern obtained by considering one or two characteristics alone.

INFLUENCE OF EDUCATION ON FERTILITY THROUGH AGE AT MARRIAGE

Education influences fertility in more than one way. It leads to the awareness of birth control measures and thus directly affects fertility. Also, it modifies the age at marriage, and the exposure to sexual unions. As the institution of marriage is still strong in India, the higher the age at marriage, which is positively linked to schooling, the lower the fertility. This may not be true, where education is religion-based, as is the case for Catholics. In general, education in India is secular, and therefore, we do not expect any relationship to the contrary.

Considering the women who had children in 1970 as a sample, the mean and the median age at marriage may be computed. These are shown in Table 5 by religion, education, and residence. The pattern which emerged is:

(*a*) The mean age at marriage increases as the level of schooling increases, irrespective of religion and/or residence.

(*b*) The Christians have the highest mean age at marriage and the Hindus, the lowest, flanking the Jains and the Muslims.

(*c*) The mean age at marriage is higher in the urban as compared to the rural areas.

TABLE 5 Median (M) and Mean ($\bar{X}$) Age at Marriage by Religion, Residence and Education, Gujarat, 1970

Residence	Education	*Hindu*		*Muslim*		*Christian*		*Jain*	
		M	$\bar{X}$	*M*	$\bar{X}$	*M*	$\bar{X}$	*M*	$\bar{X}$
Urban									
	Group 1	16.5	17.0	17.9	19.1	17.6	18.6	17.2	18.0
	Group 2	17.3	18.1	18.8	17.9	19.7	19.3	18.2	17.0
	Group 3	20.0	19.8	20.3	20.2	21.5	22.0	20.6	20.6
	Group 4	22.1	22.9	22.2	23.0	24.9	24.5	22.9	24.2
	Overall	18.7	19.5	19.6	20.1	20.6	21.1	19.4	20.0
Rural									
	Group 1	16.6	17.1	17.9	16.7	17.1	18.0	16.7	17.3
	Group 2	16.7	17.3	18.4	17.4	18.0	16.8	17.6	18.7
	Group 3	19.6	19.2	20.2	20.1	20.4	20.3	19.8	19.4
	Group 4	21.5	22.0	NA	NA	NA	NA	NA	NA
	Overall	16.7	17.9	18.4	18.1	18.0	18.4	17.6	18.5

SOURCE: Computed from Census data

Legend: Group 1—Illiterate; Group 2—Literate but less than Matriculate; Group 3—Matriculate and above but less than Graduate; Group 4—Graduate and above

Similar to these are the differentials in fertility, as corroborated by the findings cited earlier.

FERTILITY BY AGE AT MARRIAGE, RELIGION, AND EDUCATION

Table 6 is complementary to Tables 3 and 4 showing total fertility (hence standardised for the present age distribution) by religion, age at marriage, education and residence. Owing to the paucity of data, only two education groups are shown. The differentials by religion, residence, and education with one of the two characteristics at a time are in the expected direction. The age at marriage brings in a curvilinearity to the relationship. Those married between 13–17 years of age have higher fertility than those married at other age levels.

This is true regardless of the group one is considering, with one, two or three characteristics controlled. Curvilinearity may arise from physiological (sub-fecundity in those who marry before 13 years), and sociological (non-exposure to sexual union due to strict social control) factors.

Log-Linear Model

A log-linear analysis[3] is developed to obtain a description of the relationship between the factors shown in the table (indicated by the variables used) either by forming a model for the data, or by testing and ordering the interactions between the factors.

A SIMPLE MODEL

This is an unconventional approach to model building.[4] But our objective is to look for patterns in the data. The TFR values of rural and urban Gujarat by religious and educational groupings have been considered (see Table 4). The variables used are R—religion (four levels), E—education (two levels), and U—residence (two levels). The frequencies in each cell are the number of births per 100 women standardised for age distribution. The census data, it is assumed for the purposes of the statistical analysis, are a sample drawn from a super population.

First we fit the saturated model to obtain the following (after necessary literations):

Effect	*Degrees of Freedom*	*Probability*
R	3	0.0000
E	1	0.3317
U	1	0.0000
RE	3	0.0001
RU	3	0.0164
EU	1	0.0055
EUR	3	0.0002

[3] For a good overview of log-linear analysis, see Fienberg (1980). The BMDP and the new SPSS packages also include good write-ups on this methodology. The BMDP programme has been used here for computation.

[4] A more detailed log-linear model is presented in the Appendix.

TABLE 6 Total Fertility by Age at Marriage, Religion, Residence, and Education, Gujarat, 1970

Residence	*Age at Marriage*	*Hindu*		*Muslim*		*Christian*		*Jain*	
		Group 1	*Group 2*	*Group 1*	*Group 2*	*Group 1*	*Group 2*	*Group 1*	*Group 2*
Urban	Less than 13	5,274	4,339	5,956	5,810	NA	NA	NA	NA
	13–17	5,741	4,432	6,291	5,528	3,582	4,571	4,571	4,476
	18–22	5,433	4,415	5,815	5,279	1,349	4,705	4,705	4,299
	23–27	4,027	3,613	4,662	4,286	NA	NA	NA	NA
	28+	2,634	2,280	3,621	2,011	NA	NA	NA	NA
Rural	Less than 13	5,914	4,831	5,876	5,903	5,122	4,745	NA	NA
	13–17	6,165	4,980	6,810	6,327	5,022	6,411	5,984	5,350
	18–22	5,949	4,966	6,291	6,367	6,724	5,204	4,398	4,498
	23–27	4,623	3,325	5,216	6,937	NA	NA	NA	NA
	28+	972	1,693	4,459	NA	NA	NA	NA	NA

SOURCE: Computed from Census data

From the fit, we see that the main effect E is not significant, whereas all the others, including the third order interaction, are significant. Therefore, we will omit E in further analysis, even though most of what we know in demographic literature goes against this finding.

Next, we fit the model (UR). It is found to be significant. The models (UE), (RE), (UR, UE), and (RE, UR, UE) are all significant. Since the probability values are all close to zero, and less than the probability parameter (– , 0.05), the lambda parameters are not computed by the programme.

We can rank order the various models on the basis of the size of the Likelihood Ratio Chi Square, or the Pearson Chi Square. These are:

Model	*LR Chi Square*	*Pearson Chi Square*
Saturated	0	0
(UR, RE, UE)	19.60	19.56
(UR, RE)	27.30	27.22
(UR)	49.35	49.12
(RE)	119.30	118.39
(UE)	155.07	156.27

The parsimonious model seems to be (UE)—residence and education together determine the fertility levels of the population groups considered here.

Discussion and Conclusions

We have presented the results from a demographic and a statistical analysis of the data on human fertility in Gujarat. We find that the fertility differentials by religion, education, and residence are in the expected direction. Surprisingly, the education variable does not appear to be statistically significant, even though substantively those women with lower levels of schooling had higher birth performance. It may be mentioned here that in the analysis the higher levels of schooling (high school and above) were omitted as the TFRs could not be satisfactorily computed. We presume that, if these groups were included, the education variable would have been significant.

Although education is not, religion, residence and the various interactions are statistically significant and hence sociologically meaningful. In fact, the best model to explain the data is seen as (UE)—residence, education and their interaction. Religion does not seem to carry much weight at all. This leads one to conclude that the modernisation elements flowing through education and residence are most influential in determining the levels of fertility of this population. Further analysis with data on the age at marriage, income, use of contraception, participation in the labour force, may bring out the fertility dynamics in Gujarat and therefore in India, better and clearer.

APPENDIX

Findings from a More Detailed Log-Linear Model

The results of a more detailed analysis are highlighted here. Instead of the TFR values, births experienced a year before the census are analysed here for Gujarat. The variables are: A—age, M—age at first marriage, E—education, R—religion, U—residence.

A falls into three categories—13–22 years, 23–37, and 38+; M is trichotomous—less than 13 years, 13–22 and 23+; E is dichotomous—illiterate and others; U is also dichotomous—rural and urban. Finally R has four categories—Hindus, Muslims, Christians and others.

Since second and higher order interactions are not easily interpretable, we shall not fit models involving higher order interactions.

First we fit (A, M, E, U, R). Then we add first order interaction terms to the above. The programme picked (A, M, E, UR) as the best. Now let us delete one or more main effects and ask the computer to select the best. The best model after such deletions was seen as (A, M, E, U). The model (AM, AE, AU, AR, ME, MU, MR, EU, ER, UR) was fit and after deletions, the most parsimonious model turned out to be (AM, AE, AU, AR, ME, MU, MR, EU, ER). Other models, by adding or deleting third and higher order interactions, have been tried but are not reported here.

We can make some sociological sense of these statistical manipulations. From the various 'best models' involving main effects and first order interactions, we see that religion as such is not an important factor in fertility in India. Factors such as the age at marriage, education, residence and its interaction with religion have greater significance in determining fertility differences.

REFERENCES

AGARWALA, S.N. (1967). *Population*. New Delhi: National Book Trust.

BALAKRISHNA, S. (1971). Family planning: Knowledge, attitude and practice. Hyderabad: National Institute of Community Development.

BECKER, G.S. (1960). An economic analysis of fertility. *In Demographic and economic change in developed countries*. National Bureau of Economic Research. Princeton: Princeton University Press.

DAVIS, K. (1951). *The population of India and Pakistan*. Princeton: Princeton University Press.

DAVIS, K. and J. BLAKE. (1956). Social structure and fertility: an analytic framework. *Economic Development and Cultural Change*, 4, 211–35.

DRIVER, E.D. (1963). *Fertility differentials in Central India*. Princeton: Princeton University Press.

EASTERLIN, R.A. (1969). Toward a socio-economic theory of fertility: survey of recent research on economic factors in American fertility. *In* S.J. Behrman *et al.* (Eds.), *Fertility and family planning, a world view*. Ann Arbor: University of Michigan Press.

FIENBERG, S.E. (1980). *The analysis of cross-classified categorical data*. Cambridge, Massachusetts: The MIT Press.

FREEDMAN, R.A. (1961–62). The sociology of human fertility: a trend report and bibliography. *Current Sociology*, 10–11, 1–121.

HILL, R., J.M. STYCOS and K.W. BACK. (1959). *The family and population control*. Chapel Hill: University of North Carolina Press.

INDIA. (1978). *Social cultural tables*, Vol. I, Series I, Part II C (III), Census of India 1971. New Delhi: Registrar General and Census Commissioner.

JAIN, S.P. (1975). *Demography—a status study of population research in India*. Vol. II. New Delhi: Tata McGraw Hill Publishing Company.

KARKAL, M. (1977). *The Parsees*. Doctoral thesis submitted to Bombay University. Bombay: International Institute for Population Studies.

MAHADEVAN, K. (1979). *Sociology of fertility: determinants of differentials in South India*. New Delhi: Sterling Publishers.

MAJUMDAR, D.N. (1960). *Social contours of an industrial city*. Bombay: Asia Publishing House.

MUKHERJEE, S.B. (1961). *Studies on fertility rates in Calcutta*. Allahabad: Bookland Ltd.

NAMBOODIRI, N.K. (1972). Some observations on the economic framework of fertility. *Population Studies*, 26, 185–206.

TIEN, H.Y. (1968). The intermediate variables, social structure, and fertility change. *Demography*, 5, 138–57.

UNITED NATIONS. (1961). *Mysore population study: report of a field survey carried out in selected areas of Mysore State, India*. New York: United Nations.

WYON, J.B. and J.E. GORDON. (1971). *The Khanna study*. Cambridge: Harvard University Press.

9

Cost of Rearing Children and Fertility

D. Usha Rani

Introduction

The cost and value of children to parents is a major and valuable research area within the general realm of the determinants of fertility. The importance of this concept and its application in the field of population studies was increasingly realised only during the 1970s. The concept of the cost and value of children is usually associated with a number of disciplines—economics, sociology, demography and psychology. While discussing the micro level concepts and assumptions towards a broader theoretical view regarding these themes, Fawcett (1976) pointed out the differences and similarities among the various approaches—sociodemographic, socio-structural, socio-psychological and microeconomic. Though the first and second approaches appear to be substantially similar, the former emphasises urbanisation and industrialisation with concomitant changes in the value and cost of children as factors in demographic transition, while the latter focuses on the institutional structure of a society as a determinant of the utility and cost of children (Blake, 1972). The third approach states that children serve different functions in different kinds of societies, as well as in the different strata within a society. Knowledge about the perceived costs and value of children can be interpreted within this framework to motivation, and can provide an additional dimension of knowledge to test the economic decision-making approach to fertility. The microeconomic approach based on the conventional theory of consumer behaviour, views the individual as trying to maximise satisfaction, given a range of goods, their prices and his own tastes and income. This approach views children as special kind of goods and focuses on the definition and measurement of the income and price variables and on the role of tastes and preferences in childbearing decisions. The third and fourth approaches are highly relevant for the present study.

The socio-psychological approach to the study of the value and cost of children is fairly exhaustive since it attempts to incorporate all the dimensions of the value and cost of children relevant to fertility decisions. Hoffman and Hoffman (1973) provided an appreciable theoretical rationale for undertaking research in this field. In this model, the dimensions of the value of children are

Acknowledgement The paper is based on the doctoral work of the author under the guidance of Professor K. Mahadevan, Department of Population Studies, Sri Venkateswara University, Tirupati.

viewed as a set of variables intervening between socio-demographic factors and psychological and social orientations on the one hand and fertility and family planning on the other (Arnold *et al.*, 1975). The interplay between the value of children and family size decisions is further summed up by Chung (1974):

> On a conceptual level, the value of children can be regarded as a balance between two opposing forces and it is assumed that child bearing decisions depend on which of the opposing forces dominates. As long as positive values of children outweigh the negative values, parents will continue having children and will stop having children only when the negative and positive values of children are at a balance or when negative values outweigh positive values.

The microeconomic approach, on the other hand, asserts that the differences in fertility are the result of family decision processes whereby households choose a family size that maximises satisfaction, given the relative prices of goods, tastes and total income. Early application of neo-classical price theory to fertility (Becker, 1960) maintained that the demand for children is a direct function of family income, the direct costs of children and the prices of other goods and services. Mincer (1963), further elaborating the same concept, suggested that the cost of children should also include the opportunity costs of parents, because additional children reduce the capacity of women to work in the market. Mullineaux (1977) also elaborated on this point: 'several studies have found that a higher value of the mother's time is associated with a lower number of children in the family. These studies typically use women's wage or number of years of education as a measure of the value of time.'

Subsequently, a modification of the economic theory of fertility has been suggested by Easterlin (1975). According to him, as urbanisation and industrialisation proceed, the excess demand for children in subsistence agricultural societies decreases and eventually becomes an excess supply as child mortality falls and children become increasingly less productive and more expensive to raise. When a household has produced enough children, then the net economic and psychic cost of every additional child exceeds the economic and psychic cost of fertility control, the parents begin to practise contraception and the theory of fertility becomes operative. Accordingly, the theory that economic factors are relevant to household fertility behaviour was supported by findings from several studies (Rainwater, 1965; Caldwell, 1967; Freedman, 1969; Nag, 1978).

It is clear from the preceding discussion that children are valued for both economic and non-economic reasons. However, from the point of view of policy intervention, economic dimensions seem to outweigh non-economic dimensions, particularly for the developing countries. The economic benefits and costs of children are more amenable to policy manipulations than most of the deep-rooted non-economic benefits and costs of children (Mueller, 1972; Ridker, 1976). Additional support for this proposition comes from empirical evidence: 'almost without exception in the studies relating costs and values to fertility, economic aspects of children were shown to bear a stronger statistical relation to measures of family size than did socio-psychological dimensions' (Espenshade, 1972). Mueller (1976) emphasises that the economic value of children deserves separate study because of its bearing on the pace of economic development.

Thus, the influence of the value and cost of children on fertility has been more or less well established, based on several empirical findings relating to other countries. However, the fact remains that all such studies were conducted not in India and they deal mostly with macro level variables and as such are not fully convincing with respect to the influence on childbearing motivations. Hence, further research on the cost and value of children can help to fill this gap in knowledge. On the basis of the above assumption and partly stimulated by the practical problem of rapid population growth and the failure of family planning programmes to control the high fertility rate, and further encouraged by the need for alternative programmes of population control, the present study was undertaken in South Central region of India.

Knowledge about the costs of rearing children has policy implications in at least three ways—to improve the effectiveness of population education, information and persuasion programmes; to suggest various kinds of incentives and disincentives that may directly influence motivations; and to understand the effect of development programmes that may influence motivations indirectly (Fawcett, 1978). Hence, there is a need to identify and assess the range of alternative policy options for reducing fertility that could be adopted and implemented by governments to slow down population growth as soon as possible. In this context, the study of the costs of rearing children assumes great importance. Research on the costs of rearing children to parents is also relevant for its own sake representing as it does a fundamental aspect of the social behaviour of human beings.

Empirical Findings from Central India

A study was undertaken in South Central India with the main objective of studying the actual and perceived costs of rearing children, which are particularly relevant to the National Health and Family Welfare Programme. Conceptually, the study was based on the model of Hoffman and Hoffman (1973). As suggested by Fawcett (1976), and Mueller (1972), a monetary and non-monetary approach was adopted to assess the financial and perceived costs of rearing children. The survey was carried out in the villages of Vayalpad Community Development Block in Chittoor District, Andhra Pradesh during the year 1980. For the present investigation, a total number of 600 couples in the reproductive age span were selected using the stratified proportionate random sample technique. Out of the total sample, 300 couples belonged to the Scheduled Caste* (SC) and the other 300 couples belonged to the non-Scheduled

* Scheduled Caste population constitute the economically and socially most backward communities in India. They are locally known as the Harijans. They fall outside the realm of the major

Caste (NSC) population. Their proportion in the general population was more or less equal and they belonged to the major segment of the population in Central India. The couples in each group were further sub-divided into two groups and were drawn in such a way that there were 150 representatives in each sub-group based on the current age of the last child (that is, 1–4 years and 5–9 years). The data was collected using a schedule and was processed with the help of a computer. Primary data was supplemented by secondary data, wherever necessary.

In the Indian setting very little information is available about the actual financial costs incurred in rearing children from different socio-cultural backgrounds and about various other perceived costs involved in rearing children. The paucity of data regarding the economic costs and benefits of children has made it more difficult to relate perceptions about desired and actual family size to the costs and benefits of children. The following discussion highlights the policy implications of the direct costs of rearing children, the indirect costs of rearing children and their influence on fertility behaviour.

Direct Costs and Fertility

The direct financial costs involved in rearing children include the cost of food, clothing and other such essential services. The effect of these financial costs on the fertility behaviour is considered to be highly significant. The hypothesis that the higher the direct costs involved in rearing children, the lower will be the fertility has been confirmed in this study.

The couples' decisions regarding family size were significantly influenced (significant at .01 level) by the costs of rearing children (see Table 1). According to the findings the mean live births for the respondents as a whole at the total cost of Rs 200 and less per annum, per child was 4.3 as against 2.4 at the cost of Rs 1,000 and above per annum. Considering the social classes in the SC and NSC groups separately, it can be seen that in all the groups, the mean number of live births decrease with increasing costs. Though the percentage of parents spending Rs 201 and above was more in the NSC group as compared to the SC group, both the groups were equally influenced by the costs of rearing children in their family size decisions.

A close examination of the analysis reveals the influence of the direct costs of rearing children on the fertility behaviour of the couples. Respondents were very much aware of these costs. These costs were cited as the primary reason for restricting the family size. Most of the respondents further stated that their additional and ideal family size decisions were also influenced, to a great extent, by the direct costs of rearing the present living children. The preceding

---

traditional caste hierarchy in India, whereas non-Scheduled Caste population have a relatively more or less prosperous socio-economic status. They belong, of course, to different caste hierarchy but have many common socio-cultural characteristics. Therefore, the sample population is grouped into two major categories of Scheduled Caste (SC) and non-Scheduled Caste (NSC) in the analysis. They are considered here as two distinct classes of population on the socio-economic lines.

TABLE 1 Mean Live Births of the Scheduled and Non-Scheduled Castes by Total Direct Costs of Rearing the Child

Total Cost Per Annum (In Rupees)	*SC*		*NSC*		
	Mean Live Births of Respondents having a Child in the Age Group 1–4 Years	*Mean Live Births of Respondents having a Child in the Age Group 5–9 Years*	*Mean Live Births of Respondents having a Child in the Age Group 1–4 Years*	*Mean Live Births of Respondents having a Child in the Age Group 5–9 Years*	*Total (SC + NSC)*
≤ 200	3.9	5.3	4.5	4.0	4.3
	(21)	(10)	(2)	(15)	(48)
201–400	3.6	4.7	4.0	3.9	3.9
	(57)	(26)	(25)	(19)	(127)
401–600	2.6	3.4	3.5	3.2	3.1
	(36)	(35)	(29)	(32)	(132)
601–800	2.5	2.7	3.0	3.0	2.7
	(12)	(28)	(39)	(15)	(94)
801–1000	2.6	2.6	2.4	2.6	2.6
	(10)	(17)	(20)	(14)	(61)
1001+	2.3	2.4	2.4	2.4	2.4
	(14)	(34)	(35)	(55)	(138)
Total	3.1	3.5	3.3	3.1	3.1
	(150)	(150)	(150)	(150)	(600)

Note: Figures in parentheses indicate the actual number of households

trend clearly shows that as the direct costs of rearing children increase, the fertility level decreases. Hence, the findings of this research study support the new policy options related to the direct costs of rearing children as a measure to control the growth of population.

However, the study also revealed that some of the respondents expected to obtain long range economic benefits through a big family comprising several children. When the costs of rearing children are considered in relation to the social, emotional and cultural motivations for having children, it appears that an increase in costs would have to be very substantial to bring about a significant decline in fertility. In order to manipulate this determinant of fertility, the government may have to provide suitable substitutes for the economic benefits accrued from having children, such as, credible and adequate old age pension programmes as well as programmes providing housing, physical care and old age financial support for the elderly. Another policy approach could be to popularise the idea that the quality of children is assured through a small family that ultimately ensures a greater earning potential and ability to support parents in their old age. This fact could be brought home to people by educating them.

However, parents were only vaguely aware in general, of the direct costs of rearing children and they lacked specific and detailed knowledge when asked to estimate the amount of money spent on their children in the past three months. Since the availability of such data could be useful in community education programmes, further studies are needed in this field. This study also suggests that though the effect of direct costs is generally recognised, it does not become a crucial factor until people are bestowed with certain optimum number of children. While normative influence and emotional needs largely operate strongly for the first and second child, economic costs become highly salient only for the subsequent children. Therefore, it becomes imperative to educate people about the higher cost involved in bringing up children properly to ensure their bright future. Such education is relevant to people before and immediately after getting married and not after producing several children.

Indirect Costs and Fertility

Decisions regarding family size also involve many non-economic subjective considerations. Therefore, anticipated costs are apt to have a subjective component. In this context, on the basis of recent empirical evidence, Fawcett and others (1974) have confirmed that perception about the satisfaction and costs of children may be a more significant determinant than the actual costs in understanding fertility behaviour. Further, the cross national study on the value of children (Arnold *et al.*, 1975) is an example of the successful application of perceived values. The following analysis deals with the indirect costs of rearing children encompassing the different perceived costs of rearing children.

TABLE 2 Mean Live Births of the Scheduled and Non-Scheduled Castes by Index of Indirect Costs

	SC		*NSC*		
Response Categories	*Mean Live Births of Respondents having a Child in the Age Group 1–4 Years*	*Mean Live Births of Respondents having a Child in the Age Group 5–9 Years*	*Mean Live Births of Respondents having a Child in the Age Group 1–4 Years*	*Mean Live Births of Respondents having a Child in the Age Group 5–9 Years*	*Total (SC + NSC)*
Do not agree	4.2 (25)	4.3 (28)	3.6 (31)	3.2 (45)	3.7 (129)
Uncertain	3.1 (38)	3.3 (47)	3.0 (46)	3.2 (31)	3.2 (162)
Agree	2.8 (87)	2.9 (75)	2.6 (73)	2.6 (74)	2.8 (309)
Total	3.3 (150)	3.5 (150)	3.0 (150)	3.0 (150)	3.2 (600)

Note: Figures in brackets indicate the actual number ot households

To study the impact of the indirect costs of rearing children on the fertility behaviour of the sample population, the respondents were asked to indicate whether having four or more children involved any indirect cost to them namely, inconveniences to parents, problems between couples, financial burden to the family.

The overall findings confirm that the various aspects of the indirect costs of rearing children depress the fertility behaviour. More than half of the sample population with 2.8 mean live births agreed that the indirect costs affect their fertility behaviour relative to only 25 per cent of the sample population with 3.7 mean live births who disagreed with the idea of the influence of indirect costs in depressing fertility. Considering the two caste groups separately, it was observed that the fertility levels of the SCs was more than that of the NSCs, even for those who admitted the negative influence of the indirect costs on fertility. The mean live births of the SCs who agreed with the influence of the indirect costs on fertility with children in the age groups 1–4 years and 5–9 years have 2.8 and 2.9 mean live births respectively as against 2.6 mean live births each for similar groups of the NSC population (significant at 1 per cent level). The mean live births of the SC women with children in the 1–4 years and 5–9 years age groups who disagreed with the influence of the indirect costs on fertility have 4.2 and 4.3 live births respectively as against the NSC women who had 3.6 and 3.2 mean live births.

This data indicates that these indirect costs are extremely important in influencing the fertility of the respondents among both the caste groups. Further, the impact of the indirect costs on fertility behaviour confirms the hypothesis that the higher the perception about indirect costs, the lower will be the fertility. Most of the parents felt that large families were a nuisance and that too many children (4 +) created worry and emotional strain for the parents. Many of the parents believed that 'a child who has a close relationship with his parents and fulfils their aspirations' is likely to belong to a small family. The trend in the sample area is towards the small family size norm and most of the respondents believe that 'A small family is a happy family'.

These findings have many policy implications. Hence, the population policy should aim to increase the awareness and importance of these indirect costs of rearing children particularly for parents in the early childbearing years. It is likely that the indirect costs of child rearing are not fully anticipated by prospective parents. The study, however, shows that these indirect costs are very important to couples with children and these costs play a part in their family size decisions. It is clear from the evidence provided by this study that the indirect costs have an inverse influence on the fertility behaviour of the couples. The government should, therefore, take immediate steps and implement policies to increase the awareness of the consequences of the indirect costs of rearing children on the life of parents. Delay in the birth of the first child can be encouraged through population education using various themes on the indirect costs of child rearing and encourage the newly weds to establish themselves financially and emotionally before taking on the additional burden of marrying early and rearing children.

Multiple Determinants of Fertility Behaviour

In order to examine the individual contribution and the combined influence of major linear variables on the dependent variable, a multivariate analysis on the lines of regression equations has been attempted for major economic and sociodemographic variables. The results of these regression equations strongly support the observations already made, besides confirming the importance of a few more new factors.

TABLE 3 Determinants of Fertility Behaviour of the Scheduled and Non-Scheduled Castes as Denoted by Proportion of Variance explained (R^2) by Multiple Regression Analysis

Prediction	*Proportion of Variance Explained (R^2)*			
	NSCs with a Child in Age Group 1–4 Years	*NSCs with a Child in Age Group 5–9 Years*	*SCs with a Child in Age Group 1–4 Years*	*SCs with a Child in Age Group 5–9 Years*
Direct Costs of Rearing Children	10.44**	19.95*	6.63*	31.79*
Indirect Costs of Rearing Children	7.62*	3.38*	8.48*	9.98*
Socio-economic Status	24.81*	1.89*	12.89**	2.16
Age of the Mother	3.41	1.63	14.07*	0.94
Family Size Norms	1.97	5.42	7.62*	2.10
Education of Wife	2.22	1.93	0.65	5.61*
Total (per cent)	53.13	40.39	54.66	53.18

* Significant at 0.01 level; ** Significant at 0.05 level

Note: The significance of each independent variable in explaining total variance was computed using an F-test with the formula:

$$F = \frac{R^2 \text{ change}/M}{(1 - R^2_{Y1,2.....K}) / (N-K-1)}$$

where: R^2 change represents a squared part correlation that measure the incremental variance explained with the addition of each new variable

K is the total number of independent variables

M is the number of independent variables for which the significance test is being made, and

N is the total number of cases

SOURCE: Norman, H. et al. (1975). SPSS: Statistical Package for Social Sciences, N.Y. In Arnold et al., *The value of children, a cross national study*. E.W. PI, E.W.C. Honolulu, Hawaii: 1975, p. 76.

The total variance in fertility behaviour explained by all the thirty-one variables considered for the regression analysis was 53 per cent for the NSCs with children belonging to 1–4 age group, 40 per cent for the NSCs having children in 5–9 age group, 55 per cent for the SCs with children in 1–4 years age group and 53 per cent for the SCs having children in 5–9 years age group. The costs of rearing children account for the single large percentage of fertility variance in all the groups (Table 3). Hence the findings evidently confirm the hypothesis that the costs of rearing children is the most important economic factor in explaining the fertility behaviour of the Scheduled and non-Scheduled Caste population. Nevertheless, the costs of rearing children explain significantly more variance in the fertility behaviour of the poorer sections of society (SCs) as compared to the higher socio-economic status groups (NSCs); this is obviously true because the former groups undergo a lot of strain even to raise the minimum required funds to rear their children. Thus, we may conclude that the costs of rearing children—direct and indirect costs—strongly influence fertility behaviour independent of other socio-economic factors. Further, these costs together with other socio-economic factors explain an overwhelming proportion of the variations in fertility behaviour.

Among the SC group with younger children, the costs of rearing children explained 15 per cent of the total variance in fertility whereas socio-economic status accounted for 13 per cent. On the other hand, in the NSCs group with younger children, the age of the mother influenced the fertility variance significantly and accounted for 14 per cent of the variance while family size norms accounted for 8 per cent of the variance in this group. However, the education level of the wife was not found to be significant in this category. But the education level of the wife significantly explained the fertility behaviour of women with older children in this caste group.

The above analysis clearly shows that the costs of rearing children are equally important in predicting the fertility behaviour of diverse populations like the non-Scheduled and Scheduled Castes in India. Similarly, other factors like socio-economic status and the age of the mother also influenced the NSC and SC groups with children in the 1–4 age group. On the other hand, other variables differentially influenced the population belonging to the non-Scheduled and Scheduled Castes and with young and old children. The other influential variables in the order of importance include socio-economic status, the age of the mother, family size norms and the education level of the wife. For instance, the education level of the wife significantly influenced only the NSCs with younger children and the SCs with older children. Family size norms, on the other hand, were significant only for the NSCs with children in the age group 5–9 years and the SCs with children in the age group 1–4 years. The costs of rearing children emerged as a strong variable which accounted for the maximum variance in fertility across a large proportion of the four strata of society.

Thus, the findings fully confirm the hypothesis that 'fertility progressively decreases with increase in the costs of rearing children'. The direct financial costs of rearing children cross tabulated with the fertility behaviour of the two castes confirmed the hypothesis that couples tend to have fewer children as it becomes more expensive to rear them. The indirect costs of rearing children confirmed the hypothesis that the higher the perception of the indirect costs of rearing children, the lower the fertility. Regression analysis also revealed that the costs of rearing children independently depress the fertility behaviour significantly among the rural population in general, though they differently affect the fertility behaviour of the population across different social strata.

These results, thus, raise a number of issues for formulating policies in the field of fertility control, not uniformly for the country as a whole, but differentially for the several cultural and sub-cultural groups in various parts of the country. Similarly, the fertility control programmes may have to be suitably modified in view of the diversity of factors and their influence on family size decisions across different socio-economic groups.

Micro level research on such topics as the costs of rearing children can facilitate the development of population policies that take into account the needs of the people, as well as the needs of the society. Policies developed on this basis will not only be more acceptable but also more effective. The costs of rearing children could be used to bring personal needs and social needs into congruence, so that demographic events will contribute to, rather than detract from, the good of improving the common welfare. The findings also implicate that to achieve reductions in fertility, policy-makers should attempt to educate the people about the higher cost involved in bringing up children for their better future and quality of life. Thus, the findings of the present study are unique in many respects and have far-reaching theoretical, methodological and policy implications in the population control programme.

To conclude, these are only some of the explicit items for programme and policy that deserve serious consideration by administrators and scientists for the betterment of our society. Many other such issues may further be extracted from the findings. Since the present study is one of the few efforts made in this direction in India, there is a need for replicating the same in different regions and cultural groups in our country.

REFERENCES

ARNOLD, F. *et al.* (1975). *The value of children: a cross national study.* Vol. I. Honolulu: East-West Population Institute.

BECKER, G.S. (1960). An economic analysis of fertility. *In Demographic and economic change in developed countries.* National Bureau Committee for Economic Research, Special Conference Series No. 11. Princeton: Princeton University Press.

BLAKE, J. (1972). Fertility control and the problem of voluntarism. *Scientists and world affairs.* Proceedings of the Twenty-second Pugwash Conference on Science and World Affairs. London, September 1972.

CALDWELL, J.C. (1967). Fertility attitudes in three economically contrasting rural regions of Ghana. *Economic development and cultural change*, 15. Chicago: University of Chicago Press.

CHUNG, B.J. (1974). Dimensions of the value of children in Asian societies. *In Report and Papers of the Expert Group Meeting on Social and Psychological Aspects of Fertility Behaviour.* Asian Population Studies, Series No. 26. Bangkok: United Nations Economic and Social Commission for Asia and the Pacific.

EASTERLIN, R.A. (1975). An economic framework for fertility analysis. *Studies in family planning*, 6.

ESPENSHADE, J. (1972). *Estimating the cost of children and some results from urban US.* Berkeley: International Population and Urban Research.

FAWCETT, J.T. (1976). *The value and cost of first child.* Paper presented at the conference on the first child. Asilomai, California, 23–26 March.

———. (1976). The value and cost of children: converging theory and research. *In* L.T.

Ruzicka (Ed.), *The economic and social supports for high fertility*. Proceedings of the conference held in Canberra, 16–18 November 1976.

———— (1978). The value and cost of the first child. *In* Warren B. Miller and Lucile F. Newman (Eds.), *First child and family formation*. Chapel Hill: Carolina Population Centre, University of North Carolina.

FAWCETT, J.T. *et al.* (1974). The value of children in Asia and the United States: comparative perspectives. Honolulu: East-West Population Institute, Paper No. 32.

FREEDMAN, R. (1969). *Family planning in Taiwan*. Princeton: Princeton University Press.

HOFFMAN, C.W. and M.L. HOFFMAN. (1973). The value of children to parents. *In* J.T. Fawcett (Ed.), *Psychological perspectives in population*. New York: Basic Books.

MINCER, J. (1963). Market prices, opportunity costs, and income effects. *In* Carl Christ, *et al.* (Eds.), *Measurement in economics: studies in mathematical economics and econometrics in memory of Yehuda Grunfeld*. California: Stanford University Press.

MUELLER, E. (1972). Economic motives for family limitation: a study conducted in Taiwan. *Population studies*, 26 (3). London: Population Investigation Committee, London School of Economics.

————. (1976). The economic value of children in peasant agriculture. *In* R.G. Ridker (Ed.), *Population and development: the search for selective interventions*. Baltimore: The Johns Hopkins University Press.

MULLINEAUX, D.J. (1977). An economic approach to family size. A new perspective on population growth. *In* J.F. Crawford (Ed.), *Readings in modern economics*. Philadelphia: Saunders.

NAG, M. (1978). An anthropological approach to the study of economic value of children in Java and Nepal. *Current anthropology*, 19 (2), 63.

RAINWATER, L. (1965). *Family design: marital sexuality: family size and contraception*. Chicago: Aldine Publishing Company.

RIDKER, R.G. (1976). Perspectives on population policy and research. *In* R.G. Ridker (Ed.), *Population and development: the search for selective interventions*. Baltimore and London: The Johns Hopkins University Press.

10

Child Labour and Fertility

D. A. Naidu

Introduction

Children are valued primarily as a source of labour or income besides old age security. Leibenstein (1957) postulated that family size desires depend on the relative utilities of the economic benefits and the costs of children to parents. A vast body of research evidence shows that the demand for child labour was both an antecedent and a consequence of fertility trends in the industrialised countries (UN, 1973; Coontz, 1968; Lebergott, 1960). By contrast, in the developing countries, econometric analyses suggest that child labour is positively related to fertility (Kasarda, 1971; Schultz, 1972; Rosenzweig, 1978; Leiberman, 1976; Davanzo, 1972). However, these econometric studies appear to be limited in scope with respect to policy relevance as they are based on secondary data and can study only one dimension of child labour in its relation to fertility. The economic potentialities of child workers, a decisive factor in the determination of the economic value of children, depend on age, sex, type of occupation and earning opportunities. In the light of preceding evidences and taking clues from the earlier studies an effort was made by this author to study the influence of child labour on fertility behaviour.

The main purpose of the study is to compare the labour value of children in two traditional sectors, namely, agriculture and household industry, and to examine the sectoral difference in the relationship between child labour and fertility. In this study not only current child labour but also 'past' child labour and the perceived labour value of children are considered in relation to fertility.

Methodology

Child labour merits attention on numerical grounds alone. The International Labour Organisation estimates reveal that not less than 52 million children under fifteen years of age are working throughout the world, of which, perhaps, the largest contingent is from India. Another interesting aspect as far as the relevance of the present study is concerned is that it is carried out in a state in India where child labour participation is the highest.

For the present study a sample of 400 male respondents were selected—200 from the household industrial sector and 200 from the agricultural sector. For the selection of respondents from the household industrial sector, two villages from the universe (Chittoor District in Andhra Pradesh)—Papanaidupet and

Narayanavanam—were purposively selected, because a unique glass beads industry is located in the first village while the second is predominantly a weavers' village. Other household industrial occupations could not be considered for the study as they were not available in sufficient number at a reasonable distance.

For the selection of respondents from the agricultural sector, Srikalahasti Community Development Block from the same district was purposively selected since it was similar to the other villages in terms of culture and development. All the villages in this Block were listed out and from this list ten villages were selected at random in order to get a sufficient sample population.

From these twelve villages, one male respondent from each household was randomly selected on the basis of the following criteria: he should be currently married and should be in the reproductive age group with at least two living children. Each selected household should have at least one working child in the age group between five to fourteen years. The term 'child worker' was used in a broader sense; the category child workers included children engaged not only in directly productive work, but also in household maintenance work.

Discussion

Demand for child labour depends, among other factors, on the type of occupation. A noteworthy difference can be made between the household industrial sector and the agricultural sector with respect to child employment and the economic value of children. While the working children in the household industrial sector appear to be busy round the year their counterparts in the agricultural sector seem to be busy for only two-thirds of the year. Moreover, the proportion of wage-earners is higher in the former than in the latter sector. Based on these observations, the following differences can be made between the household industry and agriculture with respect to the employment pattern of children:

HOUSEHOLD—INDUSTRIAL SECTOR	AGRICULTURAL SECTOR
Employment is mostly perennial	Employment is mostly seasonal
Work is mostly indoor	Work is mostly outdoor
Less exposure to hazards of nature (sun, wind, rain)	More exposure to hazards of nature (sun, wind, rain)
Household work and productive work are more compatible	Household work and productive work are less compatible
The practice of female seclusion may not affect female employment since work is mostly indoor	Non-participation of females in outdoor work (female seclusion) is often considered as an order of high social status

The above differences in the work situation between the household industry and agricultural activities, therefore, supports the hypothesis that the labour value of children is higher in the household industrial sector as compared to the agricultural sector.

Current Child Labour and Fertility

As stated earlier, children are relatively economically valued more in the household industry than in agriculture. It may be of interest to know how child labour is related to fertility in these two traditional sectors of occupation. One simple way of understanding this relationship is, perhaps, to cross tabulate the cumulative fertility with the occupational background of the currently working children. Table 1 shows the occupational background of the older child workers

TABLE 1 Mean Live Births, Marriage Duration, Sex and Occupation of Working Children in the Household Industrial Sector and Agricultural Sector

Sectors and Occupation of Older Child Workers (10–14 age group)	*Sex of Child Workers, Marriage Duration of Wives and Mean Live Births*					
	Male Child Worker		*Female Child Worker*		*All Child Workers*	
	Marriage Duration		*Marriage Duration*		*Marriage Duration*	
	Below 25 yrs	*25 yrs & Over*	*Below 25 yrs*	*25 yrs & Over*	*Below 25 yrs*	*25 yrs & Over*
Household Industrial Sector						
Bead Workers	6.0 (16)	7.8 (4)	6.1 (23)	7.9 (15)	6.1 (39)	7.9 (19)
Weavers	5.9 (23)	7.2 (19)	5.6 (54)	7.0 (29)	5.7 (77)	7.1 (48)
Other Workers	4.7 (10)	6.8 (8)	5.3 (12)	7.0 (3)	5.0 (22)	6.9 (11)
Non-Workers	5.6 (82)	6.8 (38)	5.4 (42)	6.5 (22)	5.5 (124)	6.7 (60)
Total	5.6 (131)	7.0 (69)	5.6 (131)	7.0 (69)	5.6 (262)	7.0 (138)
Agricultural Sector						
Agricultural Labour	5.1 (10)	6.6 (10)	5.4 (6)	6.9 (6)	5.2 (16)	6.7 (16)
Herdsman	6.1 (29)	6.7 (21)	5.5 (13)	6.0 (13)	5.9 (42)	6.4 (34)
Bead and Other Workers	—	6.2 (4)	5.5 (15)	6.2 (12)	5.5 (15)	6.2 (16)
Cultivators	6.0 (10)	6.1 (11)	—	—	6.0 (10)	6.1 (11)
Household Helpers	—	—	4.9 (38)	5.7 (16)	4.9 (38)	5.7 (16)
Non-Workers	5.2 (60)	5.9 (35)	5.7 (47)	6.2 (34)	5.4 (107)	6.1 (69)
Total	5.5 (109)	6.2 (81)	5.4 (119)	6.1 (81)	5.4 (228)	6.2 (162)

Note: Figures in parentheses denote number of women

(10 to 14 years age group) and the fertility differentiated for the sex of the child worker and the wife's duration of marriage. It is interesting to note that fertility is highest for occupations with high child earnings, namely, bead workers and agricultural labourers. For these two occupations, not only are the earnings high but the proportions of child earners are also high. The lower fertility of the weavers as compared to the bead workers in the household industrial sector, is consistent with their lower earning potentialities. It is also consistent that the older wives of the respondents of non-working children show the lowest fertility in the household industrial sector, although a similar trend is not observed in the agricultural sector. However, in the agricultural sector, women whose children are engaged in the least economically useful occupation, namely, household duties, exhibit the lowest fertility. Above all, in comparison with the agricultural sector, the respondents of the household industrial sector show higher fertility which corroborates the earlier observation that the labour value of children is higher in the household industrial sector than in the agricultural sector.

'Past Child Labour' and Fertility

In the preceding section, the influence of current child labour on fertility was analysed. It is also interesting to explore the relationship between 'Past Child Labour'—childhood economic activity of the illiterate adult sons and daughters—and fertility. In this analysis, it is presumed that all the illiterate adult sons and daughters would have worked full-time in their childhood. Accordingly, the literacy status of adult sons and daughters is cross tabulated with the cumulative fertility as shown in Table 2. The date confirms the

TABLE 2 Mean Live Births, Sectors, Sex and Literacy Status of the Working Adult Sons and Daughters

Literacy Status of Adult Sons and Daughters		*Adult Children, Sectors and Mean Live Births*					
		Adult Sons		*Adult Daughters*		*All Children*	
		Household Industrial Sector	*Agri-cultural Sector*	*Household Industrial Sector*	*Agri-cultural Sector*	*Household Industrial Sector*	*Agri-cultural Sector*
No Living Adults		5.3 (100)	5.2 (96)	5.4 (112)	5.3 (127)	5.4 (212)	5.3 (223)
Literate Adults		6.8 (80)	5.7 (58)	6.6 (27)	5.9 (12)	6.7 (107)	5.7 (70)
Illiterate Adults:	One	6.9 (18)	6.2 (28)	6.6 (40)	6.4 (39)	6.7 (58)	6.3 (67)
	Two	7.5 (2)	7.2 (18)	8.0 (21)	6.6 (22)	8.0 (23)	6.9 (40)

Note: Figures in parentheses denote number of women

hypothesis that the higher the portion of illiterate adult sons or daughters, the higher is the level of fertility. Stated conversely, the hypothesis that 'Past Child Labour' relates positively to fertility is confirmed. With respect to the overall inter-sectoral differences in fertility, the agricultural sector continuously and consistently maintains lower fertility than the household industrial sector reflecting the inter-sectoral variations in the demand for child labour.

Perceived Labour Value of Children and Fertility

Another approach to understanding the labour value of children is to measure the respondents' attitude and perception relating to child labour. In this respect direct measurement of the net economic value of child labour is a difficult task since some of the dimensions are not amenable to quantification. Moreover, child labour represents a fraction of the working lifespan. Therefore, an assessment of the perceived labour value of children may be more realistic than the measurement of actual child labour alone. These perceptions may be more meaningful as they are based upon the past life experiences of the individuals.

To measure the perceived labour value of children, an index was constructed based on the responses to the four variables described in the Appendix. The score value for each respondent is the sum of points assigned to each response. Individual scores, ranging from 8 to 32 points, were classified into four levels.

The respondents in the two sectors differ significantly (at 0.01 level) with respect to the perception of the labour value of children. The proportion expecting the highest value of labour from children (Level 1) account for 51 per cent in the household industrial sector as against only 35 per cent in the agricultural sector. Further, it may be observed that only in the agricultural sector about 7 per cent of the respondents have perceived the least value of child labour (Level 4). On the basis of these findings, it may be concluded that the perceived labour value of children is higher in the household industrial sector as compared to the agricultural sector.

One of the major hypotheses of the economics of fertility is that the labour value of children is an important correlate of high fertility. The survey data presented in Table 3 provide the most formidable evidence in support of the above hypothesis. The perceived labour value of children is inversely related to the cumulative fertility in both the traditional sectors. It is also noteworthy that, with the exception of Level 3, the respondents in the agricultural sector show lower fertility than their counterparts in the household industrial sector which is consistent with the inter-sectoral differentials in the labour value of children as shown earlier. The results of the study, therefore, clearly demonstrate that not only the actual child labour but also the perceived labour value of children is significantly related to fertility behaviour.

This finding also corroborates the studies conducted in Taiwan (Mueller, 1972) and Thailand (Arnold and Pejaranonda, 1977) showing a positive relation between the perceived utility of children and the ideal family size.

The findings of the study may have several policy implications. First, the results of the study are of particular interest to educational planners since the

TABLE 3 Mean Live Births, Distribution of Respondents, Sectors and the Levels of Index of Perceived Labour Value of Children

Index of Perceived Labour Value of Children	*Sectors, Respondents and Mean Live Births*					
	Household Industrial Sector		*Agricultural Sector*		*Both Sectors*	
	Respondents Per cent/No	*Live Births*	*Respondents Per cent/No*	*Live Births*	*Respondents Per cent/No*	*Live Births*
Level 1 (High labour value)	50.5 (101)	6.5	35.0 (70)	6.6	42.7 (171)	6.5
Level 2	37.0 (74)	6.0	40.5 (81)	5.5	38.8 (155)	5.7
Level 3	12.5 (25)	4.4	17.5 (35)	5.1	15.0 (60)	4.8
Level 4 (Low labour value)	—	—	7.0 (14)	4.0	3.5 (14)	4.0
Total	100.0 (200)	6.1	100.0 (200)	5.7	100.0 (400)	5.9

Note: Figures in parentheses indicate number of respondents

achievement of the goal of universalisation of primary education is linked to the decline in child employment. Second, employment policy-makers may find these results useful in their attempts to realise the goal of full employment, because it is said that child workers deprive the adult workers of employment opportunities. Further, these findings may provide some clues for effective manpower planning and human resources development. Third, a proper understanding of the various dimensions of the value of children is a prerequisite for the formulation of a comprehensive population policy. This study, therefore, identifies the various dimensions of the labour value of children in relation to fertility behaviour. For further details one can read the thesis presented to Sri Venkateswara University, Tirupati.

REFERENCES

ARNOLD, FRED and CHINTANA PERJARANONDA. (1977). *Economic factors in family size decisions in Thailand*. Bangkok: Chulalongkorn University.

COONTZ, SYDNEY H. (1968). *Population theories and the economic interpretation*. London: Routledge and Kegan Paul.

DAVANZO, JULIE. (1972). *The determinants of family formation in Chile, 1960: an econometric study of female labour force participation, marriage, and fertility decisions*. R—830—AID. Santa Monica: Rand.

KASARDA, J.D. (1971). Economic structure and fertility. *Demography*, 8: 307–17.

LEBERGOTT, STANLEY. (1960). Population change and the supply of labour. *In Demographic and economic change in developed countries*. Princeton: Princeton University Press, pp. 377–422.

LEIBENSTEIN, HARVEY. (1957). *Economic backwardness and economic growth*. New York: John Wiley.

LIEBERMAN, SAMUEL S. (1976). *An economic approach to differential demographic behaviour*. Doctoral dissertation, Harvard University.

MUELLER, EVA. (1972). Economic motives for family limitation: a study conducted in Taiwan. *Population studies*, 26: 383–403.

ROSENZWEIG, MARK R. (1978). The value of children's time, family size and non-household child activities in a developing country: evidence from household data. *In* Julian L. Simon (Ed.), *Research in population economics*. Grunwich: Jai Press, pp. 331–47.

SCHULTZ, T.P. (1972). Fertility patterns and their determinants in the Arab Middle East. *In* C.A. Cooper and S.S. Alexander (Eds.), *Economic development and population growth in the Middle East*. New York: American Elsevier, pp. 399–500.

UNITED NATIONS. (1973). The determinants and consequences of population trends. *Population studies*, 1, 50. New York: United Nations.

APPENDIX

Index of Perceived Labour Value of Children

The index of the perceived labour value of children was constructed by combining data based on the responses to the several questions:

Question/Statement	*Score*	*Response category*
A. How valuable is child work to you? Is it of great importance, less importance or very little importance?	1	Great importance
	2	Less importance
	3	Very little importance
	4	No value at all
B. If children were not doing this work, what would you have done?	1	Would have hired some one
	2	Work could be done by adult with qualification
	3	Work could be done by adults in the family
C. (If would have hired help) What might be the cost per year?	1	Rs 700 and over
	2	Rs 600–699
	3	Rs 500–599
	4	Rs 400–499
	5	Rs 300–399
	6	Rs 200–299
	7	Rs 100–199
	8	Below Rs 100
	9	Not applicable

D. *Statements*

(i) 'Child labour deprives children of educational opportunities'

(ii) 'Child labour stunts physical growth'

(iii) 'Child labour hampers intellectual development'

(iv) 'Child labour renders a worker unskilled with low wages during the entire production life'

(v) 'Child labour deprives employment opportunities for adults'

1 Disagree
2 Uncertain
3 Agree

11

Factors in Fertility Decline in Kerala

P.K.B. NAYAR

The objective of this paper is to examine the various factors in the rapid fertility decline in Kerala in the second half of the twentieth century and to explain the process by which these factors influenced fertility. An attempt is made to relate Kerala's fertility behaviour to the fairly wide coverage of social welfare and social security programmes obtaining in the state and contributing to a more egalitarian societal base facilitating better accessibility to, and more efficient use of, family planning services. It is argued that the two major factors in fertility decline in Kerala in recent times, namely, mass education and massive health care programmes, were able to make their full impact on fertility when superimposed on a near-egalitarian base which supplied the supporting inputs required for the quick absorption of these programmes by the people. It is further argued that the existence in Kerala of a sound social infrastructure, compared to the other states in India, is largely the contribution of its ecology. Thus, in the case of Kerala, it may be argued that ecology contributed to the establishment of a good social infrastructure, including social welfare and security programmes; these, in turn, contributed to better distributive justice which enabled a fuller use of the family planning programmes by the target groups. However, it is not the argument presented in this paper that ecology is a necessary or sufficient factor in demographic behaviour. But, in the case of Kerala, its influence cannot be minimised.

Kerala's population increased by 24.76 per cent during 1951–61, 26.29 per cent during 1961–71 and 19 per cent during 1971–81. The corresponding figures for India as a whole were 24.80, 24.80 and 24.75 respectively. During the three decades birth rates for Kerala were 38.9, 31.9 and 25.9 per thousand. The rates for India were 41.7, 37.8 and 33.1. The death rates for the same decades were: Kerala 16.9, 9.2 and 6.9; India 23.0, 18.0 and 13.9 respectively. The net reproduction rate for Kerala during 1971–81 was 17.5 against India's 22.2. Only one state other than Kerala, namely, Tamilnadu had a net reproduction rate lower than that of Kerala (16). But in Tamilnadu both birth (28.9) and death rates (12.9) were higher than those in Kerala.

Another area in which even wider differences are observed is in the field of infant mortality. For the three decades under review Kerala's achievements were 120, 66 and 44. The all-India infant mortality rates were 160, 140 and 125. Clearly, there is a significant difference in the demographic characteristics of the two populations.

Many hypotheses have been formulated to explain the dramatic change in the demographic characteristics in Kerala over the last three decades, especially the steep decline in fertility. P.R.G. Nair (1974) argues that the decline in birth

TABLE 1 Selected Demographic Indicators—Kerala and India (1961–81)

Decade	*Population Change*		*Birth Rate*		*Death Rate*		*Infant Mortality Rate*	
	Kerala	*India*	*Kerala*	*India*	*Kerala*	*India*	*Kerala*	*India*
1951–61	24.76	24.80	38.9	41.7	16.9	23.0	120	160
1961–71	26.29	24.80	31.9	38.3	9.2	18.0	66	140
1971–81	19.0	24.75	25.9	33.1	6.9	13.9	44	125

rate in Kerala began well before the intensification of the family planning programmes. While these programmes gathered momentum only by the second half of the 1960s, crude birth rate had begun to fall in the late 1950s as indicated by the fall in enrolment in the first year of the primary school in the early 1960s. Nair attributes this phenomenon to the fall in infant and child mortality rates during the latter part of the 1950s which, in turn, was due to the extension of primary health centres and other health measures over a period of time. T.N. Krishnan (1976) believes that the decline in Kerala's fertility is due to a change in nuptial rates and the consequent rise in the age of effective marriage brought about by continuous and sharply increasing rates of female literacy. According to Leela Gulati (1976), the rise in the age of marriage of women in Kerala was the major factor in bringing down the birth rate. The other contributory factors include the spread of female education and improvement in medical and public health which improved child survivorship which, in turn, influenced the number of children a woman wanted to have.

However, Kurup and Cecil (1976) have attributed the decline in the birth rate mainly to increased acceptance of family planning programmes. They argue that the major factor contributing to the fall in the birth rate between 1961–75 was the decline in marital fertility rates as a result of the increasing acceptance of birth control. Pillai (1980) also supports the view that changes in the age at marriage and marital fertility were preceded by more fundamental changes in attitudes and motivation.

Radcliffe (1978) discounts both the family planning programmes and the mortality explanation as contributing to the fall in the birth rate in Kerala. He contends that the role of linkage between education and fertility has not been properly established in the earlier studies. He argues that these studies exclude Kerala's unique economic and political context in the demographic transition. Radcliffe propounded his social justice hypothesis. He states that the observed shifts in Kerala's demographic rates are the consequences of identifiable shifts from less to more social justice in Kerala's political and economic policies and development strategies. Moni Nag (1981) has found some support for this hypothesis in his comparative study of Kerala and West Bengal.

Panikar (1981) disagrees with the arguments advanced by Radcliffe. According to him, while Radcliffe's argument that the role of the state government's policies in ensuring that the benefits of education and medical care reach the majority cannot be disputed, there still seem to be some missing links in the phenomenon. He identifies these links as the emergence of the nuclear family system consequent to the break-up of the Nayar Tharwad, fostering the attitude towards family limitation, low proportion of children in the labour force, higher status of women and modernisation of the rural masses. However, the spread of education has played a crucial role in bringing about family nucleation, improvement in health, reduction in child labour, rising status of women and modernisation. It also changed the attitudes and motivations of couples so as to make them willing to accept family planning techniques. 'These developments', he admits, 'were unfolded over the first half of the present century. The ground was thus prepared for the germination of the idea of family limitation and when simple and effective methods became widely available, people readily accepted them.'

Panikar's argument thus points to the presence of several contextual factors which were in operation long before the observable shift in fertility change occurred. As stated earlier, the variables identified by Panikar, which Radcliffe and the others have endorsed, were the products of a peculiar social structure prevalent in Kerala for centuries. This social structure, in turn, was the contribution of Kerala's ecology which not only evolved and nourished certain institutions but also paved the way for several others, all of which provided the infrastructure for Kerala's demographic transition.

The influence of ecology on fertility reduction has not received adequate attention of demographers and population experts. While the case of Kerala is amenable to the application of the equity-fertility hypothesis put forward by Radcliffe, it will be simplistic to argue that social equity *per se* has achieved fertility decline in this state. Various antecedent and intervening factors (for example, ecology and social structure) have to be identified and properly located in any scheme of explanation of the Kerala phenomenon in terms of the equity-fertility hypothesis. For, these have not only contributed to the egalitarian social structure but have generated other factors which have accelerated the downward trend in fertility.

Ecology and Social Structure

The low birth rate and other demographic features peculiar to Kerala have to be explained in terms of a variety of factors in which ecology, social structure, welfare needs, political activism, social legislation, welfare programmes and social equity assume great prominence.

Kerala's ecology is characterised by the striking absence of the village system which obtains in the rest of India. Households in Kerala are scattered far and wide; one cannot see in this state the clustered habitats of the average Indian village. Kerala is favoured by both the North West and South East monsoons which provide its lands with a plentiful supply of water during the greater part of the year and especially for the two major crops of the area. A comparatively fertile land and an abundant supply of water for agricultural operations enabled the Keralites to dispense with the village system and the community chieftain to regulate the water supply and administer the village land as had been the case in traditional societies. It enabled them to settle down anywhere they liked. The concept of the village chieftain as a mediator between the individual and the community and between the community and the state did not develop in Kerala. Naturally, the individual looked directly to the state for services and the state, in turn, was obliged to render these services to its subjects. Scattered dwellings also promoted individualism and competition in political and economic pursuits. It is in this context that the state developed a very good social infrastructure as compared to the rest of India.

The influence of ecology is also reflected in the religious composition of the population. Kerala has a long sea coast (425 kilometres) facing the west and this acted as a 'window on the west' and exposed the people to the influence of western culture and ideas from the time of recorded history. Both Christianity and Islam found their way to Kerala during the century of their founders and were not only encouraged by the rulers but also welcomed by the people. St. Thomas, one of the twelve disciples of Christ,is believed to have established a church in Kerala—considered to be the first church in India. The last ruler of Kerala in the earlier times (Cheraman Perumal) is believed to have embraced Islam, renounced the throne and gone on a pilgrimage to Mecca. The Zamorins of Calicut (now a part of Kerala) had their entire navy manned and commanded by the Muslims. As a result of the benevolent attitude of the state towards foreign religions and peoples, the population of Kerala is 21 per cent Christian and 20 per cent Muslim. For the whole of India, Christians form only 3 per cent and Muslims 14 per cent of the population. This is indicative of the cosmopolitan and innovative outlook of the people of Kerala which was largely promoted by the ecological setting of the state.

Kerala has a long tradition of benevolent and paternalistic rulers who spent a major part of their revenue on public welfare—roads, irrigation works, schools, hospitals and the like. As a matter of fact, the princes who ruled over Travancore and Cochin (now parts of Kerala) had earned a reputation among the Indian princes as the most enlightened and progressive rulers of the country. Naturally, when India gained independence in 1947, the state was much ahead of the other Indian states in the field of social welfare. The simplest evidence of this is the predominance of the service sector in the state's economy. The contribution of this sector to the state's net domestic product was one-fourth in 1950. In the next two decades, it rose rapidly to one-third. As a result, a large part of the state's revenue has to be earmarked every year for meeting the social service needs of the people. Besides these, the Five-Year Plans of the state have also allotted sizeable proportions of their total outlay for direct social welfare schemes (average of 22 per cent for all plans—1951–78).

The ecology of the state has also led to the construction of a large network of roads in the state. Scattered type housing requires greater length of roads than cluster type housing to link together the different centres of living. Notwithstanding the undulated nature of the terrain, roads have criss-crossed throughout the length and breadth of the state linking all the centres of human habitation in the state. Naturally, Kerala has the largest road length per 100 square kilometres in India—232 against 49 for all India (1978–79). Even Kerala's nearest competitor, West Bengal, had only 158 kilometres of roads per 100 square kilometres in that year. The importance of roads in fertility decline has been tremendous. Roads accelerated communication, enabled newspapers (in which the state has a record circulation compared to the rest of India) to reach every nook and corner of the state, helped in the quick delivery of the government's health care and family planning inputs and made the medical staff more inclined to serve in the remote villages. This network of roads also contributed to the willingness of the paramedical staff of the family planning centres to visit more houses for their promotion work.

Slant Towards Egalitarianism

Kerala was the first state in India to abolish the caste system. The famous Temple Entry Proclamation issued by the Maharaja of Travancore in 1936 threw open the portals of Hindu temples to people of all castes and creeds. Thus, in Kerala, untouchability was abolished two decades before it was abolished in India (1956).

Kerala has a long history of women's freedom. The Nair 'Tharavad' following the 'Marumakkathayam' system of inheritance gave women primacy over family property. The total impact of 'Marumakkathayam' on women's status has not been fully appreciated. According to Woodcock (1967), 'probably the most important effect of the matrilineal (Marumakkathayam) system is that it had left Malayali women with an influence and an independent outlook which one will not find anywhere in India.' Under this system, all children born on the female line had an equal right to family property while only the male members, not their children, enjoyed this right. Thus, a woman with children (even minor children) commanded more claim to the wealth of the family than a man with or without children. This claim over wealth gave women command over family matters. The Tharavad catered to the physical and intellectual development of its members. Each Tharavad maintained a 'Kalari' where its members—both men and women—were trained in physical fitness and in the arts of traditional wars, and especially in fighting duels. According to Nagam Aiya (1906: 474),

> No (Nair) girl is permitted to grow up to womanhood without a fair knowledge of reading and writing. Education is a *sine qua non* of every (Nair) household. Curriculum for the well-to-do go beyond the 3 'rs', to Sanskrit for reciting Kavyas, Natakas, etc., and to the practise of musical instruments and of music and dancing.

Kerala has been, by and large, a land of small farmers. It did not have the Zamindari system. The state had abolished the Jenmi system (a distant variant of the Zamindari system) in the early part of the twentieth century. Partitioning of the 'Tharavad' was permitted by law during the same period. The enactment of the Cochin Nayar Act in 1920, the Nayar Regulation in Travancore in 1925 and the Madras Marumakkathayam Act in 1933 were legislations in this direction. Thousands of partition deeds were registered in the wake of these laws. During the 1950s, 92 per cent of the farmers in Kerala had less than two hectares of landholdings and 83 per cent had only less than one hectare. Following the emergence of the Communist Party to power in Kerala in 1957, there had been great tension between landlords and tenants including the agricultural workers; and many landlords in anticipation of the land ceiling laws had reduced their landholdings through sale, benami transfers and in numerous other ways. The Kerala Agrarian Relations Act (1969) imposed a fifteen acre ceiling on land-ownership, abolished tenancy of all types, gave right of purchase of land to the tenant at a rate less than the market price and on easy instalments and conferred proprietary right on the 'tiller'. It gave entitlement to all hutment dwellers on 0.4 hectare of land around their hutments. The Act virtually wiped out the category of landless households in the state. The enactment of this legislation was preceded and followed by large-scale encroachments on government lands and lands belonging to the big landlords by the members of the Marxist Communist Party (CPI-M). This practice was later continued on an even larger scale through the so-called 'land-grab agitation' of the early 1970s by which members of the CPI-M openly grabbed the 'excess' land belonging to the landlords whenever they suspected that the state had not moved effectively in the direction of implementing this clause of the Act (according to the Act, an individual was permitted to hold only a maximum of fifteen standard acres of land and land in excess of this was to be taken over by the state for distribution to the 'landless'). The psychological impact of this agitation was much more than its material contribution. In many cases, the hutment dwellers were able to keep in their possession more than the area conferred on them by the law. A large number of encroachers who had no legal title to the encroached land also became *de facto* owners of the land.

The Agricultural Workers Act (1974) provided security of employment and fair wages, and also introduced a provident fund scheme for all the agricultural workers in the state. In view of the fact that the agricultural workers constitute the lion's share of the total workers in the state, the effect of this legislation was far-reaching. The two Acts brought distributive justice to the millions of small farmers and farm labourers who together constituted more than 45 per cent of the population of the state as per the 1971 Census.

There are a large number of other legislations enacted by the state in recent times which further broadened the egalitarian base of society. Kerala has the most extensive coverage of the Minimum Wages Act (1948) and it extends to even small groups and odd jobs like the care and handling of elephants. Kerala had pioneered a Maternity Benefit Act (1957) four years before such a measure was passed in the country. The Kerala Act provides for three months' leave with full pay besides other benefits for maternity. The Kerala Destitute Pension Scheme (1960) grants state pension to old destitutes above sixty-five years of age, to destitute widows and to the physically and mentally handicapped persons. These measures which reflect strong biases towards social equity and distributive justice were the natural and necessary outgrowths of a strong and expanding social welfare infrastructure. Flegg (1979) has collected data covering several countries to show that lower inequality of wealth and income, especially wealth, was associated with the fertility level. This supports Radcliffe's hypothesis on Kerala.

Political Climate

Ecology has made its impact on the political climate as well. The high degree of individualism,which scattered type habitats nourished,made people politically active and innovative. The state was the first in India to vote an all-Communist ministry to power in 1957. As in the case of religion, the attitude of the people towards political ideologies of all persuasions has been one of tolerance and even encouragement. Hence, political parties in the state have a tough time wooing people and retaining their loyalty. As a result, these parties have to offer the most fastidious programmes to the people and welfare programmes with a tremendous mass appeal dominate their manifestoes. Part of the explanation for the proliferation of welfare services in the state after independence has to be found in this situation. The high level of literacy along with the ever vigilant and critical press compelled political parties not only to make tall promises to the people but to attempt to live up to them when elected to power. These factors have also compelled the bureaucracy to carry out the welfare programmes more vigorously and thoroughly than otherwise. Left-oriented parties have tutored the weaker sections about the benefits and services available to them from the state and party leaders and ranks are always around to help them to secure their dues from the government,even against administrative lethargy.

Mass Education

Perhaps the largest single factor that contributed to fertility decline in Kerala in recent times has been education—especially education of women. Kerala is way ahead of the rest of India in the field of education. According to the 1981 Census, the literacy level of Kerala is 69 per cent against the all-India achieve-ent of 36 per cent. As far as women's education is concerned, the disparity between the state and the nation is even wider (64 per cent against 25 per cent). The high status of women in Kerala and the educational facilities available to them in traditional times have already been mentioned earlier. It was customary in those times for wealthy families to hire teachers ('Asans') to educate their children; some of these teachers imparted knowledge in their own homes to those who could not hire their services on a personal basis. There were a number of such one-teacher schools (known as 'Palli' in the local language) in the state before the state dealt with the issue of education in a systematic manner. Besides, there were also schools run by various temples, though admission to these schools was restricted to children of the upper castes.

However, the democratisation of education began only in the beginning of the nineteenth century when the state not only undertook the opening of schools on a large-scale but made them available to children of all castes and creeds. Women, Muslims and the untouchables received special encouragement in these schools. There were not only special schools for women in many places, but special schools were also opened for training women teachers. Lord Curzon, Viceroy and Governor-General of India, visited Travancore in the first decade of the twentieth century and 'The sight that pleased him most was the vast array of bright eyed student girls who were called upon together to receive him; and of the several acts of administration which he commended much, one was the generous help given by His Highness in the cause of female education' (Aiyar 1923: 235).

These measures had contributed to the rapid growth of education in Travancore and Cochin. According to the Census of 1901, Travancore was foremost among the Indian states in both general education and women's education. It was the first state in India to introduce free and compulsory primary education (1946). After independence, popular governments continued the tempo of education; it was an irreversible process. When the district of Malabar (formerly part of the Madras province) was merged with the newly formed state of Kerala in 1956, education was promoted as a top priority item in that backward district as well.

These achievements have been possible at great cost to the exchequer. The educational investment as a proportion of the total investment of the state is now about 7 per cent. Public expenditure on education was Rs 185.5 million in the 1960–61 budget and it rose to Rs 2,021.9 million in the budget of 1980–81—a more than ten-fold increase in twenty years. Per capita expenditure on education in 1979–80 was Rs 72.65 whereas it was Rs 27.42 for the whole of India. Primary education absorbs 57 per cent of the total outlay on education whereas secondary education accounts for 26 per cent.

An important point to note from the point of fertility decline, as mentioned earlier, is the high literacy of women in Kerala. The factors contributing to this have already been mentioned. Table 2 gives the position of India and Kerala in the area of female literacy.

TABLE 2 Literacy Level—India and Kerala (1961–81)

	India		*Kerala*	
Year	*General*	*Female*	*General*	*Female*
	(Per cent)		*(Per cent)*	
1961	24.20	12.95	48.80	33.00
1971	29.30	17.32	60.40	54.30
1981	36.20	24.88	69.16	64.50

Education of women has contributed in several ways to fertility decline. It has raised the age of marriage of women in Kerala (twenty-three years against the national average age of eighteen years). Kerala has consistently maintained a high age at marriage of women in India as shown in Table 3.

TABLE 3 Mean Age at Marriage of Women—Kerala and India

Period	*Kerala*	*India*
1901	17.13	13.2
1911	17.35	13.6
1921	17.80	12.6
1931	19.66	15.0
1941	19.35	15.4
1951	19.85	16.1
1961	20.64	17.2
1971	20.88	18.0

Late marriage coupled with education enabled women to have a better choice of husbands and a better understanding of family responsibilities, family planning and child care methods. Women became more health-and-hygiene conscious, better equipped for pre-natal and post-natal care including child rearing and its associated problems. These, in turn, reduced child mortality (Nayar, 1983). As a result, Kerala women are better adapted, more ready and, in many cases, very anxious to avail of the family planning services (Nayar, 1977).

Another contribution of education to fertility control is that the economic value of children became less and less as more and more children began to spend their days in school,where education up to Standard X is free and socially backward children are given financial incentives to remain in school. Kerala ranked first in the proportion of children in the 0–14 years age group attending school full-time in 1961 (43.58 per cent boys and 37.42 per cent girls against 25.26 and 9.84 respectively for India as a whole). Similarly, only 3.37 per cent of the girls in this age group in Kerala were engaged in household work in 1961 against the Indian average of 8.86 (Panikar, 1981).

Comprehensive Health Care System

The second factor in fertility reduction in Kerala in recent years has been the extensive coverage of public health by a comprehensive health care system. Following our earlier focus on welfare, it may be pointed out that good health is not only a prime objective of all welfare policies, it is also necessary for the enjoyment of welfare programmes. The native rulers of Travancore and Cochin took an active part in encouraging public health programmes even in the nineteenth century. They enforced sanitary measures in temples, bazars and other public places and developed a paramedical public health and sanitation infrastructure in the second half of the nineteenth century. A free hospital was opened in Trivandrum by the Maharaja of Travancore in 1865. Similar hospitals were later opened in the important towns. Besides, the state gave grants-in-aid to physicians to augment the hospital services.

Thus, with an already existing medical and public health infrastructure, it was easy for the state to build a comprehensive health care system on modern lines under popular ministries. When the Government of India wanted the state governments to expand their health care facilities in a big way, Kerala had no difficulty—administrative or operational—in implementing the Government of India directive *in toto*. The additional inputs were quickly and effectively absorbed by the system. This has not been the case in other parts of India where the absence of a suitable infrastructure worked as a limiting factor in their health promotion programmes. The figures in Table 4 are indicative of this.

TABLE 4 Growth of Medical Facilities—India and Kerala

Region	*Period*	*Hospi-tals*	*Rate of Growth (Per cent)*	*Dispen-saries*	*Rate of Growth (Per cent)*	*Beds*	*Rate of Growth (Per cent)*
India	1972	3863	43	10409	25	375839	32
	1977	5405		12435		495170	
Kerala	1972	122	623	471	60	23391	114
	1977	760		742		48294	

Table 4 shows that growth in medical facilities in Kerala during the five year period 1972–77 (for which comparable data are available) was several times more than the all-India increase. This is especially true of hospitals where the growth rate in Kerala was fifteen times more than the all-India growth rate. This is expected as Kerala spent a huge part of its budget on health care programmes. The state's per capita expenditure on health was Rs 20.63 in 1979–80. This is the second largest component of the state budget (the first being education). Kerala held the third place in India in 1979 in the amount of expenditure per capita on health. However, the other two states holding the first and second ranks (Haryana and Jammu & Kashmir) spent only slightly higher sums than Kerala (Haryana—Rs 27.71 and Jammu & Kashmir—Rs 21.25). These two states launched their massive programmes only recently. They have smaller populations and lower population densities and higher per capita incomes than Kerala. In spite of this they have not been able to come anywhere near Kerala in the provision of health facilities to their citizens. For example, as far as hospital needs are concerned, while Kerala has one bed for every 445 persons, the figures for Haryana and Jammu & Kashmir are 1,375 and 1,000 respectively.

With a highly literate population and a sound and extensive health care system, it was relatively easy to promote programmes of population planning. The chronological details of Kerala's family planning programme follow the general course of events common to all states in India as the initiative, directive and a major part of the funds were given by the Government of India to all the states in a uniform manner. The Government of India intensified the family planning programmes only in the late 1960s and MTP (Medical Termination of Pregnancy) was legalised only in 1975. The family planning drive in Kerala gathered momentum only in the second half of the 1960s; the progress achieved thereafter has been creditable. In 1961–62, the percentage of eligible couples protected was 0.82; in 1965–66, the percentage was 4.34; but in 1971–72 this figure was 19.62. It shot up by more than 11 points in 1978–79 (30.95). Similarly, the number of births averted in 1961 was 4,345; 19,455 in 1965; however, in 1971, the figure was 1,07,672 and five years later (1975) it increased to 1,72,428.

In short, it is clear that in fifteen years (1961–75) the percentage of eligible couples protected rose from 0.82 to 24.88 (more than twenty-five-fold increase) and the number of births averted rose from 4,345 to 1,72,428—nearly forty-fold increase.

It is certain that these results were achieved by massive programmes of family planning and they have profoundly influenced the state's fertility rate. But their success was, in no small degree, due to the excellent infrastructural build-up present in the state.

Role of Infant Mortality

The low rate of infant mortality prevailing in the state has also greatly contributed to the fertility behaviour. Kerala has the lowest rate of infant mortality (44 per thousand) in India. Over the last three decades, the fall in infant mortality has been rather steep and spectacular. Several studies have shown that there is a close relationship between the infant mortality rate and the fertility rate (see especially J.C. Caldwell, 1979; A.T. Flegg, 1982). High child mortality induces couples to have more children than they actually desire in the hope of having the desired number of children alive after providing for the mortality of some of them. When the probability of child survival increases with a fall in infant mortality, couples need to have only that number of children which they actually desire, especially when childbearing involves both physical and mental strain and child rearing is expensive. By achieving a drastic reduction in infant mortality, a similarly steep fall in fertility rate has been possible in Kerala.

Table 5 gives the infant mortality rate of Kerala for the last three decades. It can be seen that the rate was 120 during the 1950s but was halved in ten years and by the end of the 1970s it had fallen to 44.4. A factor worth mentioning about Kerala's infant mortality is that while infant mortality (that is, death during the first year of the child) accounts for 30 per cent of the total deaths in India, it accounts for only 7 per cent of the deaths in Kerala. A major reason underlying this phenomenon is the wide network of hospitals in the state and the excellent roads which facilitate quick medical assistance when needed. 53 per cent of the total deliveries in the state are attended by medical personnel.

The practice of depending on the untrained 'dhai' for child birth has almost died out in Kerala. Education,together with the wide circulation of newspapers, journals and other mass media,have enabled women to abandon their traditional beliefs and practices regarding pregnancy, child birth, child rearing, food habits and daily routine associated with these activities and to adopt modern practices without much mental adjustment or emotional strain.

TABLE 5 **Infant Mortality Rate in Kerala—1951–1979**

Year	*Infant Mortality Rate*
1951–60	120.0
1965–67	74.0
1971	60.9
1975	57.3
1976	56.3
1977	50.1
1979	44.4

In this paper, an attempt has been made to identify and describe the factors responsible for the significant fertility decline in Kerala in recent times. Two types of factors have been identified—direct causal factors and indirect enabling factors. Massive programmes of educational and health care promotion (the latter includes family planning programmes as well) have directly contributed to the fall in fertility. But the success of these programmes has been largely due to the state's ecology and culture which produced a unique social structure which, in turn, supplied the necessary infrastructure required for the quick absorption of the educational and health care inputs.

REFERENCES

AIYA, NAGAM. (1906). *Travancore State Manual*. Vol. 2. Trivandrum.

AIYAR, S. RANGANATHA. (1923). *Progressive Travancore*. Trivandrum.

CALDWELL, J.C. (1979). Education as a factor in mortality decline: an examination of Nigerian data. *Population Studies*, 33, pp. 395–413.

FLEGG, A.T. (1979). Role of inequality of income in the determination of birth rate. *Population Studies*, 33, pp. 457–77.

———. (1982). Inequality of income, illiteracy and medical care as determinants of infant mortality in under-developed countries. *Population Studies*, 36, pp. 441–58.

GOVERNMENT OF KERALA. (1981). *Kerala Economic Review 1980*. Trivandrum: State Planning Board.

———. (1981). *Census of India, Series 10*. Kerala Paper I Provisional Population Totals. Trivandrum: Directorate of Census Operations.

GULATI, LEELA. (1976). Age of marriage of women and population growth: the Kerala experience. *Economic and Political Weekly*, Special Number.

KRISHNAN, T.N. (1976). The demographic transition in Kerala: facts and factors. *Economic and Political Weekly*, Special Number.

KURUP, R.S. and S. CECIL. (1976). *A note on birth and death rates in Kerala, sample registration report*. DRC Series No. 11. Trivandrum: Bureau of Economics and Statistics.

NAG, MONI. (1981). Impact of social development and economic development on mortality: a comparative study of Kerala and West Bengal. Working Paper of New York Centre for Policy Studies, New York: The Population Council.

NAIR, P.R.G. (1974). Decline in birth rate in Kerala. *Economic and Political Weekly*, Annual Number.

NAYAR, P.K.B. (1977). Socio-demographic characteristics of minilaparotomy acceptors in Kerala. Paper presented at the IUSSP General Conference, Mexico.

————. (1981). Social security, welfare and basic needs policies: a study in Kerala, India. Monograph prepared for ILO, Geneva. Trivandrum: Centre for Social Research.

————. (1983). Impact of welfare measures on mortality: a study in Kerala, India. Proceedings of the Seminar on Social Policy, Health Policy and Mortality Prospects, Paris. IUSSP Publication, 1984.

PANIKAR, P.G.K. (1981). Towards demystifying Kerala's fertility experience. Working Paper 122. Trivandrum: Centre for Development Studies.

PILLAI, K. DIVAKARAN. (1980). Components of fertility decline in Kerala. Population Research Centre, Paper No. 98. Trivandrum: Bureau of Economics and Statistics.

RADCLIFFE, JOHN. (1978). Social justice and the demographic transition: lessons from India's Kerala state. *International Journal of Health Services*, 8, pp. 123–44.

WOODCOCK, GEORGE. (1967). *Kerala: A portrait of the Malabar coast*. London: Faber & Faber.

III

Social Change and Fertility

12

Modernisation and Fertility Change: A Review of Theoretical Developments

K. SRINIVASAN

Introduction

The term 'modernisation' is used to denote the collective set of processes of economic development and social change that are operating in any developing country. As Karl Marx noted over a century ago in his preface to *Das Kapital*, 'The country that is more developed industrially only shows, to the less developed, the image of its own future.' Development economists, as typified in the words of Arthur Lewis (1955: 9), define modernisation as 'the growth of output per head of population'. Modernisation is viewed as a process of social change in which development is the economic component. Economic development, as reflected in the per capita GDP, was viewed as the principal goal of modernisation with the other social changes as necessary ingredients for such a growth. In recent years a number of articles have been published on the concept of modernisation and in many writings the concept has been used rather synonymously with other terms such as westernisation, industrialisation, urbanisation, development and social change. In simple terms 'modernisation' connotes the process by which a tradition-bound agrarian society of very low levels of per capita income, health, education, status of women and welfare, gets transformed into an industrial society with a large proportion of the population living in the urban areas, with a high level of income, expectation of life, literacy and education and other concomitant social changes specially relevant to the improvement of the status of women in the society.

When we assume that the following five dimensions of modernisation, namely, planned development improvements in educational levels, industrialisation, urbanisation, provision of widespread health facilities and improvement of the status of women are crucial factors of modernisation, then these changes tend to be accompanied by concomitant demographic changes in terms of reduction in the mortality levels, increased migration from rural to urban areas, reduction in fertility levels and changes in the rate of growth and composition of the population.

This article attempts to review the frameworks that have been developed during the past three decades for studying the nature of interrelationships between modernisation and fertility change, particularly from the point of view of their relevance to the understanding of the nature of changes that are taking place in India. The theories that have been developed relating the process of modernisation and fertility change can be grouped into three categories:

(*a*) Theory of demographic transition and modifications thereon;
(*b*) Theory of intermediate or proximate variables affecting fertility and elaborations thereon; and
(*c*) Socio-psychological and synthetic framework of fertility.

Though there may be many other theories which may not fall into any of these three categories, only these are chosen for discussion in this paper in view of their relevance to understanding the nature of modernisation and fertility change currently taking place in the developing countries in general and in India, in particular.

Theory of Demographic Transition with Modifications Thereon

The theory of demographic transition, in terms of three phases of fertility and mortality trends of a population subject to the process of industrialisation, seems to have been first postulated in 1929 by Warren Thompson, but an elaborate and mature version of this theory can be attributed to Notestein in 1945. The theory is a general exposition of the fact that mortality and fertility levels remain high in pre-modern agrarian societies, that mortality declines first, with modernisation, that fertility starts declining after a time lag with the continuing process of modernisation, and that fertility and mortality levels stabilise at relatively low values at late stages of modernisation. Notestein also postulated that the prevalence of high fertility in pre-modern agrarian societies is a societal response to high mortality, necessary for the preservation of the species. This high fertility has been maintained by a number of sociological and cultural supports for high fertility, 'props' if we can use the term, through religious doctrines, moral codes, laws, education, community customs, marriage habits and family organisations all focused towards maintaining high fertility levels. With declines in the mortality levels caused by development, such props for high fertility become unnecessary, which is felt by the society only after a time lag, and fertility starts to decline.

Thus modernisation is presumed to lead inevitably first to a decline in mortality, followed by a decline in fertility. In this theory there is a certain amount of inevitability in the nature of demographic response of a population subject to the theories of industrialisation-modernisation. There are various modifications and suggestions that have been made from time to time in this demographic transition theory delineating the phases of transition in greater detail, and primarily based on the experience of Western societies, but the theory seems to retain its validity in the broad explanation of the demographic changes even in developing countries. However, there are many major gaps in the theoretical framework. The theory does not indicate the stage of modernisation at which mortality decline commences, the pace of decline in mortality in relation to modernisation, the stage at which the onset of fertility decline commences and the subsequent pace of decline in fertility in relation to development. A detailed analysis of historical demographical data in a number of European communities have recently revealed that fertility decline has commenced under varying mortality situations. The major gap seems to be a lack of understanding of the determinants of fertility decline in stage III.

Thus, the theory seems to be a very general one in the sense that fertility decline follows after a time lag of the mortality decline in a population undergoing development, which is based on the experiences of Western societies and the theory does not seek to explain the factors underlying the decline in mortality or fertility in terms of social, cultural or economic conditions of the population. The usefulness of the theory, either for understanding the demographic transition currently taking place in developing societies or for predicting their future course, is thus extremely limited. Further the demographic transition in Western societies occurred in situations where there were no governmental interventions with regard to mortality or fertility control in their population. Mortality declines occurred before the organisation of any large-scale public health programmes by the governments and in none of the developed countries were there any national policies which either reduced the birth rates or curtailed the growth rates. The observed demographic changes are aggregate values of individual choices on matters of hygiene, health, family size preferences, practice of birth control, etc. On the other hand, in the developing countries, since their political independence in late forties, most of the national governments felt themselves responsible for the social and economic development of their populations and launched national programmes of public health and family planning to achieve declines in their mortality and fertility levels. These programmes were undertaken as a part of their strategy for the economic development and social transformation of their populations. The public health programmes have had unprecedented success in a number of countries in reducing the mortality levels drastically. While the demographic phenomena in the West have occurred essentially as unplanned effects of their industrialisation, urbanisation and accompanying social changes (which were rather uncontrolled), in a number of developing societies the governments are vigorously popularising public health measures and norms of small family size to achieve rapid reductions in their mortality and fertility levels. With rapid declines in the mortality level, the growth rates of population have touched unprecedented high levels and these are considered as deterrents to development. Also, the traditional demographic transition theory did not recognise the wide range of variations in the fertility and mortality levels that existed in pre-modern societies, particularly in their fertility levels. In a number of developing societies of Asia, particularly in the larger countries such as India and Indonesia, the fertility levels in the past in the pre-transition stage appear to have been far lower than those that existed in the pre-transition European societies. For example, around 1950 before the introduction of any official family planning programme, the total marital fertility rate in India and Indonesia was around 6.0 as compared to the total fertility rate of 8.1 in Norway and 10.8 in Canada in the nineteenth century.

The major contribution of the demographic transition theory seems to be that fertility decline and low levels of fertility are linked to earlier mortality declines and low levels of mortality. Many studies carried out in recent years in

India and other countries both at the macro level (using countries or states within a country as units of analyses), and at the micro level (using families as units of analyses) reveal that period and cohort fertility measures are strongly and consistently associated with mortality levels, especially with infant mortality rates. For example, there is no country in the world which has a crude birth rate of 25 or less, an infant mortality rate above 60, or expectation of life below 65. The elasticity of IMR on crude birth rate appears to be of the order of +0.33 (McGreevey and Birdsall, 1974). Various explanations, valid at the micro or family level, such as child replacement theory, child insurance effect, advanced fecundability of the mothers after child birth,have been empirically tested as to why mothers who experience infant mortality have higher fertility than those who do not, even after controlling for other factors such as social status and occupation. The linkage of fertility to previous experience of infant and child mortality is one of the established facts of demography.

The decline in fertility during the process of modernisation from the stable high levels prevailing in the pre-modern period, postulated in the theory, has also come to question in recent years. Romaniuk (1969, 1978) has found empirical evidence indicating a substantial increase in fertility among Zaire and Canadian Indians in the early stages of modernisation. Nag (1980) has made a comprehensive review of increases in the fertility level recently observed in a number of developing countries. In India, for example, in a comparison of the changes over a generation in the marital fertility rates in Karnataka state (data obtained from two large-scale surveys in the state), Srinivasan, Reddy and Raju (1978) found that the marital fertility rate recorded a substantial increase between 1951–75, in spite of modernisation and an increase in the use of contraceptives, both in rural and urban areas. The increase was greater in rural than in urban areas. For example, the mean number of children born to women increased from 4.8 in 1951 to 6.3 in rural areas, whereas the number increased from 5.3 to 5.7 in Bangalore City. Similarly, by comparison of the data on age specific marital fertility rates between 1959 to 1972 compiled from a published source, Srinivasan and Jeejeebhoy (1980) found that in rural areas in nine out of eleven states investigated in the study the total marital fertility rates were higher in 1972 than in 1959.

This increase in the marital fertility rates of a population in the early stages of modernisation could be due to one or more factors such as (*a*) reduction in the duration of breastfeeding of children by mothers, which, in turn, leads to earlier resumption of menstruation after child birth leading to possibilities of next conception, (*b*) increased coital frequency because of nucleation of families and breakaway from traditional or cultural taboos on intercourse for specific reasons, (*c*) increased fecundity of couples because of improved nutritional status and health of couples, (*d*) reduction in period of widowhood or separation of the spouses within the reproductive span because of reductions in mortality levels, increased prevalence of widow remarriage, etc., (*e*) decrease in involuntary sterility owing to availability of increased medical facilities, and (*f*) earlier onset of menarche, later onset of menopause and reduced risks of spontaneous abortion because of improved nutrition, health condition, prevention and treatment of communicable diseases which impair the female reproductive system.

Such a rise in the potential fertility of a population in the early stages of modernisation can, of course, be offset by the adoption of modern methods of contraception. But it has to be realised that the observed fertility level is the net balance of these two opposing sets of forces, namely, fertility and mortality. The demographic transition theory did not envision a potential rise in fertility during any of the stages of demographic transition. This appears to be a serious limitation of the theory as it is applicable to the demographic transition currently taking place in the developing countries.

In a recent study of fertility levels and trends in selected developing countries of Asia during the period 1960 to 1975, Srinivasan and Pathak (1981) observed a wide variation in the pretransition fertility levels in ten developing countries, even in the absence of the adoption of contraception. The fertility levels in all these countries, however, recorded a declining trend during this period. The fertility transition in each country appears to take place in two stages: the first stage occurs when the marital fertility levels of older women (over age 30) decline because of adoption of family planning, without a decline in the fertility levels of younger women; and the second stage occurs when the marital fertility levels of younger women decline. It has been seen that while the first stage of decline in the fertility level of older women could occur without any major socio-economic changes—with declines in the mortality level and the introduction of family planning programmes together with the availability of modern methods of contraception—the second stage of decline in the level of fertility of younger women appears to be strongly correlated with socio-economic development. Thus, fertility transition in developing countries seems to be partly linked to the widespread availability of modern methods of family planning, and partly and sequentially to the social and economic changes. Most of these countries have already experienced substantial declines in their mortality levels.

Theory of Intermediate Variables

An analytic framework that has been very influential in fertility studies is the one formulated by Davis and Blake (1956), which identifies and classifies a set of 'intermediate variables' through which societal changes have their impact on fertility. In this scheme, fertility change is the focus of the theory and the intermediate variables posited immediately before fertility are the proximate determinants of fertility. This is an exhaustive classification of the intermediate variables in the sense that any change in fertility must be effected through changes in one or more of these intermediate variables, but it is not a mutually exclusive set in that the intermediate variables are not independent of each other. The eleven intermediate variables were classified into three categories: (*i*) factors affecting exposure to intercourse (intercourse variables); (*ii*) factors affecting exposure to conception (conception variables); and (*iii*) factors affecting gestation and successful parturition. This classification assists in identifying the appropriate intermediate variables for explaining any observed fertility change in a society. Sociological elaborations of the model have attemped to include the effects of norms about family size and about the intermediate variables themselves on each of these intermediate variables (Freedman, 1961–62).

In recent times a major contribution in this regard has been made by John Bongaarts who has tried to quantitatively assess the effects on fertility of some of the key intermediate variables. In seeking an explanation for the observed differentials in fertility among various populations in both developed and developing countries, Bongaarts found that a major portion of the variability in the fertility levels in the population can be explained by only four out of eleven intermediate variables proposed by Davis and Blake. These were found to be nuptiality variable (age at marriage and proportion of non-marriage), period of lactation following child birth, incidence of foetal wastage, and prevalence of contraceptive practice (Bongaarts, 1978). Two populations with the same level of contraceptive use of modern methods of family planning can have substantially different fertility levels when the levels of the other three proximate variables (or intermediate variables), namely, lactation, foetal wastage, and nuptiality are different. Bongaarts developed a method of estimating the effect of each of these four intermediate variables in reducing the biologically maximum fertility levels that could otherwise prevail in a population. He also observed that the impact on fertility of traditional practices (such as, post-partum abstinence and lactation) are considerable and that fertility levels could increase substantially during the process of modernisation when these traditional child spacing practices are abandoned or loosened. As an illustration, he studied the fertility increasing effects of decreased duration of breastfeeding in a traditional society and found that a high prevalance of contraceptive use is required to offset this rise in fertility. Bongaarts' work suggests that in any study of fertility change in the developing countries, attention should be focused, in the first instance, on the analysis of the extent to which the intermediate variables (such as, breastfeeding, abstinence, foetal wastage and contraceptive practice) influence the fertility levels, and subsequently an investigation should be made to assess the influence of various social and economic factors among the selected intermediate variables and on fertility, in order to determine the nature of relationships between socio-economic factors and fertility. This means a study of the determinants of fertility. The initial levels of some of the selected intermediate variables in the developing countries, such as India, the nature of changes likely to take place in these variables during the process of modernisation and the nature of effect that such change is likely to have on fertility are illustrated in Table 1.

Studies on Determinants of Fertility

During the past two decades there has been a spate of studies aimed at an understanding of the nature of relationship between various economic factors and fertility. Such studies have been undertaken both at the macro level using

TABLE 1 Likely Changes in the Intermediate Variables in India and their Effect on Fertility during Process of Modernisation

Intermediate Variable	*Level before Modernisation*	*Likely Change during Modernisation*	*Effect on Fertility (− ve reduce) (+ ve increase)*
1) Age at marriage	Low	Increase	−ve
2) Per cent remaining single throughout life	Low	Increase	−ve
3) Broken spells of marriage by divorce or separation of spouses	Low	Increase	−ve
4) Widowhood during reproductive span	Relatively high	Decrease	+ve
5) Voluntary abstinence because of taboos, social customs, etc.	High	Decrease	+ve
6) Coital frequency	Relatively low	Increase	+ve
7) Fecundity during menstruating intervals	Relatively low	Increase	+ve
8) Post-partum infecundity due to breastfeeding	High	Decrease	+ve
9) Contraception and sterilisation	Low	Increase	−ve
10) Spontaneous abortion	Medium	Decrease	+ve
11) Induced abortion	Low	Increase	+ve

countries (or areas within a country) as units of analyses or at the micro level using the household or a woman of reproductive age as the analytic unit. The analysis is carried out usually using a single stage linear multiple regression model, a multi-stage regression model or other more complex multivariate techniques. One of the major problems in the application of multiple regression methods seems to be the presence of large biases in the estimates of the regression coefficient of any predictor variable on fertility because of the presence of strong multi-collinearity among the predictor variables.

Table 2 shows the findings from a recent correlational analysis carried out for the years 1960 and 1978, providing the values of correlation coefficient between the crude birth rate and a number of socio-economic variables using countries as the units of analysis. The correlation coefficients are computed separately for 52 developing, 27 developed and for the total group of 79 countries. The table reveals that the crude birth rate is very strongly associated with the infant mortality rate (positively), the per capita GDP (negatively), percentage of economically active population engaged in non-agriculture (negatively), literacy levels (negatively), urbanisation (negatively), and the population per doctor or hospital bed (positively). The correlation coefficients appear to be very high and significant in each case, and if all the variables are considered together, more than 90 per cent of the variability in crude birth rate among the countries could be explained. The presence of such high correlation between various fertility measures and each of these socio-economic factors at the macro level has been documented time and again through a number of studies. An illustrative summary can be seen in a review by the Research Triangle Institute (1971), *Social and Economic Correlate of Family, Fertility: A*

TABLE 2 Correlation Coefficients of Crude Birth Rates with Selected Socio-Economic Variables with Countries as Units of Analysis

Independent Variables	*Developing Countries n = 52*		*Developed Countries n = 27*		*All Countries n = 79*	
	1960	*1978*	*1960*	*1978*	*1960*	*1978*
Infant mortality rate	+0.71	+0.82	+0.43	+0.63	+0.84	+0.91
Expectation of life at birth (. . .)	−0.73	−0.91	−0.43	−0.48	−0.89	−0.92
Per capital GDP	−0.50	−0.65	−0.42	−0.65	−0.81	−0.78
Per cent GDP from agriculture	+0.54	+0.56	+0.40	+0.67	+0.64	+0.73
Per capita energy consumption	−0.54	−0.67	−0.40	−0.17	−0.76	−0.78
Per cent economically active in non-agri-cultural occupation	−0.70	−0.75	−0.37	−0.66	−0.77	−0.86
Enrolment in primary school	−0.69	−0.84	−0.26	−0.39	−0.83	−0.88
Percentage urban population	−0.63	−0.58	−0.21	−0.43	−0.66	−0.71
Population per physician	+0.35	+0.57	+0.27	+0.37	+0.56	+0.54
Population per hospital bed	+0.28	−0.39	−0.08	+0.56	+0.35	+0.49

Note: Computed by G. Rama Rao, International Institute for Population Studies, Bombay.

Survey of Evidence. The type of correlation coefficient given in Table 2 can be viewed as an empirical validation of the demographic transition theory that with continuing modernisation, as reflected in the above variables, fertility is bound to fall. However, the mechanisms through which the above factors influence the reproductive behaviour of the couple are not yet clearly understood. Do couples desire a smaller family size with modernisation? If so, why? Does modernisation help couples to realise their desired family size more efficiently? If so, how? The answers to such questions are possible only through micro level studies and analyses wherein the fertility of the couples is correlated with their social and economic conditions, and, if possible, changes in these conditions are sequentially related to subsequent changes in their fertility behaviour. Such type of longitudinal studies at the micro level are very few but there have been a large number of cross-sectional studies at the micro level. Unfortunately, the evidence from these studies correlating fertility with the socio-economic variables at the individual or at the family level indicates that not more than 15 to 20 per cent of the variability in the fertility level could be explained by these factors as compared to over 90 per cent of the variability in fertility that could be explained at the macro level. For instance, in two large-scale surveys covering approximately 10,000 and 6,000 households respectively in the states of Bihar and Rajasthan (Srinivasan and Kanitkar, 1981), the completed family size of women in the age group 45 to 49 years were regressed on a number of socio-economic conditions of the woman and the family. Some of the results are furnished in Table 3. It can be seen that not only are the correlation coefficients small but the correlations are smaller in rural areas than in the urban areas and even the nature of relationship is in an unexpected direction. The correlation coefficients are almost close to zero in some cases. Thus more than 80 per cent of the variability in the fertility levels of women remain unexplained by the micro level analyses. Such a major gap in the explanatory power of the variables at the macro and micro levels is one of the puzzling areas of fertility studies.

TABLE 3 Correlation Coefficients of Children Born to Married Women between 45–49 Years with Selected Socio-Economic Variables (Micro Level Analysis)

Variables	*Less Developed Villages (I)*		*More Developed Villages (II)*		*Urban Areas (III)*	
	Bihar $n = 149$	*Rajasthan* $n = 74$	*Bihar* $n = 454$	*Rajasthan* $n = 260$	*Bihar* $n = 143$	*Rajasthan* $n = 93$
Education of wife	−0.08	− 0.12	−0.01	−0.14	−0.06	−0.33
Education of husband	−0.07	−0.16	−0.04	−0.22	−0.14	−0.32
Index of standard of living of the family	0.01	−0.04	0.002	−0.07	0.05	−0.06

Note: Data obtained from the surveys in Bihar and Rajasthan, India, 1981.

Synthetic Framework

Human fertility behaviour is a complex phenomenon which cannot be understood by any one discipline and there is a need to use the essentials from various disciplines to have a holistic understanding of the phenomenon. One such interesting development in recent years is the 'synthetic framework' developed by Easterlin (1975), wherein he combines some of the basic principles of micro economic theory with the sociology of tastes and preferences for children. Easterlin's model characterises the phase of fertility decline during a demographic transition, as a shift from natural fertility to deliberately controlled fertility as well as a change from initially high to eventually low levels of fertility. In this context, factors affecting fertility change can be understood by a study of factors affecting fertility regulations. The regulation of fertility is again viewed as a function of the level of 'motivation' of couples to control fertility and the perceived and objective costs of regulation. 'Motivation' is considered to be a function of the surplus children a couple would have, that is, the excess of potential supply over demand. Thus, Easterlin conceptualised that the fertility levels of a population, in a developing country, are determined by the following three variables:

(*a*) desired family size (C_d—being the number of surviving children) parents would want if fertility regulations were costless;

(*b*) the potential supply of children (C_n—being the number of surviving children) parents would have in an unregulated fertility regime; and (*c*) the costs of fertility regulation (C_r—which includes subjective, psychological and objective costs), namely, the time and money required to learn about and use specific techniques. The first factor, C_d, represents the demand for children. The second factor, C_n, symbolises the supply of surviving children, which is a function of the natural fertility and the proportion of children surviving to adulthood. It represents the supply side of the equation. The excess of supply over demand ($C_n - C_d$) is considered to be an indicator of the level of motivation on family planning in a community to limit the family size. The actual practice of family planning is a function of 'motivation' and the cost of fertility regulation. The observed fertility can, thus, be considered as a function of the level of motivation ($C_n - C_d$) and the costs of regulation (C_r). Motivation for fertility control in a population can increase if C_n increases and/or C_d decreases. C_n could increase if the natural fertility of the population increases and/or the child mortality level decreases. Similarly, the desired family size can decline during the process of modernisation because of increasing education and higher costs of raising children. The costs of fertility control (C_r) may also decline because of national family planning programmes which offer free or subsidised birth control methods and services, and as the level of education increases couples can be expected to acquire greater knowledge of fertility control. This framework seems to be particularly relevant to the developing countries around the stage of fertility transition. A good validation of this synthetic framework has been made from an analysis of the data compiled for Karnataka state, on the basis of two large-scale surveys carried out in 1951 and 1975, and data for ten states in India through large-scale surveys carried out in 1971 (see Crimmins *et al.*, 1981 and Srinivasan *et al.*, 1984). These analyses revealed that the framework developed by Easterlin is generally applicable for an explanation of various interesting observations on Indian fertility, for instance, why the fertility levels in some states are higher than in others; why marital fertility increases in the early stages of modernisation even in the context of increased use of contraceptive methods, and what is the extent of use of contraception in relation to the level of motivation. The analysis of the data from ten states revealed that the factor of motivation is as crucial (if not more) as the cost of regulation in the explanation of differentials in contraceptive use among the states. The synthetic framework of Easterlin offers a plausible explanation for the hitherto apparently low impact of the family planning programme on the fertility levels in India and indicates the possibility that fertility levels in many developing countries could drop substantially in the near future, due to the combined effects of increasing motivation and reducing costs of fertility regulation.

Another school of thought attempts to explain the recently observed dramatic declines in the fertility levels in some of the developing countries (such as China) and in some states of countries (such as Kerala in India) in terms of reduction in the disparities in wealth and income within a society. This theory, which can be called 'the social justice theory of demographic transition,' attempts to link demographic levels and trends, particularly the fertility levels, with the political and economic institutions of a society which promote social justice (William, 1973; Repetto, 1981) and postulates that the better the social and economic equity, the lower the fertility. In a cross-country analysis of the fertility levels regressed on a number of socio-economic variables, Repetto found that the association between fertility and income inequality is much stronger than that between fertility and female literacy, infant mortality, or per capita income. The declines in the fertility levels in China and Kerala are largely attributed, by some scholars, to their Left-wing political institutions which promote equity in wealth and income. These observations have, however, been criticised by a number of scholars (Birdsall, 1977; Nag, 1983). Nag compared the fertility levels and trends in two states of India, Kerala and West Bengal, which have been governed for quite some years, at the state level, by duly elected Left-wing communist governments emphasising land reforms and a decrease in inequalities in wealth and income. He concluded that the relatively more rapid decline of fertility in Kerala can be attributed to the more equitable distribution of public health and family planning services, and improved education in the state, than to the equity in the distribution of economic assets and income. Between the two states, West Bengal has fared better in terms of equity in the distribution of wealth and income compared to Kerala, but in terms of the availability of public health services and educational facilities Kerala has consistently fared better than West Bengal and as such Nag attributes the more rapid decline in fertility in Kerala to the equity of social welfare services than to the equity in the distribution of economic assets and income. Social justice theory of fertility decline seems to support the need for providing universal education and primary health care as the basis for a decline in the fertility level.

Conclusion

The fertility level of a population observed at any given time is the outcome of a complex set of factors—biological, social, cultural, economic, demographic and political—that operate on the population and these factors interact with each other with regard to their effects on fertility. The people of India, have strongly held social and cultural values with regard to marriage, fertility of a couple and particularly the preference for a son. It cannot be expected that the desired societal or national goals of fertility, expressed in terms of desired crude birth rates, should be congruent with the goals for family size at the micro level. Only in an extremely controlled condition, existing as in China, can couples be expected to conform to a nationally specified family size of one or two children for each couple. Within a democratic framework, as is prevailing in India, every couple has the freedom to choose the number of children they want, and when they want them. Under these conditions it can hardly be expected that any targetted path of fertility decline would be realised with precision by national programmes on family planning. The processes of modernisation that have already been ushered in the country have begun to erode many of the social and cultural values that placed an emphasis on a larger family and the desired family size can be expected to go down sharply in the future. Programmes directed at providing education for children, especially for the females; reduction in infant mortality rates; providing employment opportunities for women outside the household sector; and improving the economic conditions of the poorest section of the society by appropriate subsidies and labour intensive employment (such as the employment guarantee scheme in Maharashtra), can be expected to accelerate the pace of decline in the desired family size, and reduce the demand for children. It must be recognised that a large percentage of children, currently being born, are considered as unwanted births—unwanted by the parents themselves as revealed in various surveys and in comparison to their desired family size. In this context, it is equally important to prevent such unwanted births by providing family planning services in a socially and psychologically acceptable manner. Offering increased and sustained incentives to acceptors of family planning methods can be justified on grounds of social equity and care for the children already born to those who have resolved to limit their family size. Provision of efficient family planning services with care and concern for the acceptor and his family can, in themselves, be expected to generate further demand for family planning. In the present state of knowledge of factors influencing fertility, there appears to be no unique, all pervasive theory of fertility decline which can be invoked to realise desired fertility goals; neither it appears that such a theory could ever exist. The varying political, social and cultural conditions under which fertility has started declining during the past decade, indicate that with the sweep of modernisation it is becoming universally felt that the small family norm (of one or two children) could become a universal goal in all countries before the turn of this century. High fertility levels could as well become memories of the past, even for the next generation.

REFERENCES

BIRDSALL, NANCY. (1977). Analytical approaches to the relationship of population growth and development. *Population and Development Review*, 3(1 & 2).

BONGAARTS, J.A. (1978). A framework for analysing the proximate determinants of fertility. *Population and Development Review*, 4(1), pp. 105–32.

CRIMMINS, E., R.A. EASTERLIN, S.J. JEEJEEBHOY and K. SRINIVASAN. (1981). New perspectives on the demographic transition: a theoretical and empirical analysis of an Indian state, 1951–1965. *In* K. Srinivasan and S. Mukerji (Eds.), *Dynamics of population and family welfare*. Bombay: IIPS, 1981, chapter 3.

DAVIS, K. and J. BLAKE. (1956). Social structure and fertility: an analytical framework. *Economic Development and Cultural Change*, Vol. 4, pp. 211–35.

EASTERLIN, R.A. (1975). An economic framework for fertility analysis. *Studies in Family Planning*, Vol. 6, pp. 54–63.

FREEDMAN, R. (1961–62). The sociology of human fertility. *Current Sociology*, 10/11 (2), Oxford: Basil Blackwell.

LEWIS, WILLIAM ARTHUR. (1955). *The theory of economic growth*. London: Allen and Unwin, 1963.

MARX, KARL. *Capital*. Translated from the third German edition by Samuel More and Edward Aveling, and edited by Friedrich Engels. Moscow: Foreign Languages Publishing House, 1959.

MCGREEVY, WILLIAM P. and NANCY BIRDSALL. (1974). *The policy relevance of recent research on fertility*. Washington: Smithsonian Institution Press.

NAG, MONI. (1980). How modernization can also increase fertility. *Working Papers*. New York: Centre for Policy Studies, The Population Council.

————. (1983). The equity-fertility hypothesis as an explanation of the fertility differential between Kerala and West Bengal. *Centre for Policy Studies Working Papers*, No. 96. New York: The Population Council.

NOTESTEIN, FRANK W. (1945). Population: the long view. *In* Theodore W. Schultz (Ed.), *Food for the world*. Chicago: University of Chicago Press.

REPETTO, R. (1981). *Economic equality and fertility in developing countries*. Baltimore and London: Johns Hopkins University Press.

RICH, WILLIAM. (1973). *Smaller families through social and economic progress*. Washington, D.C.: Overseas Development Council.

ROMANIUK, A. (1969). Increase in natural fertility during the early stages of modernisation: evidence from an African case study, Zaire: *Population Studies*, Vol. XXIII, No. 1, March.

————. (1978). Evidence of increase in natural fertility during the earlier stages of modernisation: two case studies. Paper presented at the annual meeting of the Population Association of America, Atlanta. 13–15 April.

SRINIVASAN, K., P.H. REDDY, and K.N.N. RAJU. (1978). From one generation to the next: changes in fertility, family size preferences and family planning in an Indian state between 1951 and 1975. *Studies in Family Planning*, 9–10.

SRINIVASAN, K. and S.J. JEEJEEBHOY. (1980). Changes in natural fertility in India, 1959–72. Prepared for the Seminar on Determinants of Fertility Trends, sponsored by the IUSSP, held at Bad Homurg, West Germany, 14–17 April 1980.

SRINIVASAN, K. and K.B. PATHAK. (1981). Nature of stable fertility and the determinants of its destabilisation process in selected countries of Asia. Invited paper submitted to the IUSSP General Conference, Manila, 9–16 December 1981.

SRINIVASAN, K., S.J. JEEJEEBHOY, R.A. EASTERLIN, and E.M. CRIMMINS. (1984). The roles of motivation, attitudes and access in the adoption of fertility control: a cross-sectional analysis of ten Indian states, 1970. *Population and Development Review*, January 1984.

SRINIVASAN, K. and TARA KANITKAR. (1981). A survey on fertility and mortality in Bihar and Rajasthan. Bombay: IIPS (mimeo).

THOMPSON, WARREN S. (1929). Population. *American Journal of Sociology*, 34(6): 959–75.

WILLIAMS, ANNE D. (1973). Fertility and infant mortality. *Thesis Prospectus*, University of Chicago.

13

Demographic Transition in Kerala

R.S. KURUP

Introduction

The state of Kerala came into existence on 1 November 1956, under the States' Re-organisation Act of 1956. This was formed by integrating the erstwhile Travancore-Cochin state (excluding the four southern taluks of Vilavancode, Kalkulam, Agastheeswaram and Thovala and part of the eastern taluk, Shencottah) with Malabar district, excluding the Laccadive and Minicoy Islands of Madras state and Kasargod taluk of south Canara district. It has an area of 38,864 square kilometres. It is 576 kilometres in length (North-South), while its breadth (West-East) varies from 11 to 120 kilometres. The land has been divided into three natural regions—low land, midland and high land.

The two monsoons, South-West from June to September and North-East from October to December, provide rains to Kerala. Of these the former is more important as it makes plenty of water available to the people of the state. The landscape remains ever green. Even then, the temperature rises very high during March, April and May.

In this narrow strip of land, people live in very crowded conditions, more crowded than in any other state in India. According to the 1981 Census, the density of the state is 654 per square kilometre as against 221 for India as a whole. The density, which was 134 during 1881, multiplied nearly five times over a hundred years. The per capita availability of land is less than 40 cents out of which only less than 23 cents form the cultivated/cultivable area. All the other land is covered by forests or is used for non-agricultural purposes.

The state income during 1977–78 was Rs 2404 crores while it was Rs 43.2 crores during 1960–61. This records a growth of 456 per cent over a period of seventeen years. But the increase in the per capita income was only 281 per cent during this period, from Rs 259 in 1960–61 to Rs 987 in 1977–78. The gap between the increase in the state income and per capita income is 175 per cent. Even though there is a price rise at the rate of 10 per cent per year, the contribution of population growth is tremendous. At constant prices, while the state income increased by 214 per cent from 1960–61 to 1977–78, the per capita income increased by 115 per cent only. Kerala has a much superior positive rate as compared to the all-India average in respect of some of the indicators of infrastructure development like electrification of villages, road/rail length per 100

Acknowledgements The author is deeply indebted to Mrs. A. Syamala Devi, Research Officer for technical assistance and Mrs. Hymavathy, Confidential Assistant for typographical assistance.

square kilometres of area, education and health. The growth of population was much higher here till very recent times. Naturally, the factors which contribute to the growth have to be examined in order to provide a clue to the transitionary aspects of population.

Population Growth

It has been estimated that nearly 30 lakh people lived here at the beginning of the seventeenth century (Nair, 1965) and by the middle of the nineteenth century this figure had increased to 45 lakhs. The growth rate was accelerated during the twentieth century. The rate of growth steadily increased except during the two decades 1911–21 and 1931–41. However, the decadal variation during 1971–81 has shown the beginning ot a reversal of the trend. Table 1 gives figures on the growth of population of Kerala compared to the all-India figures.

TABLE 1 Growth of Population of Kerala Compared to All-India (from Censuses)

Year	*Population of Kerala (lakhs)*	*Decennial Percentage Growth*		*Annual Geometric Rate of Growth of Population*	
		Kerala	*India*	*Kerala*	*India*
1901	64.0	—	—	—	—
1911	71.5	11.75	5.73	1.19	0.56
1921	78.0	9.16	−0.30	0.90	−0.13
1931	95.1	21.85	11.00	1.98	1.06
1941	110.3	16.04	14.13	1.50	1.34
1951	135.5	22.82	13.31	2.08	1.26
1961	169.0	24.76	21.64	2.24	1.98
1971	213.5	26.29	24.80	2.30	2.24
(1 April–1 March 1981)	254.0	19.00	24.75	1.71	2.24

While the all-India decennial growth of 24.80 and 24.75 during the two previous decades are more or less the same, the Kerala figures show a decline in 1971–81. This is the first time in recent years that the decennial growth rate decreased in Kerala. There is every possibility of this trend continuing in the later decades of this century as well as in the next century.

Birth Rate

An examination of the trend in the birth rate in Kerala from 1931–40 onwards (vide Table 2 for birth rate estimates) shows that the straight line

$$Y = 44.15 - 0.38\,t$$

is a good fit. If the information from 1966 onwards is considered, the straight line of best fit is

$$Y = 29.65 - 0.16\,t$$

There is thus a reduction in the rate of decline in the last fifteen years. A projection of this recent trend leads to a birth rate of 24–25 during the next two decades. If, however, the reduction in 1956–80 is projected, the birth rate will come down to 19.8–23.6. If the average decline during 1966–80 of 76 continues, then the birth rate will decline to 19 in 1990 and to 12.5 in 2000. Apart from projecting the trend in terms of the order of birth and the number of children born, it can be seen that if third and higher orders of birth are avoided, the birth rate will be 15 and the average number of children will be 2 for each couple, so that there is near replacement in a generation or a net reproduction rate of about 1, granting that the death rate has already become very low. In order to reach this stage, with the average number of 3 children at present, it should take at least 15 to 20 years.

Death Rate

A similar analysis of the trend in the death rates leads us to two equations:

$$Y = 27.22 - .56\,t \text{ for the period 1931–1980}$$

and

$$Y = 11.08 - .35\,t \text{ for the period 1966–1980.}$$

The rate of reduction has shown a deceleration in recent years. As the death rate has already become very low by 1980, the scope for further reduction will be very limited. One can safely assume a reduction to 5 per thousand by AD 2000 and thereafter an increase in the death rate, so that there will be a severe toll of death among the aged. The trend in the expectation of life at birth shows that it has been increasing at the rate of 0.9 years during 1936–78. The line of best fit is

$$Y = 27.0 + 0.92\,to$$

If this is projected to AD 2000, the expectation will be more than 80 years, a little more than the expectations of life reached by Sweden and Netherlands currently (the figures on birth and death rates and the expectation of life at birth are given in Table 2).

Migration

In a state like Kerala where employment opportunities are limited and educated employment seekers are proportionately more,out-migration plays a

TABLE 2 Birth Rate, Death Rate and Expectation of Life at Birth for Kerala from 1931–40 to 1980

Year	*Birth Rate*	*Death Rate*	*Expectation of Life at Birth*		
			Male	*Female*	*All Persons*
1931–40	40.0	29.1	32.1	31.4	31.8
1941–50	39.8	22.3	32.5	31.7	32.1
1951–60	38.9	16.1	41.9	40.6	41.3
1966	37.4	10.1	. .	. .	. .
1967	36.3	10.4	. .	. .	. .
1968	34.3	10.1	. .	. .	. .
1969	33.3	10.0	. .	. .	. .
1970	32.3	9.2	. .	. .	. .
1971	31.9	9.2	60.6	61.2	60.9
1972	32.1	9.2	59.1	61.5	60.2
1973	29.9	8.5	60.6	62.0	61.4
1974	27.0	7.8	60.9	63.3	63.2
1975	28.2	8.4	61.2	64.4	62.8
1976	28.5	8.1	62.5	65.2	63.8
1977	26.1	7.2	63.6	66.2	64.9
1978	25.2	7.0	63.8	66.2	65.4
1979	25.9	6.9	. .	. .	. .
1980	26.7	6.8	. .	. .	. .

SOURCE: Upto 1951–61, the figures are from census calculations and those for 1966 onwards are from the sample registration scheme
'. .' indicates not available

TABLE 3 Net Migration from Kerala

Period	*State*	*Net Migration*	
		Males	*Females*
1901–11	Travancore Cochin	3152	7390
1911–12	Travancore Cochin	7111	11763
1921–31	Travancore Cochin	33210	34744
1931–41	Travancore Cochin	−8422	3395
1941–51	Travancore Cochin	−60091	−30650
1951–61	Kerala	−199536	−85517
1961–71	Kerala	−614690	(males and females together)

leading role. The volume of net migration during 1901–11 to 1961–71 is shown in Table 3.

Net out-migration assumed significance in 1961–71, and has continued to do so in 1971–81 as well. During this decade net out-migration of almost 6 lakh persons is considered to be a reliable estimate.

Demographic Explanation of the Decline in Growth Rate

The birth and death rates declined at varying rates even before the present decade, while their difference showed a widening gap. The death rate declined first along with its downward march; followed by the birth rate after a few years. During 1961–71 there was a positive net out-migration rate and in 1971–81 this rate increased. But the major proportion of the decline to the present position is due to the reduction in the birth rate; if this had not taken place the growth rate would have further increased. This aspect would become clear if the birth and death rates given in Table 2 are scrutinised,

From 1931–40 to 1966 the death rate declined to one-third of its original value: the birth rate dropped from 40.0 to 38.9 by 1961 and to 37.4 by 1966. This gap between the birth and death rates widened up to 1966, but when the death rate remained almost steady at 10 and the birth rate continued to decline, the growth rate again declined. From 1970 to 1973, the death rate was almost stable at 9, till 1976 it was stable at 8 and at 7 up to 1980, during this period there was a decline in the birth rate except during the last triennium and a break at 1974 which may be due to undue sampling fluctuations.

Factors Contributing to Reduction in Mortality

The toll of mortality which was very heavy among infants declined substantially. While a fourth of the live born children died before reaching one year of age in the early decades of this century, less than one in twenty dies early today. Control of communicable diseases and maternal mortality has been a major contributing factor in mortality decline. The major causes of death during the previous decade were heart disease, cancer, accidents and injuries and tuberculosis; with the exception of tuberculosis, the others are common in all developed societies. There are also some causes peculiar to infant mortality.

As in other states and countries, with the passage of time mortality began to decline first. The acceleration in mortality decline was brought about by the improvement in hospital services, environmental sanitation and hygiene. The per capita expenditure on health by the government is one of the highest in Kerala among the states in India, next only to Jammu and Kashmir. Literacy level also plays a part in the reduction of mortality, perhaps by making people health conscious and motivating them to utilise the health facilities available. These health facilities came into existence by a curious combination of circumstances. The progressive rulers of Travancore and Cochin initiated health programmes in the princely states. The Christian missionaries also established hospitals in the various parts of the state. This was followed by the Nairs, Ezhavas and Muslims who competed with each other in the fields of health and education. The sectoral enthusiasm contributed largely to improvements in the fields of health and education. However, the Malabar area lags behind in both these fields. Also, the per capita availability of facilities show a declining trend, if one goes from west to east or from the low land to the high land. As a result, the decline began earlier in the low land and reached its minimum here and then in the other natural divisions. Similarly, among the various religious groups—Christians, Hindus and Muslims—there is a difference in the death rates, with the Christians having the lowest rate and the Muslims the highest rate—the state has nearly 21 per cent Christians, 60 per cent Hindus and 19 per cent Muslims. The Muslims, as a whole, are backward in literacy.

A multiple regression analysis of the variation in the death rate shows that the doctor-population ratio and the expenditure on health explain a large part of the variation; similarly, the variation in infant mortality is explained by the doctor-population ratio and the birth rate. With a decrease in the birth rate and an increase in the doctor-population ratio, the infant mortality rate declines. Though there is room for much more reduction in this rate which now stands at a level just below 50, the influence on the death rate can be only slight as the latter has fallen to less than 7 per 1000 in recent years.

Factors Contributing to Reduction in Birth Rate

The main causal factors which lead to a reduction in the birth rate are the changes in age distribution, marital status and acceptance of family planning. Applying the method of standardisation, Kurup and Cecil (1976) found that of the 156 per cent reduction in the birth rate, 14.8 is due to family planning. In the later years, the effect of family planning was only 17.9 out of 21.1 per cent reduction and 21.8 out of 31.5 per cent reduction. Pillai (1979) analysed the components of fertility change in two papers and showed that 84.6 per cent of reduction in the crude birth rate and in the general fertility rate between 1971 and 1977 was due to family planning, while 90 per cent reduction in the total fertility rate was due to family planning.

A total of 1.21 lakh sterilisations as against 1 lakh targetted had been done in 1981–82 reaching the cumulative total of 12 lakhs by the end of March 1982. Nearly 35 per cent of couples have been protected by various methods of family planning. The average age and number of children living at the time of acceptance of family planning has shown significant decreases in the previous decade. The proportion of tubectomy in the total number of sterilisations is increasing; there is more emphasis on the terminal methods of sterilisation as compared to IUD and conventional contraceptives.

As regards the age at marriage of females, the average was 17 even during the early decades of this century and this increased to 21 in 1971; it is certain that the average age may have increased to 22 by 1981 (the results of the 1981 Census of population are yet to come). The number of eligible couples in the population (married females in the age group 15–44 years with husbands living) was 140 in 1971 as against 170 for India as a whole.

With the decline in mortality, there has been a slight change in the age distribution which has also contributed to a reduction in the birth rate.

As regards the contribution of agro-socio-economic variables, Krishnan (1976) used multiple regression techniques but he could not obtain any significant effect on the birth rate stemming from any variable. In 1979, Kurup analysed the variations in the birth rate among the states in India; it was found that facilities of mobility measured through road mileage was a major variable in 1961 and 1971 supported by general literacy in 1971 and by measures of cooperative movement and irrigation facilities in 1961. Variables like the age at marriage, female literacy and rate of female workers in non-agricultural occupations to those in agricultural work were also seen to operate in 1961. While this analysis considered birth rate as the dependent variable, it is interesting to find a different type of explanation when fertility rate was considered as the dependent variable. In 1961, the major variable which could explain a large part of the variation in the fertility rate turned out to be an indicator of cooperation; in 1971 the ratio of female workers in non-agricultural and agricultural occupations, supported by the age at marriage and road mileage, explain a large part of the variation. In 1976, electrification, female employment in the organised sector and irrigation were the major variables. In short, rural development, women's role and agricultural practices may have had their impact on fertility. In the case of Kerala there is no information on the fertility rates in the various districts. The sample registration scheme which has been in operation from July 1965 onwards cannot provide district-wise estimates because the scheme does not have representative probability samples for each district. However, with the available data,district estimates can be obtained by considering the probability of selection of each sample. This may lead to varying accuracy levels. Child-woman ratio based on population count alone in the half-yearly surveys will provide a better estimate of fertility in this respect rather than an estimate of the birth rate.

The variations in child-woman ratios in 1961, 1971 and 1976 have been analysed. The 1961 and 1971 data are obtained from the censuses while the 1976 data on child-woman ratio are from sample registration. Table 4 shows the contribution of the several variables in explaining the variation in child-woman ratios among the districts in Kerala.

TABLE 4 Contributions of the Variables in Explaining the Variation in Child–Woman Ratios

S.No.	*Variable*	*Contribution and Percentage*		
		1961	*1971*	*1976*
1.	Multiple cropping	23.93	0.01	20.17
2.	Rural female literacy	21.67	0.90	..
3.	Wage rate of female unskilled workers in construction	34.86	0.91	1.07
4.	Percentage of villages electrified	2.80	73.20	24.70
5.	Ratio of female workers in unagricultural to those in agricultural occupations	1.05	0.51	..
6.	Total variation explained	84.31	74.53	45.94
7.	Multiple correlation coefficient	0.91	0.87	0.68

The choice of variables may seem rather inappropriate but one should think of the variables which can possibly fall under rural development and agricultural practices especially those for which information is available in the present case for the districts in Kerala. Wage rates, multiple cropping as an improved agricultural practice and female literacy made a significant contribution in 1961 while the electrification of villages accounted for a large part of the variation in 1971 and for more than half of the variation in 1976; multiple cropping was also a very important variable in 1976.

Land redistribution would be able to effect some changes in fertility at least in some of the states which progressed substantially in this respect. Crude death rate, infant death rate and doctor-population ratio have high correlations with birth and general fertility rates.

The Transition in Kerala

Considering the most general conceptualisation of the demographic transition with high birth and death rates in the early periods followed by a decline in mortality and subsequently a decline in fertility, so that the growth of population is very rapid in the middle period only, it can be seen that Kerala has just entered the final stage. But if one considers the economic analogy and the various hypotheses including the threshold hypothesis, Easterlin's hypothesis and Caldwell's re-examination of the transition hypothesis, there is not enough conformity in the Kerala case. At the outset, it should be borne in mind that being a small part of a large land mass there is no separate existence for the state and whatever happens in the country is naturally passed on to this part also in some measure. With this limitation in mind, let us examine how the population of the state has undergone the transition.

Progress in Attaining Demographic Transition

Bogue (1969) has devised an index for arriving at the progress in attaining demographic transition.

$$I = \tfrac{1}{2}\left[\frac{235 - \text{g.f.r.}}{175} + \frac{7.5 - \text{t.f.r.}}{5.3}\right] \times 100$$

Here 235 and 7.5 are the maximum general fertility and total fertility rates obtained by examining various communities of the world; for low fertility similar values are 60 and 2.2 respectively. The differences between the maximum in high and low fertility areas are used as denominators in the above index for the two terms. Kerala showed that 49.64 per cent of the transition was completed in 1969, 60.81 per cent in 1969 and 70 per cent in 1978.

Threshold Hypothesis

Dudley Kirk (1972) pointed out the basic assumption of demography students—humanity is midstream in a revolutionary change to reach lower birth and death rates. Stolnitz stated that all the industrialised nations moved to that stage from traditional agro-based economic systems, experiencing in this long-term process enormous population increases. This used to be called the vital revolution or demographic transition.

According to the theory of demographic transition, the less developed countries may follow the experience of Western countries by reducing birth and death rates, if those countries achieve major socio-economic development. A reduction in the birth rate is widely seen among peoples from very different cultural backgrounds with different ways of life; the higher the natality rate at the time of entering the transition, the more rapid is the rate of decline. A certain threshold and mix of socio-economic development with variations in different cultures has been a requirement for initiating a strong downward trend in natality. A rough measure of the countries in transition can be thought of by considering the level of birth rate as between 25 and 35. Several of the smaller countries, especially islands, have been experiencing unusually rapid social and economic development; birth rates have been affected by the emigration of young adults. Mortality reductions are, however, seen in all cultures and in large countries as well.

There may be a deceleration in the reduction of birth rates because of the large cohorts of young people, as a result of the baby boom after the Second World War. Given the time and socio-economic development, one may expect a general diffusion of lower birth rates in the less developed areas. The process has already begun. But much depends on how fast these reductions will occur and at what level of socio-economic development they may be expected to begin.

There is a major important gap between the European countries and Japan and those countries entering the transition later. While the European countries achieved transition much before the Second World War, Japan achieved it a little after the war. There was an interval of twenty-five years during which no major country entered the transition.

The so-called threshold hypothesis was examined in a study on world mortality as of the year 1960 (UN, 1963) by considering all the less developed countries. No specific threshold could be found. However, Kirk in 1972 analysed each major cultural region and found that consistent patterns emerged. In Asia and the Far East, for cultural reasons and perhaps because the effects of over-population are more salient, people have undertaken family limitation at a lower limit of socio-economic development than in Latin America. As of 1960–64, three countries, namely, the Philippines, South Korea and Thailand, fell in the threshold area for socio-economic indicators, though their fertility had not come down as desired; later, however, South Korea could bring down the birth rate to less than twenty-five, along with Taiwan and Sri Lanka.

In the Islamic world, measures of literacy/education stand out as related largely to natality levels; a decline was seen in Albania, Lebanon, Malaya and

Soviet Islamic Republics which lead in education and other indices of socio-economic development.

Caldwell (1981) discussed the mechanism of demographic change in historical perspective. His aim in analysing fertility transition was both to present a general theory of change and to explain why, though fertility remains high in contemporary third world countries, a sustained section may well begin without having to wait for further fundamental economic change. The major basis of the argument was that, within the upper and lower limits, fertility in all societies is rational given the identity of the decision makers and the nature of the socially determined goals. High fertility was beneficial in pre-capitalist economies, the situation changed as the capitalist-labour market developed.

In 1976 Caldwell began to examine the demographic transition theory with the purpose of re-stating it in view of his experience in Africa. The economic organisation of relatively self-sufficient agrarian communities turns almost wholly upon the family and the perpetuation of the family is the main guarantee of support and elemental security. With a high death rate, the individual feels insecure and unimportant. Combined with this is the situation in which there is no facility for education. In such societies, there is scant opportunity for women to achieve economic support or personal prestige outside the role of a wife and mother and women's economic functions are organised in ways that are compatible with continuous childbearing. Fertility is high in poor, traditional societies because of the high mortality level and the lack of opportunities for individual advances and the economic value of children. All these change with modernisation or urban industrialisation and individuals, once their viewpoints become reoriented to the changes that have taken place, can make use of the new opportunities. Economic rationality has been brought in as a precondition to demographic transition by earlier writers. Similarly, industrialisation and the concomitant urbanisation are preconditions to development.

Julian Simon (1974) assessed the available evidence on fertility in relation to economic development, but avoided the question of why fertility declines set in except by pointing out that as the educational level rises, fertility drops. He argues that fertility is primarily economically motivated.

Current research on the magnitude and direction of wealth flows and potential flows, misunderstanding of the nature of family formation and too much emphasis on modernisation has helped to understand fully the nature of demographic transition. Also, with the recent experience of countries in the family planning programme, there should be a new theory of transition. A comparison of acceptors and non-acceptors show a whole interrelated set of characteristics (such as, more education, non-agricultural employment, higher incomes and lower fertility).

As regards inter-generational wealth flows in pre-modern societies, much of the flow is indirect because of the existence of extended family clans and even villages that share in the flow. There has been a change in this situation in which the net flow is towards parents and hence high fertility is rational; in the later situation the flow is towards children and hence high fertility is irrational. In general, in societies of every type and at every stage of development, fertility behaviour is rational and fertility is high or low as a result of the perceived benefit or the inter-generational wealth flow. The wealth flows, however, are dependent on various social and cultural factors. Caldwell has tried to modify the transition theory bearing in mind all these factors.

Easterlin Hypothesis

Since the sixties Easterlin and others (1973) have introduced 'taste' as a significant variable in the relation between economic factors and fertility. Economic behaviour is affected by the gap between individual aspirations and the resources available to satisfy them. This concept can be extended to tackle the tension between income level and aspirations or tastes which may also affect fertility. Couples faced with insufficient incomes to satisfy their tastes for consumption goods can make their available resources stretch further by limiting family size. Couples with more income at their disposal may not find their income constraint eased but may find it intensified if they attempt to accommodate higher aspirations. Easterlin proposes relative income, that is, the ratio of observed income to expected living standards or aspirations as the relevant variable. His hypothesis is that the changing balance between the income earning possibilities of young adults and their desired living levels influence fertility. The basic idea is that if young men—the potential bread-winners of households—find it easy to make enough money to establish homes in the styles desired by them and their actual or prospective brides, then marriage and childbearing will be encouraged. If not, deferment of marriage and limiting family size will be resorted to.

The New Economics Approach

The Becker school uses a microeconomic model of human production to explain potential decisions about family size. Couples seek to maximise the total satisfaction derived from basic commodities as good health, satisfaction from children and enjoyment of leisure time activities each of which is produced in the household. Relevant costs include both material and time inputs. The satisfaction from children depends both on the number of children and the quality of each child, the latter determined by the quantity of inputs including both material resources and the time spent by the mother in child care. The couple's total satisfaction from all sources is constrained by their available time and income.

It may be mentioned here that there is no basic compatibility between the two schools of thought, as both are necessary to an understanding of the relation between income and fertility. There are other schools like the Leibenstein school on the economic value of children and the opposite school of Stephen Enke and Paul Demony.

The Kerala Case

In the case of Kerala, the income approach at the macro level does not hold good and the threshold hypothesis in its pure form is out of place. However, the modifications encompassing social development like improvement in the level of education, and in infrastructure facilities like roads and hospitals are applicable here. As regards the Easterlin and Becker hypotheses, not much information is available to test these—probably when the results of the study on the determinants of fertility change in Kerala are available, this can be looked into.

Kerala can safely be assumed to have entered the transition, in view of the high literacy level, and the high acceptance of family planning and health facilities, resulting in a lowering of fertility and mortality. There has been a large exodus from the state to other parts of India and to the Middle East in recent years; this will again be beneficial leading to further reductions in the fertility level and improvements in the economic levels and the standards of living.

In the recent decades, Kerala did not have a society based on familial production, hence high fertility as in most countries of the third world was not a practice here. With the matriarchal system in vogue, the women enjoyed a high place in society (which continues even today). The emotional and wealth flows were from the young to the old in the lower strata which had a high fertility level. With the emergence of new and progressive ideas in the lower strata, a transformation took place and these high fertility groups utilised the facilities available for better health and family limitation. This change has been accelerated by the imposition of mass education systems of a formal and informal character. The system of production has undergone drastic changes and with the saturation in the primary sector, children are more likely to seek work elsewhere. This can be seen now. Perhaps the demographic factors of early settlement and density are also responsible. One may be tempted to say that the abundance of water is basic in so far as this keeps the people clean and keeps away most diseases; when health is ensured, other benefits follow.

REFERENCES

Bogue, Donald, J. (1969). *Principles of demography*. New York: John Wiley and Sons.

Caldwell, J.C. (1981). The mechanisms of demographic change in historical perspective. *Population Studies*, Vol. 35, No. 1.

————.(1976). Towards a restatement of demographic transition theory. *Population and Development Review*, Vol. 2, Nos. 3 and 4. New York: Population Council.

Easterlin, A. (1973). Relative economic status and the American fertility swings. *In* Eleanor B. Scheldon (Ed.), *Family economics behaviour*. Philadelphia: Lipincott.

Kirk, Dudley. (1972). A new demographic transition. *Rapid population growth—consequences and policy implications*. Vol. II, Research Papers. U.I. National Academy of Sciences.

Krishnan, T.N. (1976). The demographic transition in Kerala—facts and factors. *Economic and Political Weekly*, Bombay.

Kurup, R.S. and S. Cecil. (1976). A note on birth and death rates in Kerala. *In Demography India*, 1980.

Kurup, R.S. (1979). Rural development, agricultural practices, women's roles and fertility in India. Presented at the Workshop of Population Council, Bangkok.

Nair, N. Gopalakrishnan. (1965). Welcome speech. *In* Kurup and George (Eds.), *Population growth in Kerala, its implications*. Trivandrum: Demographic Research Centre.

PILLAI, K. DIVAKARAN. (1979). Components of fertility decline in Kerala 1961–71. Trivandrum: Population Resource Centre.

SIMON, JULIAN, L. (1974). *The effects of income on fertility*. Chapel Hill: Carolina Population Centre.

United Nations Population Division. (1963). Population Bulletin No. 7, New York.

14

Socio-Demographic Change in Indonesia

TERENCE H. HULL

In his long association with the demography of Indonesia Professor C. Chandrasekaran has witnessed at first hand one of the more outstanding fertility declines in recent history. The fifth most populous country of the world reversed a long-standing pro-natalist policy in the late sixties, and a decade later had reduced its fertility from a crude birth rate of around 44 per thousand to a level of 34. The indications are that this decline will continue though there are a number of questions concerning the future speed and ultimate stopping point of the decline.

The present paper explores the institutional changes which underlie the fertility decline in an attempt to specify those elements which appear to be unique to Indonesia, and those which may have more general applicability to other countries. The conclusion reached in this paper is that the experience of Indonesia involves much more complex social explanations than are contained in most theories or models of demographic change.

The Theory of Fertility Decline

Attempts to build theoretical frameworks to explain fertility declines in Western societies have often included extensive lists of socio-economic changes associated with modernisation which were thought to influence fertility.[1] In the mid-1950s Davis and Blake pointed out that such explanations were insufficient if they did not specify the way social changes caused reductions in fertility through a modification of sexual, contraceptive and post-partum behaviour (which they call 'Intermediate Variables').[2] Building on this principle other researchers have developed comprehensive frameworks for the analysis of fertility levels and trends.[3]

Coale has identified three general prerequisites for a major fall in marital fertility. First, the timing of childbearing must be within the 'calculus of conscious choice' by potential parents. Second, people must perceive reduced fertility to be advantageous. Third, effective techniques of fertility control

[1] See, for example, Frank W. Notestein, 'Population—the long view'. *In* Theodre W. Schulz (Ed.), *Food for the world*. Chicago: University of Chicago Press, 1945, pp. 30–40, 52.

[2] Kinsley Davis and Judith Blake, 'Social structure and fertility: an analytical framework,' *Economic Development and Cultural Change*, Vol. 4, No. 3, 1956, pp. 211–35.

[3] Ronald Freedman, 'The sociology of human fertility,' *Current Sociology*, Vol. 10/11, No. 2, 1961–62, pp. 35–121; John Bongaarts, 'A framework for analyzing the proximate determinants of fertility,' *Population and Development Review*, Vol. 4, No. 1, 1978, pp. 105–32.

must be available.[4] The first two elements combined constitute what other writers treat as the demand for children, while the third element can be characterised as the supply of fertility control.

The important principles to be drawn from the work of fertility theorists in the current consideration of the Indonesian experience are: first, social changes must be linked to fertility through their intermediate effects on sexual, child-bearing and child care behaviour; and second, an understanding of changing reproductive decisions should distinguish between the changing calculus of choice, and changes in the substantive alternative opportunities available to decision-makers. Such detailed specification of relations between social and fertility variables facilitates the validation of arguments and data analysis. It is widely agreed that Indonesia has undergone complex changes in social relations, and that the fertility level is falling. It would not be useful, though, to attribute an amorphous agglomeration of social changes as the 'cause' of fertility decline. Rather, we need to specify what aspects of social change have affected fertility, how the effect has worked through the intermediate variables and decision-making processes, and whether the decline is likely to continue.

Dimensions of Social Change in Indonesia

Modernisation of Thinking and Family Relations

Confronted with the challenge of explaining the confused array of social changes going on within their society and the need to justify planned changes associated with socio-economic development, Indonesian intellectuals make frequent references to what they call the 'people's mentality'.[5] The phrase has come to mean more than the English terms 'attitudes', 'ideas', or even 'world-view', since it contains a normative thrust arising out of its use to differentiate traditional and contemporary interpretations of the proper roles of individuals in society. Whereas Western writers sometimes describe these differences as conflicting bipolar attributes—achievement versus ascription, individual versus group or rural versus urban—much of the discussion in Indonesia involves the concept of balancing elements of traditional and modern ways of thinking and behaving. Basic principles like mutual discussion and consensus decision-making (*mus jawarahmufakat*) and mutual assistance (*gotong royong*) are cited as the proper ways of organising activities. Traditional values are enshrined in the five principles of the State (the *Panca Sila*)[6] and interpreted as the basis of the

[4] Ansley J. Coale, 'The demographic transition,' International Population Conference, Liege, IUSSP, Vol. 1, 1973, p. 65.

[5] Virtually all of this discussion is carried on in *Bahasa Indonesia* in books and national newspapers. Writers who have grappled with the concept include such prominent people as the Minister for Information, Ali Moertopo, the anthropologist, Professor Koentjaraningrat, UN University President, Soedjatmoko, and the newspaper editor, Mochtar Lubis.

[6] The *Panca Sila* has been undergoing a refinement of definition. In 1970, the official translation was 'Belief in the One Supreme God, Civilised Humanity, Nationalism, Democracy and Social

Indonesian polity and economy.[7] On an informal level, weaknesses in social structures are attributed to traditional practices such as patrimonialism (*Asal Bapak senang*—'as long as the Boss is happy . . .'), procrastination (*Alon-Alon waton kelakon*—'slow doesn't matter as long as we arrive'), or avoidance of situations which may cause a loss of face; but even such identifiable weaknesses are seen as unfortunate byproducts of a strong culture rather than singular attitudes to be rooted out.[8]

If tradition refers broadly to personal attributes and values and ways of behaving, modernity in its positive aspects refers mainly to *things*—the material culture—though necessarily this also includes some consideration of the processes of producing goods. In Indonesia two terms which are often used interchangeably are *cara modern* (modern ways) and *tingkat internasional* (international standard), both of which imply that the essential characteristic of modern goods is that they are imported. Moreover, while the goods—jeans, cars, plastic buckets—are manufactured in Asia, the technologies are largely derived from Western industrial societies. In Indonesian thinking, then, modernisation is sometimes regarded as being synonymous with Westernisation. Understandably, in a nation which fought to rid itself of Western colonialism, modernity is regarded with mixed emotions. On the one hand modern material life is pleasurable, convenient and comfortable, on the other the goods are symbolic of a form of neocolonial domination and loss of hard-won independence.[9]

The most relevant aspect of these new identities for understanding fertility decline is the changing relationship of parents and children. Old bonds are being broken and new ones forged, but whereas the traditional family was a primary economic unit, the new arrangement involves a growing emphasis on social and emotional ties. There are many factors responsible for these changes. Parents and children alike have been swayed by the dramatic rhetoric of the Revolution with its emphasis on personal freedom and equality, and its promise of new lives for an oppressed people. Young people in this new society are increasingly prone to demand control over their own life choices.

A new material culture has given children more mobility, while at the same

---

Justice', *Indonesia Handbook 1970*, Jakarta: Department of Information, 1971, p. 47. By 1978 these *sila* (pillars) were described as: '(*1*) Belief in one Supreme God; (*2*) Just and Civilised Humanity; (*3*) The Unity of Indonesia; (*4*) Democracy, led by the Wisdom of unanimity arising from deliberations amongst representatives; (*5*) Social Justice for the whole of the people of Indonesia', *Indonesia 1978: An Official Handbook*, Jakarta: Department of Information, p. 59.

[7] An indication of the importance with which this is regarded is the Government's programme of indoctrination in *Panca Sila*—principles for all Government officials and a large section of the private sector. For many officials, including all teachers, this consists of an intensive two week long fulltime course of instruction, reading, discussion and examination.

[8] An exception is Mochtar Lubis in *Manusia Indonesia* (Indonesian Identity), where a very critical approach is taken to such qualities. Jakarta: Yayasan Idayu, 1977.

[9] The most detailed analysis of this problem in Indonesia is given by one of the country's foremost sociologists, Professor Soedjito Sosrodihardjo in *Perubahan Struktur Mas jarkat di Djawa: Suatu Analisa* (The changing social structure of Java: an analysis), Yogyakarta: Karya, 1972.

time increasing their financial dependence on parents to satisfy their demands for consumer goods. Both reflecting and causing changed relationships is the modification of language patterns toward less formally polite speech. The change is slow, of course, and proceeds at very different paces and takes different forms among people of different classes, but the evidence of change is ubiquitous.

A focal point for understanding such change is the development of formal education,[10] since schools command substantial amounts of the children's time, allegiance and energy, with a resultant decline in the potential contributions children can make to the household economy. Schools also cost money, and present the average parent with a hard choice of budgeting limited resources among children. Frequently, such choices are exacerbated for the parent who, never having been to school, cannot understand what the child is learning or precisely how it is relevant for future employment. Blind faith is frequently the determinant in such cases. At times, though, the faith of parents is not blind, but based firmly on the idea that their children must be educated for the sake of the nation.

Schools are the vehicles for new ideas. Mass education is a post-independence phenomenon and apart from the basic skills of reading, writing and mathematics, it has sought to inculcate new values, a spirit of nationalism and an understanding of the world beyond each child's own village, ethnic group or religion. This is done through an increasingly centralised, standardised set of curricula and activities. Across the nation children are increasingly dressed in standard school uniforms, they perform common warm-up exercises and nationalistic rituals every morning, they join scout groups, organise formation marching and study the '*Panca Sila* Morality'.[11] Symbolically, most of this instruction is done in the national language, *Bahasa Indonesia*, so the content of children's thought is sometimes in a vocabulary many parents cannot understand since they often speak only regional languages. The school can, therefore, be seen as an agency which serves to modernise thinking, and alter relationships between children and parents.

In terms of the concepts of intermediate variables and decision-making presented in the previous section, how does the modernisation of thinking and family relations affect fertility?

There is a clear expansion in the individual's range of personal reproductive decisions. Parents have less influence in choosing marriage partners, and matches are increasingly made on the basis of personal choice.[12] Modern

[10] For a recent exposition of a similar argument see John C. Caldwell. 'Mass education as a determinant of the timing of fertility decline,' *Population and Development Review*, Vol. 6, No. 2, 1980, pp. 225–55.

[11] Uniformity is closely related to the general concept that Indonesians need to develop 'discipline' to achieve orderliness, economic growth and social stability. One side-effect, though, is an increasing social gap between the rich and the poor. The cost of uniforms and equipment prevent many children from attending school (*Tempo*, 7 February 1981, p. 18), joining scouts or participating in marching groups.

[12] A variety of aspects of this trend are of interest. There has been a rapid increase in the number of booklets and newspaper advertisements aimed at people who want to find a spouse, and these

marriages are often love matches, and the emotional tie between the partners facilitates more open communication, so that a discussion of the concepts of family size and methods of fertility control are possible.[13] Such marriages are less prone to disruption through divorce, and the concept of continuing parental responsibility for the financial needs of children becomes prominent. As children marry later and are in school longer, they sometimes constitute a significant burden on the family economy. Even if they leave school and work, many parents do not gain the benefits of their productivity. Rather, the profits are saved by the child for marriage or spent by the child on consumer goods for personal use. Parents and potential parents are learning that children can be expensive and are giving more thought to the trade-off between the numbers of children and the quality of the life style each child can enjoy. They realise that in an increasingly commercialised economy with declining employment opportunities for the very young, they will face more substantial dependency burdens over a longer part of their married life if they do not limit family size. Experience of school also prepares people for access to other institutions like clinics and hospitals, where fertility control services are dispensed. There is less fear of confronting a doctor or nurse in adulthood if teachers have been confronted in childhood. Moreover, clinic workers are happier to deal with literate clients since they 'understand' instructions better.[14] Schooling, therefore, both encourages an awareness of the possibilities of reproductive choice, and creates conditions in which children are increasingly expensive. These, in turn, are complemented by the experience of schooling which improves access to institutions dispensing the knowledge and technology of fertility control.

How are the changing patterns of thinking and family relationships related to the recent fertility decline? The answer to this question is complex and few generalisations apply equally well to the different regions, cultures and classes of the nation. However, two points stand out. First, increases in the age at marriage seem to have coincided with the post-Revolution spread of schooling, and thus a restriction of total fertility through delayed marriage is found in some regions (notably in southern Java) prior to the formulation of the policies of population control. Second, the family planning programme came into being at a time when almost half the women of prime reproductive ages (20–39 years) had been to school and were literate in Latin characters.

---

are particularly designed to appeal to the young people and not their parents. Columns in a broad spectrum of magazines are full of correspondence from young people seeking advice on ways to deal with parental opposition to sweethearts. Finally, following legal changes in 1973, it is now possible for young people to go through civil marriages without parental consent, provided they have reached the minimum age, and this is used to circumvent family opposition which can often prevent a religious ceremony.

[13] For instance, in one village in 1972 over 80 per cent of the younger generation (35 years old) husbands had discussed contraception with their wives compared with 56 per cent of the older generation men and 38 per cent of the older men in the bottom third of the income distribution. Terence H. Hull, *Each child brings its own fortune*, Doctoral thesis in Demography. Canberra: Australian National University, 1975, p. 440.

[14] Robert Repetto, 'Correlates of field-worker performance in the Indonesian family planning program: a test of the homophily-heterophily hypothesis,' *Studies in Family Planning*, Vol. 8, No. 1, 1977, pp. 19–21.

Some idea of the rapid expansion of schooling can be gained from the growth in literacy from the time of the Revolution in the late 1940s when only one in seven women of peak childbearing age was literate, compared to four out of five women at present. Following the argument set out above, this suggests that some of the changes already in motion are unlikely to stop in the near future since a rapidly increasing proportion of women has been directly exposed to the influences of schooling and changes in world-view and family relations. Moreover, as the balance shifts from the predominantly unschooled population of the 1950s and 1960s to a situation where a majority of young women in the 1970s and 1980s have been to school, the indirect effects of schooling on the unschooled should increase. Women of the 1980s who are illiterate will be in continuous contact with educated women, and will be exposed to the ideas schools spread. In all likelihood, they too will experience an increasing awareness of the potential choices available in their reproductive life, and will perceive their own children as increasingly expensive economic burdens. If, as the government has often planned, primary school education becomes universal and compulsory, these women may also find the current economic productivity of their young children falling.[15] In any case, future fertility is likely to be lower than the present levels; if stated ideals are any indicator, women, on an average, will still want three or four children.[16]

The trends in schooling, and the changing ideas and family relationships are the key elements in accounting for the changes in family formation. They imply that women have an increasing awareness of reproductive options, a decreasing demand for high parity children and improved accessibility to health and fertility control services. This raises two further questions. If a policy of fertility control had been formulated in the 1950s rather than the 1970s would it have succeeded? If the government had done nothing about fertility control in the early 1970s, would the social changes and the private sector have combined to provide sufficient motivation and means of fertility control to bring about the observed decline in birth rates?

The answer to the first question is necessarily speculative, but the issue is relevant to a consideration of the replicability of the Indonesian experience in other countries.

There can be no doubt that the world-view and social relationships of the Indonesians were substantially different in 1971 from what they had been in 1946. To the extent that schooling, revolutionary ideology and exposure to foreign ideas and life styles caused such changes, these factors can be regarded as indirect causes of fertility decline. By way of contrast, many societies in the Middle East and Africa do not educate women, espouse personal freedom or raise expectations of material well-being, and also have not experienced many of the social changes seen in Indonesia over the past twenty years. In those societies fertility is often not within the calculus of individual choice and

[15] Terence H. Hull, 'The influence of social class on the need and effective demand for children in a Javanese village'. *In* Lado T. Ruzicka (Ed.), *The economic and social supports for high fertility*. Canberra: Department of Demography, Australian National University, 1976, pp. 287–306.

[16] *Indonesia Fertility Survey 1976, Principal Report*, Volume II: Statistical Tables. Jakarta: Central Bureau of Statistics, 1978, pp. 223–74.

children are not perceived as an economic disadvantage. The strategy of planned voluntary fertility control is unlikely to make rapid progress in such settings, just as it is hard to imagine that Indonesia could have experienced a fertility decline in the 1950s when these social changes had yet to affect a majority of the community.

Reorganisation of Political Structures

The place of the village in Indonesian consciousness is similar to the concept of 'home' in the West. It is the source: the focus of strong emotional bonds, the last refuge. Popular literature—largely written by people who have left the village—abounds with descriptions of the tensions of life outside the village and the relief or disillusion, of return. For the people who stayed, and this forms the majority of Indonesia, the village community is a world in itself and they understand the place that they and their neighbours occupy in the economy, culture, and polity.

To the Central Government in Jakarta the village is the end of a long line of command stretching through departments, provinces, regencies, sub-districts, and village complexes. The officials realise the strength of the people's ties to small communities, and seek to tap these bonds to facilitate governance, but are frustrated by the problems of using vertical lines of command. Messages, orders and requests are sometimes lost or garbled as they make their way from Jakarta to the administrative villages (*kelurahan*) of the country. The villages in Indonesia are very different depending on the ethnic group, population density and ecological setting (see Table 1), thus making the relevance of many central instructions problematic.

From the earliest days of Dutch colonialism, attempts have been made to

TABLE 1 Variation of Village Sizes among Major Islands of Indonesia

Island	*Number of Villages*	*Average Size of*	
		Land (km²)	*Population*
Java	21,472	6	3,541
Sumatra	9,065	52	2,295
Kalimantan	6,155	88	837
Sulawesi	4,025	47	2,118
Bali, Nusatenggara and Maluku	4,441	33	1,736
Irian Jaya	114	3,702	7,137
Total	45,272	42	2,629

Note: 'Village' here is limited to administratively defined units, not the groupings defined by local people as their traditional village. As such the figures here understate the number of natural village or tribal units. A recent (1979) Government handbook gives the number of villages as 63,058, of which 21,570 are in Sumatra, 12,629 in Kalimantan, and 1,717 in Irian Jaya. The difficulty encountered in defining and counting villages across the archipelago is a symptom of the problems of organising vertical chains of command

SOURCE: Central Bereau of Statistics. A Search for a Better Definition of an Urban Village in Indonesia (VP 78–04), Jakarta, June 1977, Table 1, p. 20

streamline government structures.[17] Villages have been reorganised and numerous attempts have been made to ensure that leaders at every level of the government would be loyal and responsive to the superior echelons. During the 1950s and 1960s such attempts were hampered by local and factional bickerings. A myriad of parties and interests politicised all levels of the government and the attempt to satisfy the wide range of competing forces led to a bloated bureaucracy and the eventual collapse of administrative effectiveness.[18]

In late 1965 and early 1966 Indonesia witnessed a sharp upsurge of bloody conflict climaxing many years of growing confrontation between the military and Islamic extremists on one hand and the Indonesian Communist Party on the other. Tens of thousands of people accused of leftist activities or tendencies were killed or imprisoned and by the time calm returned to the nation the Communist Party had been decimated and outlawed; the self-styled 'President for Life' Sukarno had been eased from power; and a military government led by General Suharto declared a 'New Order' for Indonesia.

The political changes of the late 1960s led to significant transformations of Indonesia's social and economic structure, not the least of which was a push to consolidate power through vertical linkages to the village level.[19] While replacing the purged officials, the government often recruited active or retired members of the armed forces. By 1973, for example, 84 per cent of the Regency heads in Indonesia's most populous province, East Java, were military appointees. At the lower levels of the government there was a tendency to appoint men who had retired from active military service (at age 55) to positions such as sub-district heads, or to support their candidacies in elections for village headships. Emmerson also reports that the government sought to improve the efficiency of non-military sub-district heads by giving them military-style training to make them 'more loyal and responsive'.[20] As a result the members of the restructured bureaucracy 'tend to preserve and apply the no-nonsense leadership to which they have become accustomed'.[21]

The reorganisation of the government was a key factor in the promotion of fertility control. Prior to 1965 President Sukarno had repeatedly avoided any discussion of birth control because of the need to balance competing political interests; many of these interests regarded contraception as immoral, or population control as economically unnecessary. In forming a military-backed government President Suharto sought the advice of, and responded to, advisors from universities and foreign donor agencies who argued the necessity of fertility

[17] Selosoemardjan, *Social changes in Jogjakarta*. New York: Cornell University Press, 1962, pp. 87–99. John D. Legge, *Central authority and regional autonomy in Indonesia: a study in local administration, 1950–1960*. New York: Cornell University Press, Ithaca, 1961.

[18] Donald K. Emmerson, 'The bureaucracy in political context: weakness in strength'. *In* Karl D. Jackson and Lucien W. Pye, *Political power and communications in Indonesia*. Berkeley: University of California Press, 1978, pp. 88 ff.

[19] For one view of the process see Emmerson (*ibid.*), pp. 90–109. A more general though incomplete analysis is in Fritjof Tichelman, *The social evolution of Indonesia: the Asiatic mode of production and its legacy*. The Hague: Martinus Nijhoff, 1980, pp. 156 ff.

[20] *Ibid.*, p. 104.

[21] *Ibid.*, p. 105.

reduction. While acting with an acute awareness of potential moral outcries, he, nonetheless, had the power to commit the loyalty of the bureaucracy to ensure that the policies were implemented.[22] Beginning in 1967, moves were made to organise a national programme. An 'institute' was formed in late 1968 and in 1970 it was strengthened, expanded and renamed the National Family Planning Coordinating Board (BKKBN).

There are a number of features of the Indonesian family planning programme which have encouraged commentators to view it as a model of organisational effectiveness. First, it is not based in a department, but rather in an autonomous board (BKKBN) responsible directly to the President. It is 'young' in a gradually aging bureaucracy, and since it essentially stimulates, supports and coordinates services directly provided by the departments, the BKKBN is by its very nature dynamic. Second, in executing its role the BKKBN has been remarkably successful in organising two basic activities: the logistics of contraceptive supplies and the system of rapid feedback monthly service statistics. These are vital to the programme since it is only through the flow of information and supplies that the effective demand for contraceptives can be identified and satisfied. Third, the areas in which the programme has been less successful in developing effective activities (for example, mass media communications, data analysis and commercial distribution) are not crucial to the basic operation of providing contraceptives to satisfy a latent demand for fertility control. Finally, after the initial emphasis on the urban areas and IUDs, the programme moved aggressively to work in the rural areas, initially only in Java and Bali, relying largely on the oral contraceptive pill. The success of the programme, which has been internationally publicised,[23] refers almost exclusively to these islands. Some confusion arose when the statistics on acceptors and fertility decline in Java were erroneously attributed to the whole of Indonesia, and this was exacerbated by the fact that the World Fertility Survey data from Indonesia were actually based on a sample confined to only Java and Bali. In fact, the BKKBN is facing greater challenges in both urban Java and the other islands than it ever faced in rural Java and Bali. However, it is true that the success in Java and Bali, which comprise over 60 per cent of the population, is itself a matter of great importance.

Returning to the question of intermediate variables, and Coale's preconditions, how has the reorganisation of political structures helped to reduce fertility?

Most obviously, it can be argued that the previous policies, attitudes and laws against birth control devices were only reversed because the balance of power in the mid-1960s shifted so dramatically to a military government which was responsive to technocratic advice on economic and social issues and relatively unhampered by conflicting views from other quarters.

[22] Terence H. Hull, Valerie J. Hull and Masri Singarimbun, 'Indonesia's family planning story: success and challenge,' *Population Bulletin*, November 1977, Vol. 32, No. 6, Washington: The Population Reference Bureau, Inc.

[23] See for example, *International Family Planning Digest*, June 1976, Vol. 2, No. 2; and James W. Brackett and R.T. Ravenholt, 'World fertility, 1976: an analysis of data sources and trends'. *Population Reports*, November 1976, Series J, No. 12, pp. J-209, J-217, J-315.

Once founded, the programme grew according to the principles established by the military leadership of the country and a goal-oriented approach was adopted. While a few leaders of the programme were military men, they recruited civilians on the basis of their skills, loyalty and commitment. Their military training fitted easily with the civilians' technical expertise and both groups stressed the need for a strong vertical organisation. The vocabulary of the planners came to include terms like 'target groups' and 'special drives', and the initial strategy was derived partly from the concepts of military sweeps through areas, and partly from the health campaigns of infectious disease control. At the outset this was conducive to reaching a large number of potential acceptors and involving all levels of the bureaucracy.[24]

As time passed such tactics were seen to be of limited value, since family planning services required both continuity of supply and the regular availability of trained staff to recruit and maintain significant numbers of acceptors. As a result the role of the local bureaucracy became even more important. Whereas previously they were expected to round up potential acceptors (presumably those voluntarily desiring to limit family size) to go to a nearby clinic, or be prepared for a visit from a mobile team, now they were asked to assist the programme in organising regular service points in the village and maintain records on family planning acceptors.[25]

While some officials talk of the BKKBN fieldworkers and village volunteers as providing 'motivation' for fertility limitation, the programme functions primarily to provide supplies, and to help overcome the fears and provide assistance relating to the side-effects of the method. Only as a secondary objective does it change attitudes and it achieves this through mass media and by creating a climate of awareness of the family size issues. Perhaps the mere knowledge of contraceptive techniques serves to broaden the villagers' awareness of potential reproductive choices, but the critical point is that awareness of choice does not reduce fertility unless parents perceive children to be disadvantageous. Programme workers do little to affect this perception.

If political reorganisation was crucial in explaining the Indonesian fertility decline in the 1970s, what are the implications for the future and for other nations?

The degree to which fertility decline was due to changes in the marriage patterns probably has no direct relation to changes in the political structure. Thus, the bulk of the decline outside Java, and a substantial part of the decline in West Java, can be explained mainly with reference to the discussion on modernisation of thinking and family relations.

The portion of the decline due to falling marital fertility in Java and Bali may

[24] Geoffrey McNicoll discusses this with reference to Bali in 'Institutional Determinants of Fertility Change,' Center for Policy Studies, Working Papers, No. 59. The Population Council, June 1980, pp. 10–11.

[25] This is the Village Contraceptive Delivery Center scheme which is described by Haryono Suyono, *et al.*, 'Village family planning: the Indonesian model'. *In* Jacqueline Gardinaer, *et al.*, *Village and household availability of contraceptives: Southeast Asia, 1976*. Seattle: Battelle Population Studies Center, 1976.

be said to have been facilitated by the strengthening of vertical chains of command and the consolidation of loyalties and power at the lower levels of government. Such consolidation was not uniform throughout and subsequently, programme success was not even. Bali and East Java are regions where the local governments were responsive to the centre, and the programme has been remarkably active there. In contrast, the Sudanese of West Java have long resented rule by the Javanese and other ethnic groups in the Central Government, and have sometimes resisted directions from Jakarta. Family planning officials report more difficulties in these areas, and data show lower acceptance rates as a result.

The other islands of the archipelago were covered by the programme in two phases: first in 1974, and later in 1978, and fertility declines are unlikely to have much, if anything, to do with the programme efforts. However, in terms of acceptance rates these islands are substantially behind Java at a similar stage of programme implementation.[26] The prospects for the future of family planning are mixed. In North Sulawesi education levels are high and the religious, cultural and economic conditions in the region are very conducive to fertility decline.[27] West and North Sumatra have undergone substantial village reorganisation, have been open to socio-economic changes of great magnitude including high migration rates, and have both registered fairly substantial contraceptive prevalence rates (17 per cent and 15 per cent respectively in October 1980). Moderate fertility declines can be expected in these provinces. Aceh, West Kalimantan, Irian Jaya and Timor Timur are regions where long-standing tensions or recent civil disturbances, often combined with difficult geographic conditions, have made effective village administration difficult. Educational levels are also low and the forces of social change discussed earlier are relatively weak. The family planning programme will face difficulties organising services in such regions, parents are likely to be less motivated to limit family size much below traditional numbers, and hence fertility decline will probably be slow in coming.

Some indication of the responsiveness of provinces to the programme efforts can be seen in Table 2, which lists areas according to the recorded contraceptive prevalence rate in October 1980. Rates are substantially higher where the programme has been running longest. However, because the regions included at each stage of development were selected partially according to the perceived degree of difficulty, those with lower rates are also generally those where the programme is experiencing slowest growth.

Possibly strong, loyal, responsive local governments are not the prerequisites to success in distributing contraceptives. Perhaps programmes can bypass ineffective or antipathetic community structures to reach women directly in clinic settings or with mobile units organised at supra-village levels. However, even if this is possible the process will be more comprehensive and lasting if

[26] Hull, Hull and Singarimbun, 1977, *op. cit.*, pp 36–40.

[27] For a prediction of rapid fertility decline in North Sulawesi see Gavin W. Jones, *The population of North Sulawesi*. Yogyakarta: Gadjah Mada University Press, 1977, pp. 144–52, 159.

local governments work in a dedicated fashion to provide contraceptive services.[28]

TABLE 2 Recorded Rates of Prevalence of Contraceptive Use by Province for Indonesia, October 1980

Province		*Rate*	*Province*		*Rate*
1. East Java	A	56.1	14. Bengkulu	C	15.2
2. Yogyakarta	A	53.0	15. West Kalimantan	B	10.8
3. Bali	A	51.8	16. South Sumatra	B	9.3
4. Central Java	A	46.6	17. East Kalimantan	C	9.1
5. North Sulawesi	B	29.0	18. Aceh	B	9.1
6. West Java	A	26.8	19. Central Kalimantan	C	4.9
7. South Kalimantan	B	22.3	20. Jambi	C	4.7
8. Lampung	B	22.1	21. Central Sulawesi	C	4.4
9. Jakarta	A	22.0	22. Riau	C	4.0
10. West Sumatra	B	17.5	23. South East Sulawesi	C	3.9
11. South Sulawesi	B	16.4	24. East Nusatenggara	C	2.6
12. West Nusatenggara	B	15.4	25. Maluku	C	2.4
13. North Sumatra	B	15.3	26. Irian Jaya	C	1.7
			27. Timor Timur	C	0.0

(Date of Programme Initiation: A. 1968–70, B. 1974, C. 1978)

Note: The prevalence rate is calculated as the percentage of married women in the age group 15–44 years apparently currently using programme methods. It is based on reports of pills, injectables and condoms actually distributed and estimates of IUD and sterilisation use based on past records adjusted for drop-outs, aging and death

SOURCE: BKKBN. Monthly Statistical Summary, Jakarta. Data Date: October 1980; Run Date: 30 November 1980

TECHNOLOGICAL INNOVATION

The dimensions of social change discussed in the two preceding sections were treated as fairly independent phenomena, and the impact they have on fertility was said to be distinct. Changes in thinking and family relationships primarily affect the reproductive decision-making environment, while changes in government structure facilitate the provision of birth control. The third dimension of change—technological innovation—cuts across the other two in complex ways, and consequently has much more diffuse, but nonetheless important, implications for fertility decline.[29]

Broad Changes in Technology and the Demand for Children It can be argued that the changes in technologies of transportation, communication and

[28] See Jay D. Teachman and Pudjo Rahardjo, 'Contraceptive delivery systems: an evaluation of clinic vs. village in Indonesia,' *Evaluation Review*, 1980, Vol. 4, No. 1, pp. 75–92. However, note that while their evaluation shows the village distribution system to be more effective, even the clinic system benefits from the support offered by village leaders in identifying and organising women to receive services.

[29] Geoffrey McNicoll has suggested a variety of these relations in 'Technology and the Social

production since the turn of the century have gradually helped to transform the world-view and family relations of the large majority of the Indonesian population. As such, the points made previously about the modernisation of thinking can be brought back to a materialist logic of development as technological change.

While these goods have certainly become more widespread over the past few years, it would be incorrect to limit an understanding of the technological dimension of Indonesia's social changes to the New Order period or even to the three and a half decades of the Republic. The roots can be traced back to the colonial period, back to the times when traders brought cloth to coastal *pasars* and government officials instituted various forced labour systems. The effect was to displace labour from domestic spinning and weaving activities. Women and children entered agricultural pursuits to replace men who were obliged to devote time to various government organised infrastructure activities. The population's dependence on a commercialised economy grew continuously under such pressures.[30] The history of Indonesia's economy has been a series of technological transformations which have changed the position of families in the economic structure. White[31] and Boomgaard[32] have argued in different ways that such changes increased the demand for children as workers in the nineteenth century, and used this as an explanation of high and perhaps increasing fertility rates for the period. In contrast, various recent forms of capitalisation in production processes and the advent of schooling have reduced the demand for child labour, though the amount and direction of these effects would depend on specific cases.[33]

The differences in economic structures among regions produced differentials in fertility levels even before the advent of the family planning programme. Java, and particularly East Java and Yogyakarta, had relatively lower fertility levels than Sumatra or the other major islands. Poor people in general had lower fertility levels than the more well-to-do. In part, these differences arose from involuntary or indirect causes, such as, high rates of infecundity or high divorce rates. They were also related to purposeful attempts to space or limit births, notably through the use of post-partum or terminal abstinence. The demand for children was not unlimited everywhere, though many people were unable to have all the children they may have wanted. In 1976 this evidence was interpreted to imply that fertility decline, if it occurred at all, would be slow

Regulation of Fertility,' Center for Policy Studies, Working Paper No. 46, New York: The Population Council.

[30] Clifford Geerts, *Agricultural involution*, 1963, presents perhaps the most readable account of these changes.

[31] Benjamin White, 'Demand for labour and population growth in colonial Java,' *Human Ecology*, 1973, Vol. 1, No. 3.

[32] Peter Boomgaard, 'Female labour and population growth in nineteenth century Java,' Vrije Universiteit, Faculteit der Letteren, mimeo.

[33] Masao Kikuchi, Anwar Hafid, Chaerul Saleh, Sri Hartoyo and Yugiro Hayami, 'Class differentiation, labour employment, and income distribution in a West Java village,' *Developing Economies*, Vol. 18, No. 1, 1980, pp. 45–64.

unless there were 'substantial reforms in the economic structure and the institutions of socialisation' because 'family size desires are based on a variety of deeply felt economic, social and personal' needs.[34]

The Materialist Explanation of Fertility Decline In now considering the fact of fertility decline there is a temptation to contend that there had been a build-up of an unobserved, possibly unconscious, latent demand for fertility control over the 1950s and 1960s, but this was unfulfilled until the BKKBN provided free, safe and effective birth control measures. This 'latest demand-active supply' argument fits very well with the pattern of fertility decline. In fact, the whole explanation could be reduced to two forms of technical innovation—consumer goods and capitalised production on the one hand, and contraceptives on the other.[35] Such logic risks the loss of an important dimension of understanding. This essentially materialist interpretation alone can explain *some* build-up of latent pressures for reduced fertility, but it does not easily allow for either a major or a continuing change throughout Indonesia and across socio-economic classes.[36] It would also direct policy-makers to place greater emphasis on the economics of the production of *goods* than on the psychology of the changes of *ideas* or the sociology of the alterations of *relationships* involving non-economic exchanges of services. Recalling the earlier discussion of the importance of these dimensions in Indonesian history, it can be seen that a purely materialist approach would produce a misunderstanding of the timing, speed and replicability of Indonesian fertility decline.

The more complicated, but more likely explanation of the role of technological innovation in changing the demand for children is that it was not the advent of modern consumer goods or changed technologies of production *per se* that motivated parents to limit family size, but the conjunction of these changes with the spread of new values and ideas through schools and the mass media. The late 1960s stand out as a period of two great discontinuities. First, a post-Revolution generation of literate young women was entering the reproductive age group. Though many had not experienced the full course of formal primary education, yet they had been exposed to the major innovative forces of the school system: nationalist values, national language, extra-familial authority, individual testing, organised group activities, and substantive knowledge of reading and writing. Second, following a long period of social disruption and economic decline, this generation was given the promise of rapid expansion of educational and health facilities, and a variety of personal consumer goods. Basically, they were convinced that their lives were to be very

[34] Terence H. Hull, 'Rapid fertility decline: a comment,' *Bulletin of Indonesian Economic Studies*, Vol. 12, No. 2, July 1976, p. 113.

[35] In fact this has recently been argued as such for Iran. See S.E. Tolnay, A.H. Mehryar and S. Jamshidi, 'An Analysis of Reproductive Motivation in Shiraz, Iran,' *Journal of Biosocial Science*, Vol. 12, No. 3, July 1980, pp. 301–308.

[36] In 1978 only one-third of the families owned radios and 5 per cent owned televisions. Over a half of the Indonesian households owned none of the items of a basic list of consumer durables which included items like table and chair, clothes, cupboard, bicycle and clock. If the materialist argument has any force it must be through shaping desires rather than reflecting attainment.

different from their parents' generation because the nation and the world were different. For some the difference was not positive. Children of the pre-war elite knew that their real incomes would not equal their parents', and children of the growing class of landless labourers knew that real wages in many sectors were falling. But for a large majority of people there was hope of a better life through a government job, business or a lucky break.

The focus of a large portion of this new generation's concern in childbearing is the opportunity for their children to be educated and achieve a good place in the economy. As such the rapid expansion of educational facilities in the 1960s and 1970s created an atmosphere of ever rising aspirations, and a general awareness among all classes of the population of the social and economic sacrifices required to achieve schooling. This is probably the single most important element of conscious thought concerning family limitation. Perceptions of rapid technological change increase the anxieties about the uncertainties of future employment. In this environment, many parents seek refuge either through a search for slight educational advantages for their children ('better' schools, extra tutorials, more equipment) or by placing an emphasis on developing connections who can assist in job placement (either with family or friends). These are essentially middle class concerns, though. For the bulk of the population the question of a child's future boils down to a blind faith in schooling and to the mercy of fate. The costs make parents consider each child individually, and reflect on the potential gain of concentrating the available resources on a few, rather than spreading the resources over many children.

The technological dimension of social change, then, cannot be interpreted simply in terms of material changes *per se*, but must take into account the concomitant changes in values. Together these make parents aware of broadening decision-making potentials, and change the perception of the values of children in ways consistent with smaller family size choices. The changes are cumulative, and disproportionate to the growth in numbers of goods or attendances at schools alone. This interpretation accords with the data on educational differentials in contraceptive knowledge and use presented in Figure 1. Higher levels of education are associated with higher levels of knowledge and use of birth control techniques, but even at the lower levels of education substantial use is reported. The argument presented above would imply that this graph will shift upward as the demand for large families falls in all social classes. As time passes all women in Indonesia will be aware about the different contraceptive methods and a large proportion of each group will want to space or limit family size.

Technological Change and Future Demand for Children The question is will the demand be as low as the two-child family consistent with replacement fertility? Basic to this question are the propositions that preconditions for low fertility (in addition to Coale's preconditions for change) include greater equality of incomes, low infant mortality, and provision for a minimum attainment of basic human needs.

FIGURE 1 Knowledge and Use of Family Planning Methods* According to Completed Schooling of Currently Married Women Aged under 50. Indonesia, 1976

SOURCE: 1976 Intercensal Population Survey, Tabulation Series No. 1, p. 107

* Family planning methods include both modern (oral contraceptive pill, IUD, condom, sterilisation) and traditional (withdrawal, rhythm, abstinence, herbs, massage)

1 never heard of family planning methods
2 know of at least one family planning method
3 have ever used at least one family planning method
4 currently using a family planning method

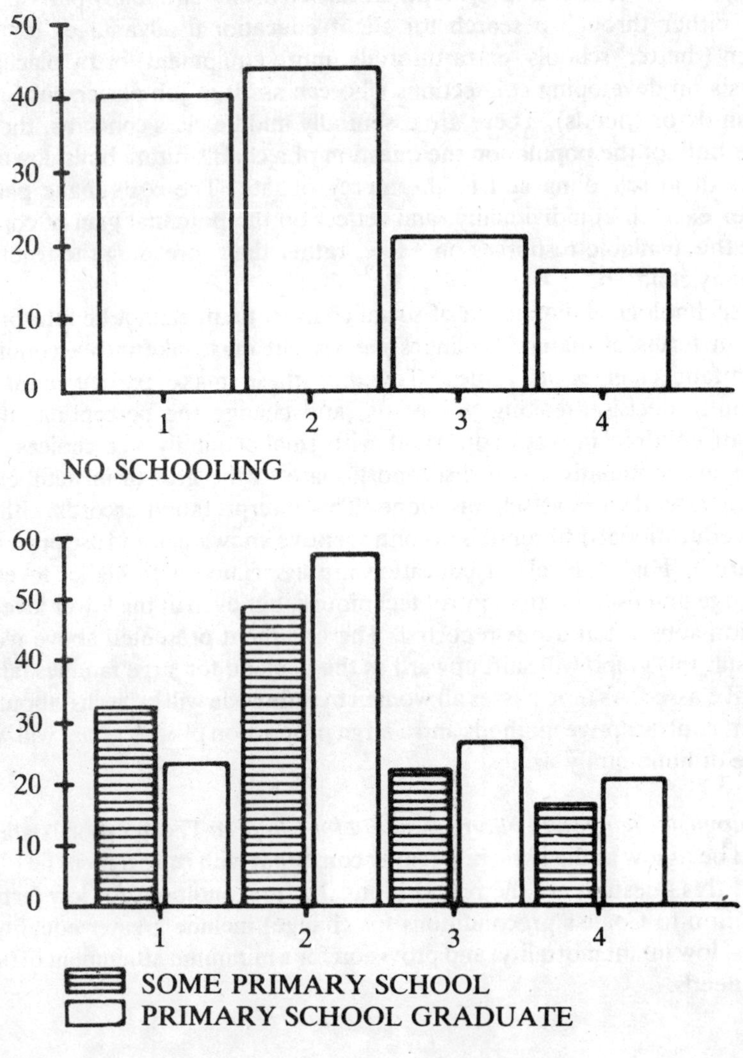

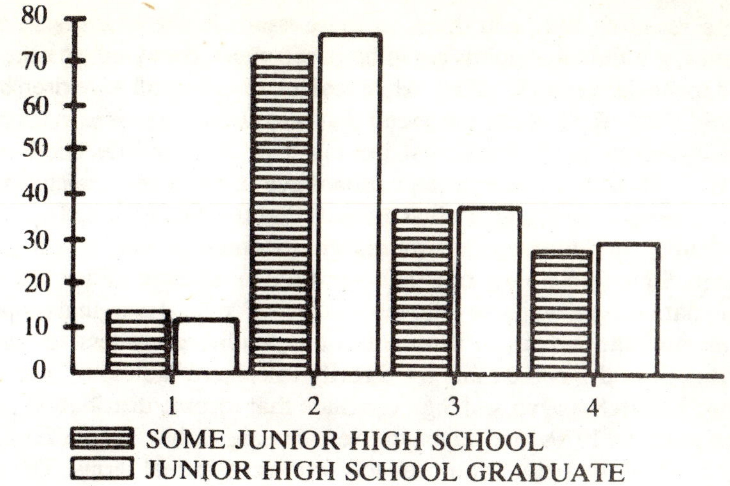
80
70
60
50
40
30
20
10
0
1
2
3
4
SOME JUNIOR HIGH SCHOOL
JUNIOR HIGH SCHOOL GRADUATE

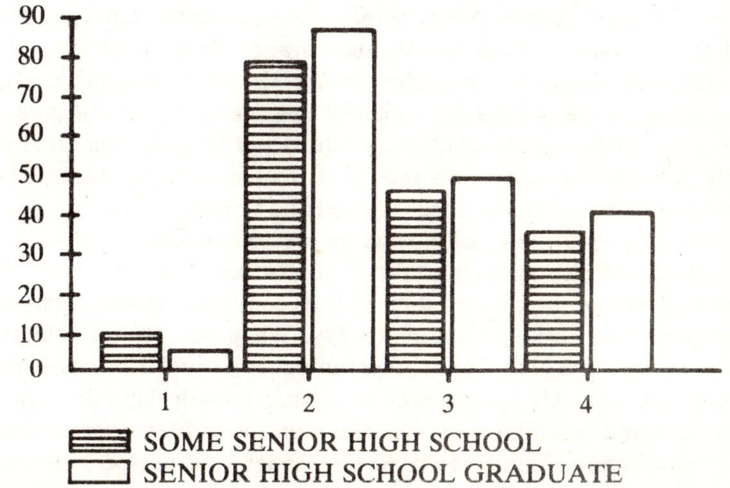
90
80
70
60
50
40
30
20
10
0
1
2
3
4
SOME SENIOR HIGH SCHOOL
SENIOR HIGH SCHOOL GRADUATE

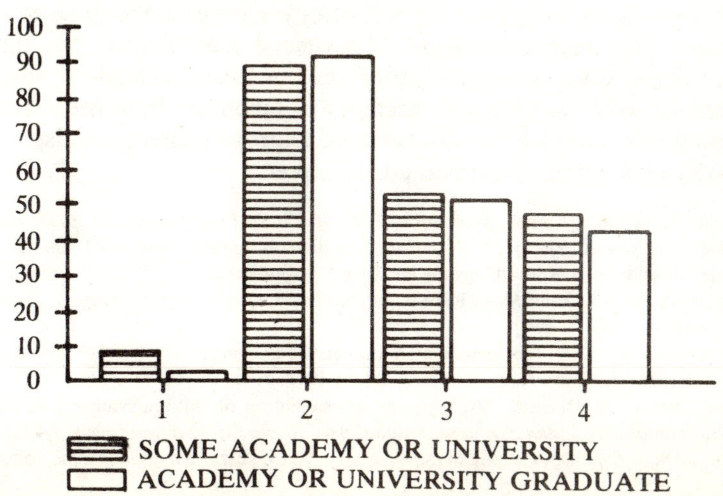
100
90
80
70
60
50
40
30
20
10
0
1
2
3
4
SOME ACADEMY OR UNIVERSITY
ACADEMY OR UNIVERSITY GRADUATE

Little research has been done on these issues in the Indonesian context. However, a number of points could be raised which could indicate the outline of an appropriate analysis. First, while real incomes overall have risen between 1970 and 1977, R.M. Sundrum found that 'inequality increased in urban areas and declined in rural areas, and increased in Java and declined in Outer Islands', with little change in the national average.[37] These conclusions have been challenged by Sayogyo who contends that the data analysed by Sundrum were faulty and understated increased incomes of well-to-do people.[38] Hananto Sigit also points out that Sundrum used expenditure data (since income data was unavailable) and because of differential marginal propensities to consume, the indices of inequality of expenditure are less responsive to increases in real income than are directly calculated indices of inequality of income.[39] Both Sayogyo and Sigit conclude that income distribution probably worsened in the 1970s, and Sayogyo goes on to argue that the lot of the poorest 40 per cent of the population probably worsened in real terms. The issue of inequality, though, is clouded by the fact of general increases. As long as most people find their income rising in real terms, moderate changes in relative distribution would not necessarily discourage rising aspirations which are consistent with the desire for smaller families. On the other hand, if inequalities can be reduced, the implication would be that the aspirations of lower income families would rise faster, and small family norms would be attained more quickly. Greater equality with growth in real income would lead to better satisfaction of basic human needs of food and housing (since the poor people spend most of their marginal income on these) and lead to rapid declines in current large differentials in mortality.[40] High infant mortality creates a climate of uncertainty and the biological conditions conducive to short birth intervals and greater than desired level of fertility. When parents are able to ensure infant survival some of the motivation for higher parity births is reduced.

While it is possible to consider economic growth without technological change in some circumstances—as when land is available for pioneering or better organisation of labour leads to efficiencies—the prospect for raising real incomes and achieving greater equality for most Indonesians involves technological innovation. However, the technology appropriate to these goals is not necessarily the 'high technology' of advanced industrialism, but may focus instead on the advances in agricultural and household technology which raises productivity with minimal displacement of labourers from the sector. Such technologies are already being developed to provide energy, transport, water, increase yields, intensify gardens and save fuel.[41]

[37] R.M. Sundrum, 'Income distribution, 1970–76,' *Bulletin of Indonesian Economic Studies*, Vol. 15, No. 1, March 1979, p. 141. Sundrum used data on reported household expenditures from consumption surveys to estimate income levels and inequalities.

[38] Sayogyo, 'Indeks yang Main Rata-Rata (Indexes that play with averages)', *Jurnal Ekuin*, 20 May 1981, p. 8.

[39] Hananto Sigit, 'Distribusi pendapatan atau distribusi pengeluaran? (Distribution of income or distribution of expenditure?),' *Kompas*, 30 May 1981, p. 4.

[40] See Peter F. McDonald, 'The equality of distribution of child mortality: Java-Bali, 1950–1976,' *Bulletin of Indonesian Economic Studies*, Vol. 16, No. 3, November 1980, pp. 115–19.

[41] The work of *Dian Desa* (Lamp of the Village) in Yogyakarta on hydraulic rams, solar cooking,

The Technology of Fertility Limitation The third of Coale's preconditions—that effective techniques of fertility reduction must be available—provides a key to understanding the timing of Indonesia's fertility decline. The family planning programme provided efficient forms of birth control to women who previously had access only to inefficient or unpopular methods. Fertility declined soon thereafter in proportion to the success of the programme in spreading this innovative technology.

The arguments presented earlier caution us not to accept these statements uncritically as proof of a kind of 'technological determinism' of supply-led fertility decline. The 1970s saw not only the spread of supplies, but also a new generation of women with changing family relationships and fertility values. The demand for birth control was on the rise. Supply systems were also changing in complex ways, not unrelated to factors influencing demand. Governmental control in villages, systems of transport and communication, and the bureaucracy of the BKKBN, were all growing in magnitude and efficiency. There are many complicating elements in the equation of supply of devices leading to a decline in fertility.

Apart from these factors there are three further issues of technology which contribute to an understanding of the spread of devices and the timing of the decline in Indonesia, and they all have to do with the oral contraceptive pill.

First, it is easy to forget how recently the pill was developed. The sixties became the first pill-decade, and its spread was rapid because of its simplicity. Pills are easily described, transported, stored, taken, and abandoned, and for these reasons are methods of preference among women and medical practitioners alike. An indirect impact of this innovation in contraception was the enormous growth of programmes of population assistance around the world.[42] In the sixties the population control industry grew many times faster than population ever had. Thus, when the New Order Government was formed in 1966, and re-established links with outside aid agencies, foundations and the United Nations, it found itself swept by a tide of commitment to fertility limitation. Six years earlier few nations were formally interested in family planning, and by 1974 the stress on family planning as an autonomous policy was taken at Bucharest. Indonesia's programme was formulated at a time when faith in a technological answer to population problems was at its peak. The forms of technology included IUDs, and condoms, but the focus of perceptions was the contraceptive pill.

Second, as the programme developed, the efficient logistics system which we have seen to be a key to the BKKBN success was designed to receive, store, transport and deliver imported contraceptive pills: it was a system of cooperation between the BKKBN and USAID. It is by no means certain that in the early

bio-gas, and the use of cheap materials in construction is an outstanding example of the potential impact of appropriate technological innovation in meeting basic needs.

[42] To get a feel for the changes a reader should look through the full series of *Studies in Family Planning*, particularly the articles of Bernard Berelson and colleagues. An interesting review of the international situation in 1969 is Berelson, 'National Family Planning Programs: Where We Stand,' *In* S.J. Behrmann, Leslie Corsa and Ronald Freedman, *Fertility and family planning: a world view*, Ann Arbor: University of Michigan Press.

1970s such efficiency would have been attainable with a more complex technology of contraception.

Third, the strength of the programme in poor rural areas is almost entirely due to the pill programme. Development of village depots, acceptor groups and many other activities rests on the lines of supply reaching as closely as possible to people's houses. The contraceptive pill, and secondly the condom, have allowed the programme to achieve wide coverage in a relatively short period of time.

Thus, the simplicity of contemporary contraceptive technology had a critical impact on creating an international framework of support for fertility control in Indonesia, establishing a foundation of efficient programme management, and spreading programme activities to isolated rural communities. The spread of the innovation in purposeful fertility control can be assessed from the fact that over half the women of Indonesia are recorded as currently using contraception.

Policies and Politics in the Analysis of Fertility Decline

Many observers have praised the BKKBN for its flexible approach to programme development.[43] At least part of this flexibility is the ability to treat formal policies as instruments to achieve informally expressed goals. At times specific laws, rules and regulations have been ignored. For example, import regulations restricting contraceptives were ignored for years on the basis of a mutual understanding between government ministers. Laws are hard to change, and an informal approach is often deemed best. Alternatively, policies may be formally announced, but not pursued with vigour, for purposes of satisfying one particular constituency or other. Thus, an analysis of formal policy alone is not sufficient to understand the experience of the family planning programme. Politics, historical accidents, and cultural and social conflicts all constitute more potent influences on behaviour than formal policies *per se*. The latter deserve critical attention, but as elements of a larger picture, rather than isolated phenomena.

The thrust of the argument in this paper was that both the formation and development of the family planning programme and the recorded fertility decline need to be understood in terms of the independent Republic of Indonesia, and more specifically the period of the New Order Government. This interpretation would not change if formal policies were regarded as artifacts rather than determinants of change.

The view that Indonesia's success offers hope to other countries wishing to achieve fertility declines must thus be tempered. There is much to be learned from Indonesia, but not through mechanical reconstruction of principles, policies and procedures. Dr. Haryono Suyono, one of the main architects of

[43] See especially Donald R. Snodgrass, 'The family planning problem program as a model for administrative improvement in Indonesia,' *Ekonomi dan Keuangan Indonesia*, Vol. 27, No. 2, June 1979, pp. 237–56.

the Indonesian programme, has pointed out that 'AID has worked closely with us to identify problems and constraints, then seek local solutions, (and) not attempted to bring to Indonesia ready made programs designed in far away places'.[44] This may be the most important lesson to be learned from Indonesia's recent fertility decline. What remains to be seen is whether the policies of the BKKBN and the trends of social and economic change will be sufficient to produce the country's goal of a 50 per cent reduction in fertility by 1990. On the basis of evidence reviewed above, all we can conclude is that things are going in that direction, but there should be caution in the speculation of likely achievements.

[44] BKKBN, 'Review of the Indonesian national family planning program,' Technical Report Series, 1979, Monograph No. 20, Jakarta, p. 12.

15

Socio-Psychological Determinants of Urban Fertility

P. J. Reddy

Introduction

Human fertility is a complex process responsible for the biological maintenance of society. The levels of fertility vary across the different societies. A high level of fertility persists among the developing countries while it is low in the developed countries. The unprecedented increase in the population of the developing countries is causing much alarm in the world because of its threat to human survival in the long run. The proportion of couples effectively planning their family size is very low in most of the developing countries. An understanding of the factors associated with high and low fertility levels is indispensable to regulate the high fertility level in the developing countries. A thorough knowledge of the determinants of fertility will help in effective implementation of the programmes facilitating the promotion of small family norm, thereby achieving a proper balance between population growth on the one hand and resources on the other.

India is one of the developing countries experiencing high rates of natural increase. Several studies on Indian fertility have been undertaken for a proper understanding of fertility behaviour. However, most of these studies have been confined to rural areas only (Dandekar, 1959; Saxena, 1965; Srinivasan, 1969; Wyon and Gordon, 1971; Mahadevan, 1979). Studies on urban fertility conducted hitherto in India are very few (UN, Mysore Population Study, 1961; Driver, 1963; Jorapur, 1967; Desai, 1970; Saksena, 1973; Rele and Kanitkar, 1980). Further, most of these studies have considered urban population as a single unit without stratifying the population into the non-slum and the slum dwellers who constitute, socially and culturally, two distinct groups within the same urban setting. Such a procedure would obscure the real differences in urban fertility. Also not much attention has been paid to the study of psychological correlates of urban fertility in the Indian context. Hence, the present study is undertaken to bridge the gaps in our knowledge on urban fertility. Here, 'urban' refers to a metropolitan city having a population of 2.14

Acknowledgements The author expresses his deep gratitude to Professor K. Mahadevan, S.V. University, under whose valuable guidance this piece of research work was carried out; and for his critical comments and suggestions in preparing this paper. He is also grateful to the ICSSR, New Delhi, for their financial assistance which enabled him to complete this work in time. The services rendered by Mr. S. Sudhakara Reddy and Mr. Nazeer are also acknowledged.

million. The major objective of the present study is to identify the social and psychological determinants of differential fertility within and between the non-slum and the slum dwellers of the metropolitan city, Hyderabad.

In this field the pioneering study on the social and psychological determinants of fertility was carried out in Indianapolis, USA by Whelpton and Kiser (1946–1958). The specific aim of this study was to test twenty-three hypotheses most of which were connected with the relation of given social or psychological factors to both the practice of contraception and the size of the planned family. The Indianapolis study could not show any significant association between the psychological factors and fertility behaviour. Even the low order of the relationships that were found between some of the psychological variables and fertility tended to fade away when socio-economic status was controlled. Only a few of the hypotheses were unequivocally supported, mainly those connected with the economic status.

The Princeton Study (Westoff *et al.*, 1961) was taken up as a successor to the Indianapolis study and explicit attempts were made to correct some of the mistakes of the earlier effort. The main personality variables included in the Princeton Study were: generalised manifest anxiety; need for nurturance; compulsiveness; tolerance of ambiguity; co-operativeness and need-achievement. Like the Indianapolis Study, the Princeton Study also did not show any meaningful association between the psychological variables and fertility indices because the relevant psychological attributes either were not chosen or were inadequately measured.

Another important piece of research in this field was the study conducted in Korea by the Korean Institute for Research in the Behavioural Sciences (Chung *et al.*, 1972). The study covered a wide spectrum of psychological factors, including value orientations, modernity attitudes, perception of environmental press, ideal family size, son preference and attitude toward fertility control. The basic finding was that psychological factors affect fertility and family planning behaviour, which, in turn, are affected by socio-demographic factors. The study confirmed that individual modernity factors are more important than the basic personality variables in explaining fertility behaviour.

The 'Value of children' forms the recent contribution as a major determinant of fertility which was studied cross-culturally by a team of scholars at the East-West Population Institute (Arnold *et al.*, 1975). According to Fawcett (1970), 'the values attached to children and functions served by children, should be a prime target for psychological research'. Hoffman and Hoffman (1973) developed a theoretical model consisting of five broad sets of variables: the value of children, alternative sources of the value, costs, barriers, and facilitators. Arnold and others (1975) found significant differences in the value of children across the three socio-economic groups: urban middle class, urban lower class and rural. The measures of expectations of economic benefits from children were found to be positively related to family size indices. Large families were observed to be associated with an emphasis on continuity, tradition, security, role motivations, happiness and affection, and social status.

Fertility desires are also influenced by parental experiences with infant and child mortality and the perception of the survival chances of infants and children. Pebley *et al.* (1979), in a study of Guatemalan women, found that experience with child mortality—own child deaths and sibling deaths—was strongly and positively related to the desire for additional children at parity three and above. Bacon and Mason (1972), on the basis of a sample survey conducted in 1966 in Ankara, Turkey, reported that '. . . those couples who have experienced an infant death have and also expect more children than do couples without such experience, even after controlling for socio-economic differences and duration of marriage.' Summarising the results of an investigation conducted in Taiwan, Heer (1972) reported that women who perceived the chances of survival of children up to the age of 15 years as 95 per cent or over, had fewer children after the birth of the second child than those who perceived their survival as 85 per cent or less. Subhadradevi (1978), in a study of rural women in Kerala, found a linear positive relationship between infant and child mortality and fertility. A recent survey conducted in rural Andhra Pradesh (Mahadevan *et al.*, 1981) also revealed a direct relationship between the number of infant and child deaths and fertility in all the three caste groups—Muslims, Harijans and Caste Hindus. There are many other variables that influence fertility. However, the preceding list of variables that are directly or indirectly linked to the psychological dimension of human behaviour, suggested the need for exploring a research venture in this field in India. Hence the section that follows discusses certain empirical findings of a study that was carried out in Hyderabad City in South Central India.

Methodology

The aim of the present study was to investigate the different and common determinants of fertility behaviour of the non-slum and slum population of Hyderabad City in South Central India on social and psychological dimensions. For the present study, a two-stage simple random sampling technique was adopted. The respondents were Telugu speaking Hindu couples married only once; the wife not exceeding 44 years of age at the time of the survey and with two or more living children. For the first stage of sampling, all the eleven 'Census enumeration blocks' of Hyderabad City covered by the Sample Registration Scheme (SRS) were selected. This was done to save time in the selection of a sample population as the area of investigation happened to be a metropolitan city. Two independent lists of the non-slum and the slum households having eligible couples of Hindu religion were prepared. From these two lists, a sample of 240 non-slum and 240 slum dwellers were selected at random. In the case of households with two or more eligible couples, only one was selected at random since all of them represented the same socio-economic status. Thus, a total sample of 480 couples consisting of an equal number of non-slum and slum dwellers were selected at random for the present study.

Conceptual Framework

For operational convenience, a simplified conceptual framework based on the models of Ronald Freedman (1975) and Fawcett (1980) was adopted for the present study. Besides these models, the 'intermediate variables' of Davis and Blake (1956) and the 'Value of children' envisaged by Arnold *et al.*, (1975) were also suitably included.

Findings and Discussion

Fertility Differentials

The non-slum dwellers showed significantly low fertility—current and cumulative—in comparison to the slum dwellers. The crude birth rate (CBR) during the year preceding the date of the survey (1980) was 20 per 1000 population for the non-slum dwellers as against 28.19 for the slum dwellers. Also, the general marital fertility rate (GMFR) for the non-slum dwellers was only 108 per 1000 currently married females as against 171 for the slum dwellers. Further, the age-specific marital fertility rates (ASMFR) were not only low among the non-slum dwellers but also the decline in ASMFRs with the advancement of reproductive age was found to be sharper among the non-slum dwellers than among the slum dwellers.

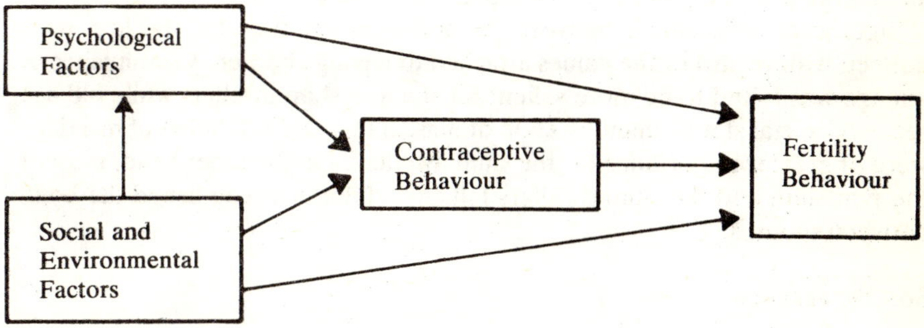

FIGURE 1 A Simplified Conceptual Frame based on the Models of Ronald Freedman and Fawcett

The cumulative fertility (children ever born) of the non-slum dwellers was significantly less than that of the slum dwellers. On an average, the non-slum dwellers had 2.96 live births in comparison to 4.49 live births for the slum dwellers. In other words, the non-slum dwellers had 1.53 live births fewer than the slum dwellers. The difference was even greater when we examine the fertility levels of the females in their late childbearing ages (35+). The non-slum dwellers with wives aged 35 years and above had, on an average, 3.80 live births as against 6.25 live births for the slum dwellers, the difference in fertility between them being 2.45 live births. The slum dwellers, besides their high fertility, also had larger additional live births as compared to the non-slum dwellers.

Some of the psycho-social, economic, demographic and modernisation variables that could significantly explain differential fertility both within and between the non-slum and the slum dwellers are dealt with in this paper. Further, a discussion is also presented here as to how much of the total variance in fertility among the non-slum and the slum dwellers could be due to each of these factors.

Psychological Correlates

Value of Children

Of the value dimensions of children examined here, only fulfilment of marriage, manifestation of masculinity and attainment of social status showed a strong positive correlation with fertility. 'Wanting children for fulfilment of marriage' accounted for 3.2 per cent of the total variance in the fertility level of the slum dwellers. On the other hand, 1.5 per cent of the total fertility variance among the non-slum dwellers was explained by 'manifestation of masculinity'.

Significant differentials between the non-slum and the slum dwellers were noticed with regard to the values attached to having children. Continuance of lineage was found to be more salient for the non-slum dwellers while old age security, social status, manifestation of masculinity and fulfilment of marriage were of greater importance for the slum dwellers. On the other hand, most of the non-slum and the slum dwellers felt that children strengthened the bond between spouses.

Son Preference

The prevalence of son preference was very high and strong among the slum dwellers in comparison to the non-slum dwellers. This is confirmed by the finding that, in the absence of a son, 81 per cent of the slum dwellers wanted to wait up to four or more conceptions till getting a son as against only 27 per cent of the non-slum dwellers expressing a similar opinion. Similarly, the VOC study (Arnold *et al.*, 1975) also reported greater son preference among the urban lower class respondents than among the urban middle class respondents of Korea, Taiwan and Thailand.

The number of sons reported as ideal for the family also lends support to the existence of strong son preference among the slum dwellers. Only 6 per cent of the slum dwellers felt that one son was ideal as against 46 per cent of the non-slum dwellers. In contrast to 17 per cent of the slum dwellers who had reported three or more sons as ideal, only 4.2 per cent of the non-slum dwellers expressed the same. About 23 per cent of the slum dwellers could not visualise the ideal number of sons for the family; they simply stated as many as given by God.

Cross-tabulation of the ideal number of sons with the fertility level of the non-slum and the slum dwellers showed a definite positive relationship between them. The non-slum dwellers stating two sons as ideal had 0.71 live births more than those who expressed only one son as ideal (significant at 0.01 level). To support this finding further, nearly 50 per cent of the non-slum dwellers actually had only one son and their fertility was lower by 0.60 live births in comparison to those having two sons which comprised 32.5 per cent of the total non-slum respondents (significant at 0.01 level). Thus, the decreasing persistence of the non-slum dwellers to have more than one son accounts for 14 per cent of the total variance in their fertility. In contrast, no such change was noticed among the slum dwellers because of the consideration of a son as the only source of economic security. Thus, son preference acts as a pronatalist force resulting in high fertility among the slum dwellers, while it is gradually losing credence as a prime determinant of fertility among the non-slum dwellers.

Infant and Child Mortality

Couples seem to adjust their fertility behaviour in response to the incidence of infant and child mortality. Several micro level studies revealed that the final family size is greatly influenced by the experience of infant and child mortality (Wyon and Gordon, 1971; Bacon and Mason, 1972; Sinha, 1975; Pebley, 1979). Harman (1970) concluded that families adjust their fertility behaviour to both their own experience with child mortality and to the mortality risks they perceive in the community. The hypothesis found true here is that fertility behaviour is shaped by personal experiences with infant and child mortality.

From the data in Table 1, it is conclusively evident that personal experiences

TABLE 1 Mean Live Births in Relation to Infant and Child Mortality

Infant/or Child Deaths	*Present Age of Wife*							
	≤ 24		*25–34*		*35 +*		*Total*	
	Non-slum	*Slum*	*Non-slum*	*Slum*	*Non-slum*	*Slum*	*Non-slum*	*Slum*
No Death	2.34	2.49	2.74	3.85	3.24	5.71	2.75	3.97
	(38)	(41)	(136)	(93)	(33)	(41)	(207)	(175)
One or More Deaths	3.50	3.40	3.80	5.21	5.45	6.97	4.33	5.91
	(2)	(5)	(20)	(29)	(11)	(31)	(33)	(65)

with infant/or child mortality have a definite positive relationship with fertility. Similarly, two of the studies conducted recently in South India (Subhadradevi, 1978; Mahadevan *et al.*, 1981) also showed a strong linear positive relationship between parental experience with infant and child mortality on the one hand and fertility on the other hand.

High tertility was significantly noticed among the slum dwellers who had personal experience with infant/child mortality, even after controlling the present age of the wife. On the whole, slum dwellers who had lost one or more children had nearly two live births higher in comparison to those having no personal experience with infant/child mortality (significant at 0.01 level). Even the non-slum dwellers were not an exception to this phenomenon although the magnitude of infant and child mortality happened to be low among them. The fertility of the non-slum dwellers who had a personal experience with infant/child mortality, as a whole, was higher by 1.58 live births in comparison to those who never had this experience. The difference in fertility between the non-slum and the slum dwellers was found to be higher in the case of those who had a personal experience with infant/child mortality than in the case of those with no such experience. Hence, high incidence of infant/child mortality serves as a motivating force for high fertility in order to ensure the survival of desired or slightly more number of children to adulthood.

Modernisation

The modernised societies tend to have less fertility in comparison to the traditional societies as modernisation brings about a change in the ways of thinking, feeling and action of the people (Inkeles, 1966). The characteristics of modernisation examined here are: general planning, aspiration for the education of children and interspouse communication.

General Planning

The propensity to plan is a characteristic of modernisation. It is generally noticed that those who observe planning in different aspects of family life also show a tendency to plan their family size (UN Asian Population Studies Series No. 16, 1974). The planning propensity of the individual is examined in relation to household budget and family size.

Economic Planning The household economic planning was assessed in terms of saving money for the future. Nearly three-fifths (57.5 per cent) of the non-slum dwellers were in the habit of saving money for the future as against only one-fifth (18.0 per cent) of the slum dwellers, which is also due to the low earnings of the latter group.

The non-slum dwellers who were saving money for the future had less fertility than those who did not do so; but the difference is not statistically significant. This may be due to the future orientation of the former group.

Though the number of slum dwellers saving money for the future was very low, they had significantly less fertility than those who did not save money (0.05 level). Thus, the tendency for future economic planning is associated with low fertility.

Planning of Family Size Significant differentials are observed between the non-slum and the slum dwellers regarding planning of family size. An overwhelming proportion of the non-slum dwellers (88.3 per cent) in comparison to only 56 per cent of the slum dwellers reported that they had thought about their family size prior to begetting children.

The findings of the study confirmed a strong negative association between the planning of family size and fertility among both the non-slum and slum dwellers. The fertility level of the non-slum and slum dwellers who had given prior thought to their family size was less by 0.70 and 1.49 live births respectively, in comparison to those who conceived children without any prior thought (significant at 0.01 level). The difference in fertility between those who had and those who had not given prior thought to their family size had steadily increased in both the groups with an increase in their present age. The fertility differential between the non-slum and the slum dwellers was found to be twice as high among those who did not plan their family size (1.59) as compared to those who planned their family size (0.80). Thus, it is evident that planning family size has a strong depressing effect on fertility.

Aspiration for Education of Children

Parents generally have high aspirations for a bright future for their children, particularly through the acquisition of higher levels of education. Parental aspiration for higher levels of education for their children is assumed to inhibit fertility. This hypothesis was tested in relation to the daughter's education rather than the son's education because the latter could not discriminate parents due to their greater aspiration for the son's education.

The data revealed that 80 per cent of the non-slum dwellers wanted to provide college education to their daughters in comparison to only 30 per cent of the slum dwellers. On the other hand, nearly 20 per cent of the slum dwellers did not show any aspiration for the education of their daughters and another 10 per cent of them desired to provide only primary education in contrast to an almost negligible percentage of the non-slum dwellers aspiring to such a low level of education for their daughters.

The findings further revealed that the fertility level of both the non-slum and the slum dwellers aspiring to higher levels of education for their daughters was significantly less in comparison to those aspiring to either 'low' or 'no education'. The non-slum dwellers who wanted to provide college education to their daughters had 0.61 live births fewer than those aspiring only secondary education (significant at 0.05 level). Among the slum dwellers, those who aspired college (30.0 per cent) and secondary education (28.2 per cent) for their daughters had one live birth less than those with 'no aspiration' for their daughters' education (significant at 0.01 level). Hence, strong parental aspiration for higher levels of education for their children (including daughters) is a determinant of low fertility because of the high cost involved in educating them.

Interspouse Communication

Interspouse communication and decision-making in the family have been reported to be positively correlated with effective family limitation (Mitchell, 1972; Chung, 1973; UN ESCAP, 1974). In the present study, 80.0 per cent of the husbands from non-slum households held discussions with their wives on the planning of family size as against only 40.0 per cent of the slum dwellers.

The negative correlation between fertility and interspouse communication regarding family size is also confirmed in the present study. The slum dwellers who discussed the planning of family size with their wives had 0.50 live births less than those who did not report interspouse communication (significant at 0.05 level). Similarly, the non-slum dwellers reporting interspouse communication had less fertility than those who did not discuss their family size with their wives; but the difference in fertility is not statistically significant. Mukherjee (1979) also found that women who had better communication with their husbands regarding family planning were more favourably disposed to the small family norm. Thus, better communication between husband and wife on family size results in low fertility as the decision regarding family size is made jointly by both the partners involved in family formation.

Socio-Economic Status

Historically, socio-economic status has been shown as having a negative association with fertility (Freedman *et al.*, 1959; Saksena, 1973). Manzoor Alan and Khan (1972), in their study of Hyderabad City, observed that the socio-economic status of the non-slum dwellers was far superior to that of the slum dwellers. A recent survey of the slum dwellers in Hyderabad and Secunderabad by the Bureau of Economics and Statistics (1980) has also confirmed their low socio-economic status. In the present study an attempt is made to determine the extent to which the existing variations in the socio-economic status have contributed to differential fertility within and between the non-slum and the slum dwellers.

An index of socio-economic status (SES) was developed to study the cumulative effect of education, occupation and household income on fertility. The respondents were divided into three socio-economic status groups: 'lower and lower-middle', 'middle' and 'upper and upper-middle'. Most of the slum dwellers (84.6 per cent) belonged to the 'lower and lower-middle' SES group while 58 per cent of the non-slum dwellers were from the 'upper and upper-middle' group.

The expected inverse relationship between socio-economic status and fertility is confirmed in this study. The non-slum dwellers belonging to the 'upper and upper-middle' SES group had 0.66 live births fewer than those of the 'lower and lower-middle' SES group (significant at 0.05 level). The fertility of the non-slum dwellers from the 'middle' SES group was also less by 0.59 live births in comparison to those belonging to the 'lower and lower-middle' SES group (significant at 0.05 level). Among the slum dwellers, the fertility of the 'middle' SES group was less by 0.50 live births as compared to those of the 'lower and lower-middle' group (significant at 0.05 level). Thus, a moderate rise in the socio-economic status from the 'lower and lower-middle' to the 'middle' level results in a significant reduction in the fertility level of both the non-slum and the slum dwellers.

Perceived Economic Capacity to Raise Children

The family size of an individual is likely to be influenced largely by one's perceived economic capacity to support children rather than the actual capacity. The hypothesis tested here is that those who perceive themselves as capable of raising a large number of children tend to have higher fertility in comparison to those perceiving themselves as able to support only a few children.

For the question as to 'how many children they could raise without becoming a financial burden to them', 84 per cent of the non-slum dwellers felt that they could afford to raise up to three children as against 49 per cent of the slum dwellers. In contrast, about 50 per cent of the slum dwellers felt that they could support four or more children as compared to only 16 per cent of the non-slum dwellers.

TABLE 2 Mean Live Births in Relation to Perceived Economic Capacity to Raise Children

No. of Children Perceived as Easy to Support	*Present Age of Wife*							
	≤ 24		*25–34*		*35 +*		*Total*	
	Non-slum	*Slum*	*Non-slum*	*Slum*	*Non-slum*	*Slum*	*Non-slum*	*Slum**
2	2.16 (19)	2.38 (8)	2.62 (69)	3.06 (16)	3.29 (14)	4.50 (8)	2.63 (102)	3.25 (32)
3	2.67 (18)	2.65 (23)	3.01 (67)	3.67 (42)	3.67 (15)	5.33 (21)	3.05 (100)	3.80 (86)
4	2.00 (2)	2.75 (12)	3.31 (16)	4.50 (44)	4.09 (11)	6.00 (28)	3.52 (29)	4.75 (84)
5 +	2.00 (1)	2.00 (2)	2.75 (4)	5.37 (19)	5.00 (4)	6.43 (14)	3.67 (9)	5.60 (35)

* 3 slum dwellers were uncertain

The data in Table 2 confirm the hypothesis that perceived financial burden acts as a depressant force on fertility behaviour. A strong linear positive relationship is observed between fertility and the number of children perceived as easy to support. This trend is seen in both the groups, particularly among the slum dwellers of young, middle and late reproductive ages and also in the population as a whole. The non-slum dwellers who perceived as capable of supporting four children had 0.47 and 0.89 live births higher as compared to those who felt that they could easily support three and two children respectively (significant at 0.05 and 0.01 levels).

Among the slum dwellers, those who had the highest fertility (5.60) also stated that they could easily raise a large number of children (≥ 5). This is most probably to rationalise their high post-fertility behaviour. The slum dwellers in this group had, on an average, 0.85, 1.80 and 2.35 live births higher in comparison to those who perceived that they could support four, three and two children respectively (significant at 0.01 level). Similarly, Arnold and Fawcett (1975) studying the value of children in Hawaii, also found that those couples who believed that they could afford to raise a large number of children actually had more children than the other couples. In each of the cells with specific number of children perceived as easy to support, the non-slum dwellers were found to have significantly less fertility than the corresponding slum dwellers. Interestingly, the difference in fertility between the non-slum and the slum dwellers has increased gradually with an increase in the number of children perceived as easy to support; thus confirming the post-fertility rationalisation among the slum dwellers.

The multivariate analysis shows that the differential perception of economic capacity to raise children accounts for 10 per cent of the variance in fertility among the slum dwellers as against only 1.80 per cent of the total variance among the non-slum dwellers. Thus, the perceived economic capacity rather than the actual capability to raise children is a powerful predictor of fertility variance, particularly among the slum dwellers.

Demographic Factors

Differentials in fertility across the various cultural and ethnic groups seem to be largely due to the variations in the age at marriage and the duration of marriage. Several studies conducted in India and elsewhere have shown that fertility is negatively correlated with the age at marriage of women and positively with the duration of effective marriage.

Age at Marriage

The age at marriage of females is early and universal in India. This is more so among the slum dwellers. As many as 65 per cent of the females from the slum areas got married at the age of 14 years or below as against only 9 per cent of the females from the non-slum areas. By 17 years of age, approximately 93 per cent of the females from the slum areas got married in comparison to only 45 per cent of the females from the non-slum areas. The mean age at marriage of males and females in the non-slum areas was 23.8 and 17.9 years respectively, while it was 20.7 years for males and 13.5 years for females from the slum areas.

TABLE 3 **Mean Live Births in Relation to Age at Marriage of Females**

Age at Marriage of Females	*Present Age of Wife* 24		25–34		35 +		*Total*	
	Non-slum	*Slum*	*Non-slum*	*Slum*	*Non-slum*	*Slum*	*Non-slum*	*Slum*
≤ 14 years	2.57 (7)	2.67 (33)	3.67 (12)	4.47 (75)	4.25 (4)	6.44 (48)	3.43 (23)	4.69 (156)
15–17	2.45 (22)	2.33 (12)	3.08 (49)	3.83 (35)	4.60 (15)	6.15 (20)	3.19 (86)	4.25 (67)
18 +	2.09 (11)	3.00 (1)	2.67 (95)	3.00 (12)	3.24 (25)	4.50 (4)	2.73 (131)	3.35 (17)

An inverse relationship between the age at marriage and fertility is found in both the non-slum and the slum areas. The fertility of the females in the non-slum areas who married at the age of 18 years or above was less by 0.46 live births in comparison to those married between 15–17 years of age (significant at 0.05 level) and 0.73 live births fewer than those married at the age of 14 years or below (significant at 0.01 level). On the other hand, the females from the slum areas who were married at the age of 18 years or above had 0.90 and 1.34 live births fewer than those married between 15–17 years of age and at the age of 14 or below respectively (significant at 0.01 level). The decline in fertility with an increase in the age at marriage of females is found to be sharp in the late reproductive ages for both the non-slum and the slum dwellers.

The findings of the study provide strong evidence for raising the age at marriage of women. The data in Table 3 show that the postponement of marriage of women in both the non-slum and the slum areas to 18 years of age results in a reduction of at least 0.5 to 1.0 live birth, on an average, per woman.

Duration of Marriage

Women who have spent more years in effective marital union are expected to have higher fertility in comparison to those with a short marital life. In the present study, 29 per cent of the slum dwellers had an effective marital union spanning 21 years and above in comparison to only 8 per cent of the non-slum dwellers. In contrast, the marital duration of more than a half of the total non-slum dwellers (51.0 per cent) was 10 years or below as against only 22 per cent of the slum dwellers. The shorter marital union of the non-slum dwellers is largely due to the relatively late marriage of females in these areas.

The findings also confirmed a strong linear positive relationship between the duration of effective marriage and fertility. Among the non-slum dwellers, those with longer marital union (≥21 years) had maximum number of children (4.26) while those whose marital life was shorter (≤10 years) had minimum number of children (2.53). For the non-slum dwellers whose marital life was of intermediate duration (11–20 years), the number of children fell between these two extremes (3.22). The fertility of the non-slum dwellers with longer marital union was higher by 1.04 and 1.73 live births respectively, in comparison to those with intermediate and shorter marital unions (significant at 0.01 level).

This is true of the slum dwellers as well. Those who spent more years in marital union had 5.80 live births as against 2.70 for those with shorter marital union and 4.17 for the intermediate group. The positive correlation between the duration of effective marriage and fertility is found to be more conspicuous among the slum dwellers, those with a marital union of 21 years or above had 1.63 and 3.10 live births higher than those with intermediate and shorter marital unions respectively (significant at 0.01 level). Thus, the high fertility of the slum dwellers is largely due to their longer marital life which, in turn, is the result of their early marriage.

The multivariate analysis also shows that the differentials in the duration of effective marriage account for 23 per cent and 41 per cent of the total variance in the fertility of the non-slum and the slum dwellers respectively. From this, it can be concluded that the duration of effective marriage is the prime determinant of fertility variance, particularly among the slum dwellers.

Closed Birth Interval

The completed family size of a couple also depends upon the closed birth interval which varies across the different socio-cultural groups. In this study, the closed birth interval refers to the interval between the last and last but one live births.

The hypothesis that females having a longer closed birth interval tend to have less fertility than those with a shorter interval is confirmed in the case of the slum dwellers. The average family size of the females in the slum areas with longer birth interval was 3.66, this is 0.81 and 1.05 live births less in comparison to those with medium and shorter birth intervals respectively (significant at 0.01 level). An inverse relationship between closed birth interval and fertility was observed for the non-slum dwellers but the differences in fertility were not statistically significant. This may be due to the effective planning of family size by a large proportion of the non-slum dwellers independent of the duration of closed birth interval.

The multivariate analysis reveals that 5.18 per cent of the total variance in fertility among the slum dwellers is due to the differentials in closed birth interval. Thus, a longer birth interval has a depressing effect on fertility, particularly for the slum dwellers among whom the practice of effective contraceptives is relatively low.

Multiple Determinants of Fertility

We have examined the causal relationship of the independent variables with fertility individually. Since fertility behaviour is determined by the simultaneous operation of a wide array of psychological, socio-cultural, economic and demographic factors at a given point of time, the relative contribution of each of these factors in the explanation of fertility behaviour cannot be assessed correctly unless they are examined together. Hence, multiple step-wise regression analysis was done to estimate the amount of variance in fertility behaviour that can be explained by each of the independent variables separately and collectively. Separate regression analyses were done to predict fertility behaviour of the non-slum and the slum dwellers.

TABLE 4 Determinants of Fertility Behaviour of the Non-slum and the Slum Dwellers, as Denoted by Proportion of Variance Explained (R^2) in Multiple Regression Analysis

Predictors	*Proportion of Variance Explained (R^2)*	
	Non-slum	*Slum*
Duration of Marriage	0.23**	0.41**
Value of Son	0.14**	–
Education of Husband	0.037**	–
Perceived Economic Capacity to Raise Children	0.018 †	0.10**
Closed Birth Interval	–	0.05**
Additional Family Size Norm	0.017†	0.038**
Wanting Children for Fulfilment of Marriage	–	0.032*
Experience of Infant and Child Mortality	0.016†	–
Wanting Children for Manifestation of Masculinity	0.015†	–
Perceived Financial Burden in Educating Children	0.015†	–

† P at 0.10 level; * P at 0.05 level; ** P at 0.01 level

Note: The significance of each independent variable in explaining total variance is computed by an F-test using the formula:

$$F = \frac{R^2 \text{ change}/M}{(1-R^2_{Y1,2,\ldots K})(N-K-1)}$$

Where: R^2 change represents a squared part correlation that measures the incremental variance explained with the addition of each new variable

K is the total number of independent variables

M is the number of independent variables for which the significance test is being made

N is the total number of cases (Nie *et al.*, 1975)

The total variance in fertility explained by all the twenty-four independent variables considered for regression is 68 per cent for the slum dwellers as against 51.9 per cent for the non-slum dwellers. Approximately 63 per cent of the variance in the fertility level of the slum dwellers is explained by five variables whereas for the non-slum dwellers eight variables have predicted only

49 per cent of the variance in their fertility. Thus, the fertility behaviour of the non-slum dwellers is more complex and relatively unknown than that of the slum dwellers.

It is quite evident from Table 4 that psychological and social factors are as important as the demographic factors in the explanation of fertility differentials, particularly among the non-slum dwellers as they account for 24 per cent and 24.7 per cent of the total variance in their fertility respectively. On the other hand, the psychological factors are found to be weak predictors of fertility differentials among the slum dwellers, since they account for only 13.2 per cent of the variance while the demographic factors account for 50 per cent of the total variance in their fertility. Thus, socio-psychological and demographic factors are of equal importance in predicting the fertility behaviour of the non-slum dwellers whereas the demographic factors are the prime determinants of fertility among the slum dwellers.

The psychological factors which predicted a significant proportion of the variance in the fertility of the non-slum dwellers are: value of son (14.0 per cent); perceived economic capacity to raise children (1.8 per cent); experience of infant and child mortality (1.6 per cent); wanting children for manifestation of masculinity (1.5 per cent); and perceived financial burden in educating children (1.5 per cent). For the slum dwellers, only two of the psychological factors—perceived economic capacity to raise children (10.0 per cent) and wanting children for fulfilment of marriage (3.2 per cent)—are found to be significant predictors of their fertility behaviour. Among the psychological factors, value of son and perceived economic capacity to raise children have emerged as the predominant determinants of fertility variance among the non-slum and the slum dwellers respectively.

Among the demographic factors, the duration of effective marriage alone has explained 23 per cent of the variance in the fertility level of the non-slum dwellers in comparison to 41 per cent among the slum dwellers. In other words, differential exposure to marriage is the prime predictor of fertility variation, particularly among the slum dwellers. Closed birth interval is the second important demographic variable accounting for 5 per cent of the variance in the fertility level of the slum dwellers. In short, the duration of marriage and closed birth interval together account for 46 per cent of the total fertility variance among the slum dwellers.

In conclusion, the study showed significant differentials in fertility between the non-slum and the slum dwellers. The non-slum dwellers, on an average, had 2.96 live births as against 4.49 live births for the slum dwellers; the difference in fertility between them being 1.53 live births. The differential fertility between them had increased linearly with increase in present age reaching the peak in the older reproductive ages. The preference for son(s) was found to be strong and widely prevalent among the slum dwellers because they consider son(s) as the only source of economic security in old age. The level of fertility was found to be significantly high among those who had personal experience with infant and child mortality in comparison to those with no such experience. Couples who discussed family planning and also planned their family size had significantly less fertility as compared to the other couples. The study confirmed a significant reduction in the fertility level of both the non-slum and the slum dwellers with a moderate rise in their socio-economic status. Perceived economic ease in raising children showed a linear positive relationship with fertility. Interestingly, those who perceived themselves as capable of supporting a large number of children actually had higher fertility than the other couples. Fertility was found to be negatively correlated with the age at marriage and positively with the duration of marriage.

The multivariate analysis revealed that psychological and social factors are as important as demographic factors in explaining the total variance in the fertility level of the non-slum dwellers. In contrast, the demographic factors were found to be the prime determinants of fertility variance among the slum dwellers. Differentials in the duration of effective marriage had accounted for much of the variance in fertility, particularly among the slum dwellers. Among the psychological factors, differentials in value of son and perceived ease in raising children were found to be the predominant predictors of fertility variance among the non-slum and the slum dwellers respectively.

The findings of the study have several implications for population policy and programme implementation. To minimise dependence on children and particularly on sons, the government should effectively implement certain old age social security schemes. People should be educated to give equal treatment to sons and daughters. Provision should be made for the expansion and expeditious implementation of infant and child welfare programmes in areas of high mortality. Improvement in the nutritional status of the mother and children along with adoption of healthy child rearing practices would go a long way in reducing infant and child mortality. The prevalence of child labour has pro-natal influence on the fertility level of the slum dwellers. To make children non-economical, laws prohibiting child labour should be strictly implemented. Such a measure would increase the cost of rearing children, thereby depressing the fertility level of the slum dwellers. The duration of marriage being the prime determinant of fertility, the best approach to depress fertility would be to cut down the reproductive period of high fecundity by raising the marriage age of females, particularly the slum dwellers. Merely passing the legislation 'raising the marriage age of boys and girls' would not serve the purpose if it is not effectively implemented. Both parents and the prospective brides and bridegrooms should be made to understand, through population education, the social, demographic, psychological and health implications of postponing marriage to later ages. Incentives such as meeting the marriage expenditure—either individually or through community marriages—of the members of the low socio-economic status group who marry at 18 years of age or above would bring about quicker social change in the institution of marriage. Provision should be made for free education and reservation of jobs for females thereby promoting gainful employment of females which reduces fertility.

REFERENCES

ARNOLD, F. *et al.* (1975). *The value of children: a cross-national study*. Vol. I. Honolulu: East-West Population Institute, pp. 5–6.

BACON, T.J. and K.O. MASON. (1972). *Turkish fertility: a review of social and economic correlates*. Rosslyn, Virginia: US Agency for International Development, p. 28.

BUREAU OF ECONOMICS AND STATISTICS. (1980). Economic conditions of slum dwellers in the twin cities of Hyderabad and Secunderabad. Reprint from the *Economic and Statistical Bulletin*, Hyderabad: Government of Andhra Pradesh, p. 4.

CHUNG, B.M. *et al.* (1972). *Psychological perspectives: family planning in Korea*. Hollym Corporation, p. 146.

CHUNG, B.J. (1973). The impact of industrialisation on fertility in Hong Kong: a social and psychological analysis. Unpublished paper.

DANDEKAR, K. (1959). Socio-economic change in rural areas during three five-year plans. Poona: Gokhale Institute of Politics and Economics. Unpublished.

DAVIS, K. and J. BLAKE. (1956). Social structure and fertility: an analytical framework. *Economic Development and Cultural Change*, 4, pp. 221–35.

DESAI, P.B. (1970). *Delhi demographic survey—tables*. Delhi: Institute of Economic Growth.

DRIVER, E.D. (1963). *Differential fertility in Central India*. Princeton: Princeton University Press.

FAWCETT, J.T. (1970). *Psychology and population: behavioural research issues in fertility and family planning*. New York: The Population Council, pp. 30–34.

————. (1980). *Cultural influence on fertility decisions: an analytic framework and some research issues*. Paper No. 6, p. 3.

FREEDMAN, R. (1975). Variables that affect fertility. *Sociology of Human Fertility*. New York: Irwington Publishers, p. 1.

FREEDMAN, R. *et al.* (1959). *Family planning, sterility and population growth*. New York: McGraw-Hill.

HARMAN, A.J. (1970). *Fertility and economic behaviour of families in the Philippines*. California, Santa Monica: The Rand Corporation, p. 46.

HEER, D.M. (1972). Determinants of family planning attitudes and practices. Cambridge: Harvard University (mimeo), p. 160.

HOFFMAN, L.W. and M.L. HOFFMAN. (1973). The value of children to parents. *In* J.T. Fawcett (Ed.), *Psychological perspectives in population*. New York: Basic Books, pp. 19–76.

INKELES, A. (1966). The modernisation of man. *In* M. Weiner (Ed.), *Modernisation: the dynamics of growth*. New York: Basic Books.

JORAPUR, P.B. (1967). Fertility study of Dharwar. *Journal of Institute of Economic Growth*, 6 (2).

MAHADEVAN, K. (1979). *Sociology of fertility: determinants of fertility differentials in South India*. New Delhi: Sterling Publishers Pvt. Ltd.

MAHADEVAN, K., M.S.R. MURTHY, P.R. REDDY, P.J. REDDY, V. GOWRI and S. SIVARAJU. (1981). *Culture, nutrition and infant and childhood mortality: a study in South Central India*. Paper presented at the General Conference of IUSSP, Manila.

MANZOOR ALAN, S. and W. KHAN. (1972). *Metropolitan Hyderabad and it's religion: a strategy for development*. New York: Asia Publishing House, p. 130.

MITCHELL, R.E. (1972). Husband-wife relations and family planning practices in urban Hong Kong. *Journal of Marriage and the Family*, 34(1), pp. 139–46.

MUKHERJEE, B.N. (1979). *Prediction of family planning and family size from modernity value orientation of Indian women*. Honolulu: East-West Centre, pp. 37–38.

NIE, N.H. *et al.* (1975). *Statistical package for the Social Sciences*. 2nd ed. New York: McGraw-Hill, pp. 339–40.

PEBLEY, A.R. *et al.* (1979). Fertility desires and child mortality experience among Guatemalan women. *Studies in Family Planning*, Vol. 10, No. 4, p. 135.

RELE, J.R. and T. KANITKAR. (1980). *Fertility and family planning in Greater Bombay*. Bombay: Popular Prakashan, pp. 68–75.

SAKSENA, D.N. (1973). *Differential urban fertility*. Lucknow: Demographic Research Centre, Lucknow University.

SAXENA, G.B. (1965). Differential fertility in rural Hindu community: a sample survey of the rural Uttar Pradesh, India. *Eugenics Quarterly*, 12(3), pp. 137–45.

SRINIVASAN, K. (1969). A perspective study of the fertility behaviour of a group of married women in rural India: design and findings of the first round of enquiry. *Population Review*, 11(2), p. 53.

IV

Analytical Framework, Models and Mortality

'. . . *the analytical framework and conceptual model on Infant and Childhood Mortality is a formidable contribution focussing on a global perspective* . . . .'

DR. C. CHANDRASEKARAN
HONARARY PRESIDENT OF IUSSP

16

Mortality, Biology and Society: Analytical Framework and Conceptual Model

K. MAHADEVAN

Introduction

Mortality research the world over gained momentum after the celebration of the International Year of the Child in 1979 by the United Nations. Several researches carried out so far, however, suffer from poor coverage of appropriate variables and lack depth. The type and number of determinants of infant and childhood mortality (hereafter referred to as child mortality) are so innumerable and diverse in nature that it is indeed difficult for a single researcher to do justice in this field. Further, for one reason or the other inter-disciplinary research does not grow to the extent needed in this field to study as many determinants required as possible. In addition to these problems, there prevails confusion regarding the nature and number of determinants, direction of causal relationship and priority of importance of different causal factors in the absence of a comprehensive and systematic analytical framework and conceptual model.

This paper presents an analytical framework and a conceptual model on infant and child mortality based on the available literature in this field. In the review of literature an effort is made to cover as many papers/reports as possible from all the continents, countries, regions and cultural groups. The discussion covers a wide spectrum of determinants on mortality, namely, ecological, cultural, institutional, socio-economic, biological, genetical, policy and programme aspects. The concept of child in this paper includes both infants and children. The discussion that follows focuses first on a systematic

Acknowledgements I acknowledge with thanks the fellowship offered by USEFI (in 1983) to visit the Universities of Columbia, Michigan, North Carolina, East-West Population Institute, U.N. and Population Council for completion of this paper. I appreciate and thank Professor Ronald Freedman, Dr. Alan Rosenfield, Dr. Joe Wray, Dr. Albert Hermalin, Dr. John Knodel, Dr. N.K. Namboodiri, Dr. Lee Jay Cho and Dr. Paul Demeny for providing me all the facilities in their respective institutions.

I thank the S. V. University administration for giving me permission to undertake this study. I thank Dr. Krishna Rao, Dr. D.A. Naidu, Dr. Samuel, Dr. Sai Prasad, Dr. P.J. Reddy for their editorial assistance. I would also like to thank the secretarial staff and others who helped me directly and indirectly in this venture.

and sequential classification of major determinants of mortality entirely within a new analytical framework, which has several sub-sections, manifesting an inter-disciplinary character. On the basis of this analytical framework, a conceptual model has been formulated. Nevertheless, the multi-dimensional and multi-disciplinary nature of this problem continues to create difficulty in completing it on a comprehensive basis, despite the efforts so far made in this direction.

Though a few analytical frameworks and conceptional models of fertility appeared nearly twenty to twenty-five years ago (Davis and Blake, 1956; Hill, Stycos and Beck, 1959; Freedman, 1961), similar developments in the field of mortality are of very recent origin (Puffer and Serrano, 1973 and 1975; The Population Council, Cairo, 1980; Chowdhury, 1982; Rohde, Hull and Hendrata (undated); Farah, 1981; Mosley, 1983; Ruzicka, 1983; Chen, 1983). Conceptualising the various determinants of infant, child and adult mortality with numerous and diverse factors is far more difficult than in the case of fertility behaviour. The determinants of infant, child and adult mortality vary across different geographical and cultural groups as well as in the developed and developing countries (UN, 1984; Chen, 1983; Mosley, 1983; Mahadevan 1983; Preston, 1976). No doubt, continuous progress has been made in the theoretical field of mortality during the early 1980s but most of the authors adopted the existing terminology on framework and models used in the field of fertility (Davis and Blake, 1956; Bongaarts, 1978), namely, 'intermediate variables' and 'proximate determinants' to explain the causal relationship of several and diverse independent variables on mortality patterns. Since the determinants of mortality are numerous and more diverse in nature than those of fertility, it was decided here to develop an altogether different analytical framework and conceptual model using new terminologies. The adoption of new terminologies to explain the various determinants of mortality may also improve the communication to explain the causal relationship in this field. This effort, however, does not preclude any other comment or limitations on the existing framework since inadequacies are likely to exist during the process of development of knowledge in any field; of course, they have their own value and merit in promoting mortality research and knowledge.

The present effort was largely constrained on account of non-availability/ paucity of data and problems in accessibility to the literature on the determinants of mortality from several sources. The diversity of the characteristics of the determinants of mortality, absence of comprehensive analytical frameworks and models, poor response to inter-disciplinary efforts continue to affect major research effort in this field even today. The present effort may be viewed in the context of these circumstances, problems and limitations.

Life-Affecting Variables (LAVs): An Analytical Framework

Though human life begins with the stage of conception, there are several antecedent and sequential factors from both endogenous and exogenous sources that affect the growth and decay of infants and children at several stages of their development. The life process is affected through preconception stages—ecology, culture, family and marriage—subsequently through parents, conception, pregnancy, perinatal and postnatal care, infant and child rearing practices, polity, policy, interventions and several other developments. Apart from these factors, on a concurrent basis morbidity, natural calamity and accidents either directly or through the preceding list of factors affect life. Each of these critical factors and progressive stages in life have several specific variables that may affect human development. There are also factors which are yet to come to light but may emerge in course of time. Whatever causal factors are already known may change their influence and several new factors may come to light through further advancement in knowledge. Besides, the same variable may effect mortality differently at different stages in life and in different contexts. For instance, minimum financial resources may be required for a peasant family to support a pregnant mother, but meeting the cost of delivery in a modern urban-oriented clinic may be a financial burden for the same family. This is how income may influence life differently at different stages of its growth. Similarly, certain cultural practices are healthy but others are fatal as can be seen from the various sections, and so is the case with respect to several other variables. Further, several specific variables individually or collectively, directly or through other factors, immediately or in a delayed manner, favourably or adversely influence life processes once or on several occasions in similar or in varying degrees at different critical periods of human development. Hence, it was decided to select a comprehensive term called life-affecting variables* for explaining the numerous and diverse determinants of mortality.

The meaning of the word 'life' here refers to 'the life of infants and children'. The literary meaning of 'affect' is 'to exert an influence' and 'to produce an effect'. A variable in empirical research is expected to have these characteristics. The influence of a variable can be either positive or negative. Most of the variables that we consider may have a positive influence in certain contexts and show a negative influence on other occasions. Therefore, the selection of the term life-affecting is very appropriate.

Life-affecting variables are, first, broadly classified under the heading of situations and sequential events from the stage of polity-cum-policy, through several other factors, institutions and stages of development of life. Second, each sequential situation and event is further classified into one or more sub-sections of life-affecting variables depending upon the requirements of clarity, communication and empirical feasibility. For the present, a total of twelve broad categories of life-affecting variables (LAVs), and under each category several other specific factors that may be relevant at each stage of life process are represented. Each specific LAV is mentioned in an already known nomenclature for facilitating their identification and better communication. Since this list of variables is not complete by any means as our knowledge of the

* The term is coined by the author.

determinants of mortality is not complete, new research may generate additional knowledge, and certain variables are likely to change their pattern of influence over time; each broad section, therefore, has the provision to add newly emerging variables in future. Any omission made involuntarily now can also be added on to this provision.

Thus, for the classification of variables here, we have adopted a new terminology, namely, life-affecting variables for the entire category of all the determinants of mortality. Of course, the same term can also be used to describe the causal factors of morbidity as well as this term is fairly comprehensive and sufficiently adequate to explain the determinants of morbidity and mortality patterns in general.

LIFE-AFFECTING VARIABLES (LAVs)

I. *Polity and Policy*
- (*a*) National and international efforts
  - Priority in investment and development
  - Change from conflict to co-operation
  - Weaponry to social development
  - Mass production and distribution of life saving drugs
  - Consultancy and material supply
  - Ideology and social justice
- (*b*) Policy
  - Strategies for programme development
  - Policy on women and child development
  - Provider-population ratio
  - Social security and health insurance
  - Rural and family centered services
  - Types of medical care
  - Law and reforms
- (*c*) Socio-economic development
  - Management
  - Infrastructure
  - Transport and communication
  - Programmes for development and modernisation
- (*d*) Research and technology
  - *Others (to be identified)

II *Ecological Variables*
- (*a*) Morphology
  - Physical environment
  - Rural-urban scene
  - Industrialisation
  - Slum-non-slum areas
  - Pollution
  - Environmental sanitation
- (*b*) Climate
- (*c*) Interaction with environment
  - Irrigation and water management
  - Cropping pattern

* Others (to be identified) may belong to any sub-category of LAVs. This provision is given to add new variables that may emerge through future research.

Animal rearing
Vector
Radio-active agents
Poisonous fauna and flora
Adulteration of food
Spurious drugs

(*d*) Housing condition
Materials used
Crowding
Ventilation
Electrification
*Others (to be identified)

III *Cultural Variables*

(*a*) Cultural patterns
Cultures and sub-cultures
Religion
Caste
Ethnicity
Tribal life
Institutions

(*b*) Culture and behaviour
Customs and habits
Belief on fatalism
Value of son(s)/children
Infanticide/homicide
Fasting
* Others (see under family, marriage, parents and also to be identified)

IV *Familial Variables*

(*a*) Family dynamics
Joint family
Nuclear family
Kinship interaction
Roles of kinship group
Family status
Primogeniture
Ultimogeniture
Needs and problems

(*b*) Economy
Labour and occupational mobility
Property and landholdings
Agricultural production and management
Resources availability

(*c*) Food
Food choice, processing and availability
Food poisoning
Storage and preservation
Pattern of sharing food
* Others (to be identified)

V *Marital Variables*

(*a*) Premarital sex and promiscuity

(*b*) Marriage patterns
Birth order of the bride and groom
Age at marriage
Monogamy
Polygamy

Affinal and consanguineous marriages
(*c*) Marital life
Marital problems
Marital dissolution
Marriage cost (dowry and other expenses)
Sex and disease
Extramarital sex
Philosophy of marriage (no baby or otherwise)
*Others (to be identified)

VI *Parental Variables*
(*a*) Biological
Age
Physical growth of parents
Height of mother
Weight of mother
Appetite
Absorption
Metabolsim
Health
(*b*) Social
Education
Occupation
Status of women
Finance
Family planning
Acceptability of medical facilities
(*c*) Life style
Alcoholism
Smoking
Drug addiction
Media exposure
Food habits and nutrition
Use of soap and disinfectant
Personal and sexual hygiene
(*d*) Psychological
Adjustability
Conflict
Violence
Tension
Frustration
Suicidal tendency
Attitude
*Others (to be identified)

VII *Conception and Pregnancy Variables*
(*a*) Bio-social
Age at conception
Pregnancy problem(s)
Nutritional status
Weight of mother and its change
(*b*) Pregnancy care
Education
Pregnancy (medical) care
Legitimate or illegitimate conception
Intention to care pregnancy
*Others (to be identified)

VIII *Perinatal variables*

(*a*) Growth
- Prematurity
- Birth weight
- Multiple births

(*b*) Delivery care
- Persons giving delivery care
- Place of delivery
- Sterilisation procedures
- Birth injury
- Materials and medicines
- Infection
- Medical check-up

(*c*) Post-delivery care
- Baby attendants
- Prelacteal feeds
- Providing colostrum
- Personal attention and care
- Style of baby care
- Post-partum psychosis

(*d*) Demographic factors
- Birth order
- Sib-ship size
- Birth interval
- *Others (to be identified)

IX *Norms on child care and socialisation*

(*a*) Neglect and abuse
- Abuse
- Discrimination
- Scolding
- Indifference
- Torturing
- Unintentional neglect

(*b*) Love and affection
- Attending to demands
- Spending time
- Parent-child kinship bond
- Fondling
- Cleaning and dressing

(*c*) Breast-feeding
- Personal hygiene
- Initiation of breast-feeding
- Quantity
- Quality
- Duration/Timeliness
- Maternal health
- Maternal environment: (*i*) Physical (*ii*) Mental posture while breast-feeding

(*d*) Supplementary and regular feeding
- Ignorance
- Preparation of food materials
- Ingredients of food (nutrition)
- Style and regularity of feeding
- Absorption and metabolism
- Hygiene of bottle and vessels
- Weaning practices

Use of proprietary baby food
*Others (to be identified)

X *Morbidity patterns and genetic factors*

(*a*) Communicable diseases
Respiratory diseases
Gastrointestinal diseases
(*b*) Parasitic infestations
(*c*) Tetanus
(*d*) Other diseases: (*i*) Aids (*ii*) Cardiovascular (*iii*) Neoplasms (including cancer) (*iv*) Sexually transmitted diseases
(*e*) Deficiency diseases
(*f*) Congenital malformations
(*g*) Immunity level
(*h*) Inbreeding
* Others (to be identified)

XI *Calamity, accidents and war*

(*a*) Natural calamity
Cyclone
Flood
Drought
Earthquake
Volcano
Pestilence
Famine
(*b*) Conflict
Accidents
Injury
Fall
(*c*) War
*Others (to be identified)

XII *Interventions*

(*a*) Strategies for health care delivery
Costs of medical service
Availability of medicines
Getting medicines to needy persons
Efforts of voluntary agencies
IEC Programmes on: Health; Nutrition; Population control; Community participation
MCH programme
Applied nutrition programme
Domiciliary and rural oriented care
(*b*) Service facilities
Clinic facilities
Medical kit for providers
Commitment and efficiency of providers
(*c*) Preventive care
Disposal of excreta and sewage
Immunisations
Control of communicable diseases
Vector control
Improvement of sanitation
Latrine promotion and use
Protected and steady water supply
Safety materials
(*d*) Promotion schemes
Treatment of diseases

Facilities for: (*i*) Oral rehydration (*ii*) Intravenous therapy (*iii*) Vitamin therapy
Periodic medical check up

(*e*) *Other service programmes (country specific schemes)

IMMINENT VARIABLES**

However, certain variables in this entire list of life-affecting variables require special mention in view of their greater and immediate influence on mortality. To clearly demarcate such variables an additional term, namely, 'imminent variable' is chosen to subclassify certain LAVs in order to highlight their prominence for urgent research and prompt attention in programme interventions. The dictionary meaning of the term imminent is 'ready to take place and near at hand'. We require the same meaning, for example, to explain terms like infanticide, injury, accident, nutritional deficiency, morbidity, breast-feeding, etc., which may cause mortality at any time and which lie very close to mortality as causal factors. Hence, this term is more suitable than any other term hitherto in use and one that could be considered to explain such variables.

The major categories of 'imminent variables' are listed here.

Mass production and distribution of life saving drugs
Provider-population ratio
Types of medical care
Environmental sanitation
Vector
Pollution
Food poisoning
Protected water supply
Crowding
Food choice, processing and availability
Nutritional status of the mother and baby
Physiological weakness
Family conflict/harmony
Sex and disease
Marriage cost
Marital problems
Maternal health and weight
Maternal education
Hygiene of persons and vessels
Use of soap and disinfectant
Age at conception
Pregnancy care
Delivery care (medical)
Birth interval
Prematurity and birth weight
Prelacteal feeds
Breast-feeding and weaning practices
Deficiency diseases
Proprietary baby foods
Abuse of baby

** The term is coined by the author.

Infanticide
Infections
Health of infants and children
Fasting
Safety materials
Respiratory and gastrointestinal diseases
Tetanus
Congenital malformation
Cardiovasculiar diseases and neo-plasms
Immunity level
AIDS
Use of medical kit by providers
Immunisation
Control of communicable diseases
Oral rehydration
Intravenous and vitamin therapies
Supplementary nutritious foods
Medical check up
Health service utilisation
Spurious drugs
Natural calamity
Accident and injury
War
***Others (to be identified)**

The preceding list of 'imminent variables' is not exhaustive since these variables may differ in their prevalence and importance in different situations. These imminent variables need special mention in view of their importance in terms of the urgent and timely attention they demand in research and programme development. The list of imminent variables and the priority of variables may change in the context of different situations, advances in societal and technological development and new knowledge. Nevertheless, all imminent variables do not fall under one category of LAVs because of the diversity in their origin and logical sequences. It is also difficult to classify and keep all of them close to the dependent variable, that is, mortality, in the conceptual model in order to magnify their special importance and priority in view of their diversity. Therefore, by mentioning them separately and highlighting their importance, we are able to project their legitimate role without obscuring them in the lengthy list of the diverse determinants of life-affecting variables.

A new classification of mortality determinants using the terms 'life-affecting variables' (LAVs) and 'imminent variables' (IVs) has become necessary to improve communication in mortality literature and to avoid overlapping and confusion with variables that influence fertility behaviour. Therefore, in the present discussion the nomenclature of intermediate and proximate variables is not adopted, although they are certainly useful terms and merit consideration in fertility literature. Further, using the term 'mortality' while referring to all the variables that either directly or indirectly, partially or fully lead to morbidity or mortality does not seem to be appropriate. Many variables, in fact, do not lead to a baby's death but they induce or affect health and life in varying degrees. Certain variables in different contexts with differing influence either induce mortality or prevent it. For example, poor education and low income may cause mortality whereas good education and high income may prevent it. Therefore, it is more appropriate to call such variables 'life-affecting variables' rather than 'mortality variables'. Hence, the terms have been coined here to describe the determinants of mortality.

The processes of development/death in a sequential manner through different stages, institutions and developmental fields form the basis for formulating the present analytical framework and conceptual model. Hence, several variables that repeatedly affect life at different stages cannot be classified on the basis of different disciplines. The present analytical framework logically begins with polity and policy followed by natural environment and culture through interventions, family, marital status, parents, conception and pregnancy, perinatal stage, norms on child care, natural calamity, infections, morbidity patterns, health and life of infants and children and ultimately leads to mortality. At a certain stage the number and diverse nature of variables will be less and at another stage it will be more. For example, the role of parents forms a distinct major category under which the number and diversity of factors is very complex. Since certain factors are likely to influence repeatedly at different stages, they need to be repeatedly mentioned to show their influence at different stages. Besides these factors, several macro and micro variables with differing influence also find their place in this scheme of presentation. Several LAVs either similarly or in a dissimilar manner may influence mortality in developed and developing countries, in same and different cultural groups, and within and across different countries. Several of these factors may apparently appear as simple or complex, important or unimportant depending upon the experience of the researchers. But identifying and listing them together, more or less, according to definite criteria may facilitate systematic consideration of their use in future studies. The present list of variables also may reduce the chances of arbitrary and random selection of variables at least in future studies.

Conceptual Model of Life-Affecting Variables

The development of a conceptual model in the absence of specific and detailed knowledge of a number of determinants of mortality pattern is a difficult task. Unlike other vital events, mortality has a large number of diverse causal factors which have their roots in several disciplines. At least some knowledge of health, nutrition, society and culture, economy, epidemiology, etc., is needed to do justice in developing a conceptual model on mortality since all these factors either directly or indirectly, positively or negatively affect mortality. Besides, no suitable term is hitherto used to classify the diverse determinants of mortality into a coherent and systematic pattern.

Till now, the practice was to borrow terms like intermediate variable and proximate variable from fertility literature to explain the causal relationship of mortality. This practice has its shortcomings for several obvious reasons. First, it may lead to a confusion in the communication of causal relationship between several independent variables separately with mortality and fertility. Further, the type of intermediate variables eminently categorised by Davis and Blake are not the type of variables that affect mortality. Most of the conceptual models on mortality, that have been developed in the eighties, have focused largely on the variables that belonged to the respective profession of the authors themselves and, therefore, have a narrow focus. Under such circumstances we have made this modest effort to bridge the gap in theoretical understanding about the causal relationship of the diverse and numerous independent variables in mortality patterns. However, certain lapses may occur which we may amend when a revision of this model is done in the future.

Conceptual Model

This conceptual model of mortality is largely based on the list of several life-affecting variables. The first level of these variables constitute the structural and macro variables such as polity, policy, culture and ecology. There are several specific variables under each of these macro variables. The second level of variables constitute programme interventions sponsored by polity and policy based on culture and ecology; parallel to these lie natural calamity, accidents and war. The first level of macro variables and natural calamity can directly influence child mortality. But some of the specific variables such as socio-economic development, may promote programme interventions, which operate through another row of variables, namely, family through norms on child care practices and these factors, in turn, may lead to morbidity and mortality. However, effective developmental programmes per se may reduce the mortality level of children. Apart from polity and policy, cultural motives of infanticide, interaction with other organisms (for example, a poisonous snake-bite in certain eco-system) may directly lead to child mortality. Similar is the case with respect to natural calamity or accident. Poor interventions based on policy and programme may also affect one or more of the LAVs, namely, family through norms on child care variables and lead to accidents, nutritional deficiency, weakness, infection and morbidity which, in turn, may lead to child mortality.

Unlike in other conceptual models, a list of six groups of variables starting from familial variables to norms on child care have several specific factors. They are listed in detail in the analytical framework. The knowledge of the causal relationship of these sequential variables and how they can be manipulated is necessary for initiating interventions by polity and policy. The causal relationship of all these variables and their individual influences are not fully known. Further, to manipulate these variables often the preceding package of interventions are also not adequate and effective in several countries. A deeper understanding of the causal relationship of various life-affecting variables is needed to make the intervention programmes more effective. However, studies have shown some understanding about the nature of relationship of a few of the life-affecting variables on mortality. Space and time do not permit one to review all the existing literature in this field. Nevertheless, we have presented the highlights of a few variables here.

FIGURE 1 A Conceptual Model on Mortality: Life Affecting Variables and Certain Imminent Variables

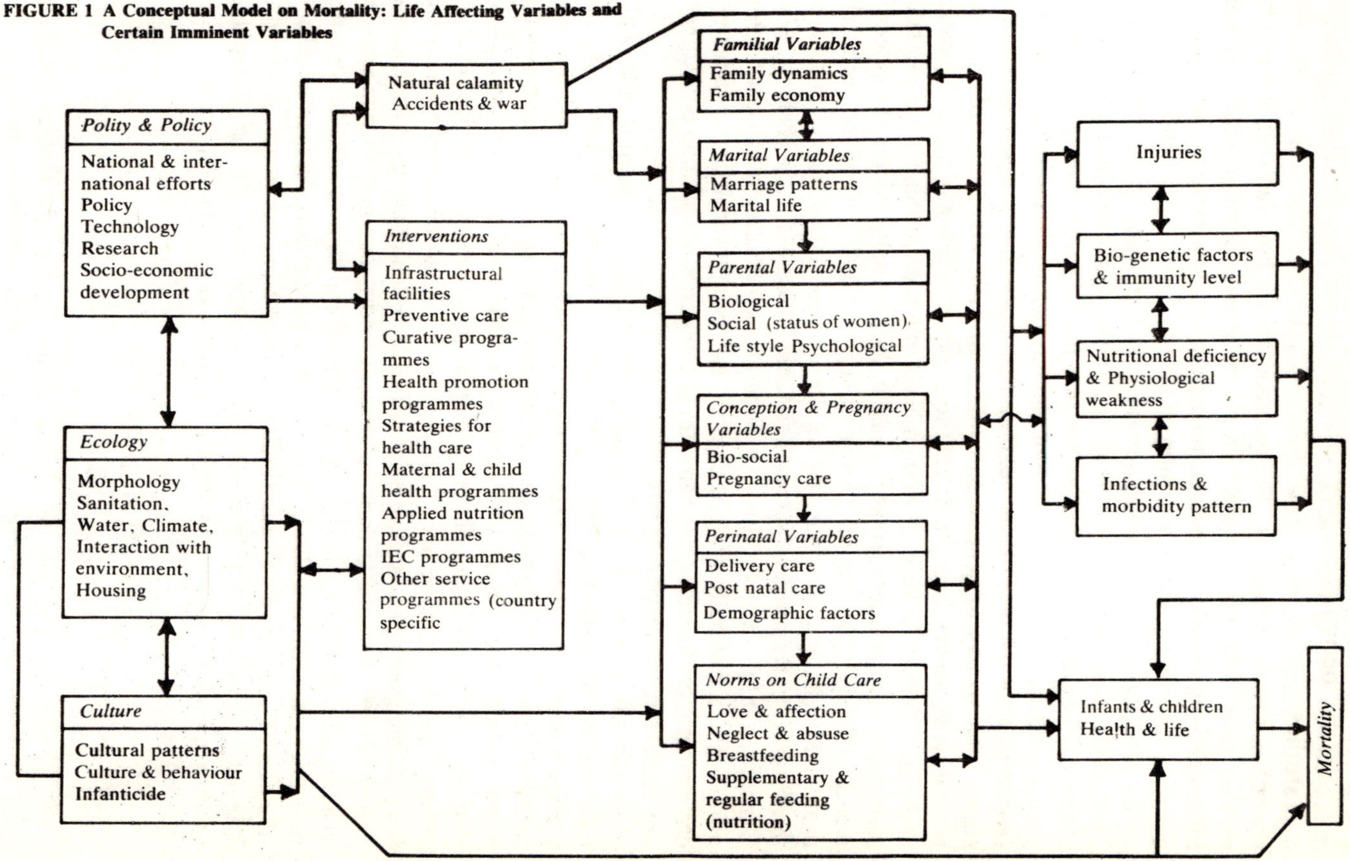

The present arrangement of variables and the direction of relationship is based on several existing studies. It is intriguing to note that the same factor, namely, occupation, may show different types of relationship in different studies in the same country. The question arises: how can such a factor be plotted on a conceptual model? Our experience, common sense and logic influenced our decision to indicate the nature of relationship of such variables (*a*) with other independent variables and (*b*) also with the dependent variable either directly or through other variables. Future research may throw some light on the pattern of such uncrystallised knowledge of causal relationship.

Several variables belonging to the first level factors, namely, polity and policy, culture and ecology cannot directly influence mortality. They are responsible for the type of interventions that may induce changes in the health status of a society in general and which, in turn, may differentially affect mortality. However, exceptions are there: rapid diffusion of higher education and particularly education of women leads to a rapid decline in the mortality level even without a concurrent increase in the per capita income as was seen in Kerala (Nair, 1980) and Sri Lanka (Meegama, 1981). Similarly, culturally motivated behaviour is responsible for the extensive infanticide noticed among the Rajputs in Rajasthan, Gujarat and Uttar Pradesh in India (Pakrasi, 1970; Barbara, 1981). Though culture, in general, cannot directly cause mortality, infanticide, a culturally evolved determinant, is a powerful causal variable of child mortality. Therefore, the first three categories of variables both directly and indirectly influence mortality.

Another feature of this conceptual model is that life is portrayed here as a continuous and sequential process and several systematic events of the life-cycle are organised here on the basis of their already known causal relationships. Life from the family through several other stages may grow or die depending upon the influence of several life-affecting variables and the interrelated factors that act at different stages in the development of human beings. However, no attempt has been made here to review in detail all the causal relationships because the existing knowledge is not sufficient for undertaking such an exercise. Nevertheless, we believe that the present effort may help to minimise the existing confusion and inadequate coverage of appropriate variables in future mortality research. It is also not expected that all the variables mentioned here need to be included in all the studies undertaken in the future. A group of relevant and related variables can be considered from the list of variables mentioned here and their individual influence on mortality patterns can be examined.

Polity and Policy

Polity and policy constitute a set of definite number of variables which find their presence and differential influence all over the world on mortality. It is difficult to imagine any civilised population, with the exception of certain tribal groups, without a political organisation and a policy of its own. No doubt, political philosophy and its complexity, and policy differences may exist in different societies. Similarly different political systems and policies produce both qualitative and quantitative changes in the level of mortality. China and Cuba adhering to a particular political philosophy, and Korea, Taiwan, Sri Lanka, Singapore and Kerala (India) with different political systems have showed a remarkable decline in their mortality levels. Of course, they pursued diverse and effective developmental and service policies which contributed to the rapid decline in their child mortality levels. Policies are, of course, formulated by the government. All other determinants have certain linkages with polity and policy which naturally vary from country to country and from region to region. These variables appear to be non-priority items but they are extremely important as already seen. Policy leads to developmental programmes and to the adoption of new technology which is a prerogative of the government. They can either be promoted with international assistance or indigenously evolved through research. For example, certain social development policies and programmes implemented in Kerala (India) and in the neighbouring country of Sri Lanka are responsible for the rapid decline in mortality in these regions, though Kerala has the lowest per capita income among all other states in India (Mahadevan, 1983). Such schemes include the Applied Nutrition Programme implemented through a chain of pre-school education-cum-recreation centres by several agencies, extensive use of mass media and rapid spread of higher education, and education of women in particular. It must be noted that the influence of education acts indirectly through the strengthening or weakening of several other variables that follow.

Socio-economic development is important for mortality control. Rapid development may reduce mortality, but sluggish growth may not reduce mortality and may even increase it. Socio-economic development may increase the per capita income, education and the opportunities for better occupation. All these are interrelated. Increased income, for example, may improve purchasing power leading to qualitative and quantitative improvements in child care, better housing, clothing, education, sanitation, personal hygiene, reduction in risk of infection, timely and appropriate adoption of modern medical services. Income, occupation and education are, therefore, composite and interrelated factors which directly and indirectly not only influence the above mentioned factors but also affect the nutritional status, age at marriage, birth interval, birth weight, foetal wastages, sib-ship size, accidents and so on.

The effects of income in changing and influencing human behaviour, particularly mortality and fertility, has been succinctly spelt out by economists Mueller and Kathleen (1983: 483), who wrote:

> . . . At the macro level, rising income is associated with the appearance of new consumption and educational opportunities. In addition, rising income promotes contacts with a larger world, make new ideas and new opportunities more widely known and raise people's sights. Assuming that consumer preferences are interdependent among families, the acquisition of watches,

bicycles, more effective health care, or greater child quality by higher-income and urban families creates a demonstration effect, inducing a felt need for these goods and services among lower and middle income families.

Existing literature has confirmed the importance and influence of most of the socio-economic variables and their influence on mortality (UN, 1984; Yang and Brian, 1983; Behm, 1980; Farah, 1981; Mata, 1983; Preston, 1975; Mosley, 1980; Omran and Standley, 1976; Brooks, 1973; Stockwell, 1966; Palloni and Wyrick, 1981). A UN report stated:

> In both developed and developing countries, the lowest level of income, shorter period of education and manual labour, occupation tend to be associated with the higher risk of death and ill health. The under privileged, trapped in poverty are disadvantaged at different stages of the disease process. They are generally more susceptible and less resistant to disease, in part due to less access to immunization services, in part due to malnutrition including protein-energy insuffiencies and imbalances in diet and in part due to deficient housing facilities that do not provide enough protection from heat, cold, wind and rain . . . injury (UN, 1984: 21).

In an analysis of socio-economic development and its influence on mortality in 124 less developed countries Yang and Brian (1983) highlighted the influence of variables like education, health services, living standards and nutritional diet. However, economic development per se did not show any direct relationship with mortality but an indirect relationship through living standards was observed.

Ecological Variables

Human beings in different environments and in relation to other organisms make many changes in the mortality patterns. The habitat of human beings, rural-urban background, slum and non-slum background, water stagnation or water scarcity, environmental sanitation, housing material, ventilation, crowding, domestication of animals, climate, and pollution are some of the ecological variables that influence mortality patterns in different ways and to varying degrees. Some of them often influence and some exert lesser influence while others exert greater influence on mortality. However, at present, we do not have enough information to precisely determine the magnitude of influence of all these factors. Universally, it has been confirmed that mortality is higher in the slum areas as compared to the non-slum and rural areas as against urban. A rare study carried out in Kerala showed that mortality is highest in hilly regions, much less in mid-lands and least in the coastal areas (Nair, 1980: 7). Though certain speculations and views have been expressed to explain the influence of these ecological variations on mortality, more specific knowledge is needed to understand their overall influence in different contexts and regions.

It is well known that building materials used for construction of houses, type of housing, and overcrowding in the house, etc., cause mortality differently (Farah, 1981; Chowdhury, 1982; Adlakha, 1970). Perennial water stagnation in low lying areas, particularly in the developing countries, acts as a breeding ground for several vectors that produce mild or severe morbidity patterns leading to mortality (Saunders and Warford, 1976). Malaria, parasitic and diarrhoeal diseases are common in such ecological niche, while similar situation in a developed country would not produce fatal morbidity conditions; on the contrary Bangladesh has become an endemic area of diarrhoea and dysentery (Chowdhury, 1982).

Housing Conditions Housing conditions directly affect health; they normally do so by leading to the incidence of infectious diseases. The more a household is congested and crowded into a small or few rooms the greater the opportunity for infection to spread by droplet or direct contact from one to the other.

> This increase in the facility of transmission of infection is even greater when bedrooms have to be shared by members of the same family or where rooms are used both for living and sleeping; it is also greater, especially for diarrhoeal diseases, where more than one family are compelled to share the same sanitary arrangements (water closets or washing up facilities). Infectious diseases particularly tuberculosis will be the major factor that will be affected with poor housing conditions (Benjamin, 1965: 42).

The poor conditions of houses with high room density, bad lighting and ventilation, absence of clean and adequate water supply and improper waste disposal together were reported as major causes of mortality in nineteenth century Europe. In Kenya (Anker and Knowles, 1977), Ohio (Adamchak, 1979), Bangladesh (Chowdhury, 1982) and India (Mahadevan, 1983) a significant association was observed between housing conditions and the incidence of mortality.

Woodbury's (1922) study of a selected area of the United States found that in households where there was an average of less than one person per room, the infant mortality rate was 52 per 1,000 live births as compared to 94.9 in households with one but less than two persons per room and 135.7 in households with two or more persons per room. Housing conditions are largely influenced by the socio-economic status of the people and the policy of the government.

Water-cum-environment has a definite influence on mortality and this has been confirmed in several studies (Saunders and Warford, 1976; Mata, 1983; Jarrett, 1970). The Pan American study revealed a definite association between water-cum-environmental factors and childhood mortality. It observed '(*a*) a strong negative correlation between childhood deaths and the use of piped water; (*b*) a strong negative correlation between childhood deaths and the use of flush toilets; (*c*) a quantifiable relationship between access to piped water and flush toilets and childhood mortality were observed' (Jarrett, 1970). Another comprehensive analysis of the influence of water and sanitation is given in a World Bank aided project by Saunders and Warford (1976). They classified water-borne diseases into five categories: (*a*) water-borne diseases (cholera, typhoid, amoebiasis, etc.); (*b*) water-washed diseases (scabies, leprosy, conjunctivitis, bacillary dysentery, etc.); (*c*) water-based diseases (guinea-worm, urinary schistosomiasis, etc.); (*d*) water-related vectors (yellow fever, malaria, etc.); and (*e*) faecal disposal diseases (hook-worm and other worms). They confirmed that differences in water quantity and quality are associated with differences in the rates of these five categories of diseases. However, they also found that local climate, geography, culture and sanitation habits are related to water-cum-environmental sanitation.

Climate The influence of climate can only be separated with difficulty from that of other environmental conditions. However, climate has a marked influence on the prevalence of a particular disease, for example, in malaria or yellow fever a high rise in body temperature is important for effective multiplication of vectors and subsequent infection. Diphtheria and scarlet fever prevail little in the tropical countries. Rainy and humid climate is conducive to the causation of diseases such as measles, pneumonia, and other viral diseases. In Sudan, climate-wise monthly distribution of infant and early childhood mortality reached its peak in July and August, that is, during the middle of summer and coincided with the wettest months. Malaria and diarrhoea are highest during this period. Mortality was at its lowest level during the winter months of January and February (WHO and Sudan Ministry, 1981). Climatic changes also showed certain variations in the incidence of mortality in the past in German villages (Knodel, 1982). But more definite knowledge is needed to confirm its specific nature of influence in several other countries.

Cultural Variables

Besides this unit of cultural variables, certain other related variables discussed under family and parents are cultural in their characteristics. Cultural variables vary in accordance with the nature of culture and therefore, their influence on mortality also may be different. However, value of children, infanticide/homicide, cultural preferences in caring persons including provision of food, seeking medical help, attending to demands, showing affection, abusing or neglecting children, fasting during pregnancy and lactation periods and life style variables are relatively more important in the light of existing knowledge which, of course, may change on the basis of emerging new knowledge in course of time.

Certain ethnic groups, wherever they may be living, manifest similar demographic characteristics. The Japanese not only in Japan, but also in Hawaii or in the USA have an extremely low mortality level and a very high longevity (Shigamatsu, 1983). Why? In the case of Japanese such intervening factors of social change either do not influence them or exert a more or less equal influence on their culture, irrespective of their habitation. Like the Japanese, the Chinese also manifest more or less similar mortality patterns, irrespective of their place of residence. A strong family structure with two to three generations of relatives living together and manifesting strong affection and rendering security are reported to be the underlying features of Japanese and Chinese cultures that minimise mortality among them. Since cultural variables were seldom studied in relation to mortality our understanding about the total influence of these variables is inadequate.

Infanticide Infanticide according to ethnographic evidence, has been prevalent since time immemorial. It was extensively practised in all the continents—Asia, Africa, America, Australia and Oceania (Freedman, 1971; Balikoi, 1967; Divale, 1970; Granzberg, 1973; Barbara, 1981). The importance of infanticide as a method of population control was stressed by Malthus (1798) and Carr Saunders (1955). But this practice was extensively observed chiefly among the lower strata of society. Infanticide was widely reported among the Polynesians in the Fiji Archipelago, among the Australian tribes, New Zealand (Westermarck, 1921); among the people of China (Tuan, 1983; Baker, 1979); among the tribal and non-tribal population all over India, particularly the Nagas of Assam (Risley, 1908), the Todas of Nilgiris (Rivers, 1906), the Rajputs in northern India (Barbara, 1981) and the Gounders of Coimbatore (Mahadevan, 1979).

There are several views about the intentions of practising infanticide. According to Carr Saunders (1955) infanticide was practised to eliminate the surplus and unwanted mouths. For different cultural groups different motives were attributed for its prevalence. These are: reduction in population pressure, economic stress, social status, to free the mothers from the excess burden of duties during adverse conditions, to save a few people from natural calamities, as an offering to the gods and goddesses, to achieve certain goals in life, for strengthening construction work, to save the people from illegitimate conceptions, and to avoid marriage expenses.

Infanticide in China was reported in ancient times and even in the contemporary period. A Chinese scholar reported the prevalence of infanticide in the past:

> I have learned of the prevalence of female infanticide in all parts of Shansi, but particularly in such Southern countries as P'ing-ting, Yu-tz'u, etc. The first female birth may sometimes be salvaged with effort, but the subsequent births are usually drowned. There are even those who drown every female baby without keeping any . . . . The poor regarded the practice as an almost legitimate means of maintaining their minimal standard of living and, in any case, as a dire economic necessity (Ho, 1959–60).

Drowning was the most common form of infanticide, but there were other ways, such as smothering, burying alive, abandoning on business, inviting the charity of the inmates and sending to orphanages. Grown up girls were sent as servants to well-to-do families once and for all or till they are married to somebody by the employer (Baker, 1979: 67).

Before the communist takeover of China in 1949, infanticide was widespread in several regions of China. It was practised primarily to reduce the economic burden on the families. Consequent to the communist revolution, land reforms, public distribution of cereals and several other reforms made life less expensive and thereby the burden of girls was eased resulting in the decline of infanticide. But according to Tuan (1983), the new concept of the one-child family advocated by the Chinese Government from the beginning of the eighties once again led to an increase in infanticide in China. The original slogan for the one-child family was 'one is the best, two squares the account, three is a mistake'. In 1970 the slogan was further changed, 'one is the best, two is the limit'. It is now being effectively implemented throughout China. After the birth of the first child of every woman, the barefoot doctors in each commune maintain the menstrual history of women of childbearing ages to advise them on the different methods of contraception and abortion. When the first baby happened to be a girl, the parents killed her hoping that the second baby would be a son. Under the one-child family system advocated in China there is a provision to go in for a second baby if the first one is deceased/dead. When this problem of infanticide assumed dangerous proportions the *People's Daily* in its editorial of 9 April 1983 warned and advised the parents not to continue this homicide practice. The *Daily* also cited the existing adverse sex ratio of girls to support this argument.

South India presents an interesting case of infanticide. The Gounders in Tamil Nadu literally kill a newly born girl by feeding the baby honey with husk which produces internal bleeding and leads to a slow and unnoticeable death; this is usually done when parents have more than one daughter. They also kill infant boys if they are unexpected and unwanted births. They do this to avoid dowry and marriage expenses, and also to prevent the division of landed property (Mahadevan, 1979). On the other hand in Andhra Pradesh weak and handicapped children are neglected because it is believed that such children will not live long. Such children are also given unpleasant and uncommon names because the belief is that in any case the child will not survive. This practice is observed even today. The continuous neglect of these babies accelerates the pace of their death (Mahadevan *et al.*, 1981).

Divale and Harris (1976) while discussing the sex ratios of over 300 societies, mentioned the magnitude of wide prevalence of infanticide in several societies. In USA the prevalence of homicide has doubled during the thirteen year period ending in 1973 (Keyfitz, 1977). The study on infanticide is a difficult and delicate issue; therefore, the extent of its influence on infant and childhood mortality cannot be fully understood now. In-depth studies in future may throw more light on this.

The demographic importance of infanticide, particularly its role in mortality has been identified in a few studies on the South Asian population (D'Souza and Chen, 1983; Gordon *et al.*, 1965; Barbara, 1981). The adverse female sex ratio prevailing in South Asian region can be, at least partially, accounted for by the incidence of infanticide. Barbara's (1981) study conclusively confirmed its extensive prevalence among the Rajputs in the northern parts of India. The same finding has been confirmed in another analysis of the data from Uttar Pradesh by Bhattacharya *et al.* (1980) who have shown that a significantly higher proportion of girls die for the later parities of high fertility couples. This can be due possibly to the infant mortality of girls. Evidences from South Asia and China support the importance of infanticide as a major determinant of infant and childhood mortality in different degrees over the years.

Other macro cultural variables like religion, ethnicity, and caste have also shown their specific nature of influence on infant and childhood mortality. In USA Negroes, in general, have a significantly higher mortality level both in rural and urban areas. In Sri Lanka Tamils of Indian origin have higher mortality as compared to the local Sinhalese (Fernando, 1981). In the multi-racial and cultural population of Singapore, between 1963–1978, higher perinatal mortality rates were observed among the Indians and Malays as compared to the Chinese (Boon, 1979). In South-Central India, the Muslims and the Harijans have higher infant and childhood mortality as compared to the caste Hindus (Mahadevan *et al.*, 1981). However, in Lucknow region of North India, religion-wise infant mortality did not show much variation though it was one-third in the upper castes as against the lower castes (Srivastava and Saksena, 1981).

Culture may also manifest its influence on mortality through fatalism, belief and value of son(s)/children and several other culturally determined variables of human behaviour subsumed under life style. Of course, these variables are again influenced by educational status and socio-economic change in society in general. However, we are yet to know precisely the nature of influence of several of these variables.

Value of son(s) Ethnographic literature all over the world, by and large, reveals that males get preferential treatment, from infancy to adulthood, in patrilineal societies and females get relatively better treatment in matrilineal societies.

As most of the societies all over the world are patrilineal, an overwhelming percentage of males in general are shown special consideration within the family. Even in a matrilineal group, the importance given to sons is no less. Several factors contribute to the value attached to son(s) in society and these have been exhaustively documented by several recent studies (Arnold *et al.*, 1975; Nag *et al.*, 1978, Mahadevan, 1979 and 1984). As a result of several values attached to sons because of their benevolent roles both within the family and in society as against the several costs involved in bringing up daughters, the latter are not valued highly in society. People who attach importance to social status, and belong to less affluent sections are largely against daughters because of a fear of financial crisis, difficulty and loss of status. People who cannot afford to bring up daughters without sacrificing their cherished desires, resort to annihilating girls right from birth mostly using indirect methods and even directly in certain communities.

The Chinese kinship system heavily stressed the importance of the male and of relationships traced through the male. Women were theoretically of little importance. They were necessary for the reproduction of the species, and in most cases for labour in the home, but in both cases they were considered to be there to serve the male and the male principle. In classical thought, women were perceived as inferiors throughout their lives, subject first of all to the men of the family into which they were born, then on marriage to the men of their husband's family, and finally on widowhood to their sons. These were known as the three 'obeyings' of women, and they seem to have first appeared in the *Yi-li*, a Confucian classic some 2,000 years ago. Even in death the fate of the woman's soul was dependent upon its worship by her descendants through males (Baker, 1979). In India according to Manu, a Hindu philosopher, a woman is a slave throughout her life, first to her parents, then to her husband and finally to her children. What we have observed in the case of China is more or less true in India as well (Kane, 1973). In the Punjab area of North India, mortality figures in the first five years of life indicated that the death rate for females was almost 150 per cent more than that for males (Wyon and Gordon, 1971).

The birth of a son was accorded extreme importance in a family not only because the son would provide for the parents in their old age, but also in connection with ancestor worship. The common expression: 'May you have a hundred sons and a thousand grandsons' may have exaggerated the issue but it faithfully represented Chinese feelings about the importance of male descent (Baker, 1979: 3–4). Data from Bangladesh also confirm the greater value attached to sons (D'Souza and Chen, 1983).

Cost of Rearing and Dowry Cost of rearing girls in many societies, particularly in several Asian countries, is considered to be an unproductive investment and also beyond the means and status of many parents. It is further aggravated by the ever increasing practice of giving dowry in developing countries. When the amount of dowry increases/and cannot be raised by the parents and other economic difficulties or needs coexist, dowry may act as a major cause for the lesser value attached to girls, and may lead to neglect and even homicide. Though information on dowry, bride price, bride wealth, etc., is available in anthropological literature, most of the authors have not examined its consequences, particularly mortality. Status consciousness and high aspirations prompt the parents to marry their daughters to individuals who are better placed educationally and professionally, which in turn leads to more expenses for the celebration of marriage and payment of dowry. In northern parts of India, particularly among the Rajputs, marriage costs have increased the incidence of female mortality. Barbara (1981: 156–59) has confirmed the hypotheses that there is a clear correspondence between marriage costs and girls with juvenile sex ratios, when marriage costs are high sex ratios are also high thereby indicating high female mortality where costs of marrying daughters are most prohibitive.

The aim of this paper is to systematically classify and visualise the direct and indirect determinants of mortality of infants and children. A theoretical development of this nature cannot be complete in the absence of adequate information about several determinants of mortality and their interrelationships. It is a continuing process. Most of the variables and the direction of causal relationship of these variables as shown in the analytical framework and conceptual model offer a number of plausible hypotheses for testing. It may not be possible to study the entire list of variables and the model in general in a given population in view of the diversity and limited number of determinants existing in a particular group. Globally, of course, most of the variables may be relevant in one country or the other or in one culture or the other. Similarly, a single researcher may not be able to test all the determinants of mortality which, in fact, require the combined efforts of an interdisciplinary team.

Familial Variables

Family is a composite variable which has several social and economic dimensions. The type of family stemming from the kinship network as nuclear, joint, and extended families exert different influences on mortality. The number of members in a family and the role of several members of the kinship group may provide different opportunities and influences on child rearing leading to a beneficial or an inimical effect on the baby. One of the secrets of very low mortality in Japan, China and Korea is largely due to the beneficial influence of the kinship network of their extended family. Even in Tokyo city 50 per cent of the households still have extended families (Shigamatsu, 1983). A similar influence of kinship may be operating in several other communities but empirical knowledge is very scant today to evaluate the extent of its influence. Similarly primogeniture, ultimogeniture, differential value attached to sons or daughters, harmony or disharmony in the family, economic conditions in a family, availability of financial and other resources, mode of distribution of food in the family may have different forms of relationship with mortality either directly or indirectly. There is evidence to show that the pattern of sharing food is based on culturally conditioned preferences for the first child or the last child or for the son or the daughter and so on (Baker, 1979). Very often the daughter is neglected even in the allocation of food in several communities. The processing of food itself affects the nutritional ingredients which may affect health in the long run. Thus the overall socialisation that goes within the familial environment may have different patterns of influence on the child (Nam and Harrington, 1983). The family's decision plays a very great role even in the case of a culturally motivated action like infanticide. Similarly, the family makes decisions regarding medication, food habits and even selection of a doctor for treatment.

Japanese familial structure is favourable for extending better child care and minimising infant mortality. Japan is noted for its stem family where parents live with their (eldest) married children. Very often marriage was a contract between families and not a love match of individual choice. All other children, except the eldest, move to new and separate households after marriage leaving the ancestral house to the eldest sibling. This 'stem type' of family structure, according to Linda and Culter (1983), was instrumental in reducing mortality of all types. The author further states that the rapid increase in the survival probabilities of Japanese at all age levels has significant implications for the living arrangements of families. It is argued that the three generational set up of families as seen in Japan is a factor leading to increased survival chances of babies and adults. The setting continues even today inspite of the rapid industrial and economic revolution that Japan witnessed during recent times. Similar research is required to study the benefits or adverse effects of the family in other countries and cultures.

Marital Variables

Age at marriage universally varies. It has tremendous implications for infant mortality, particularly the mortality of the first and even the second order of birth according to several studies (Omran and Standley, 1976; Islam *et al.*, 1982; Pethybride *et al.*, 1974; Simmons *et al.*, 1979; Mahadevan *et al.*, 1981). The physical and emotional state of the mind and the variegated experiences that accumulate with age may be responsible for the different nature of influence of the age at marriage on mortality. Besides the age at marriage , monogamy, polygamy, inbreeding, stability of marriage, marriage costs, several problems of married life either directly or indirectly and in different degrees contribute to mortality. The cultural influence of infanticide is linked with the problem of cost and dowry involved in marriage.

The National Academy of Sciences (1970) reported that mothers under 17 years of age show higher infant mortality rates, while the risk for mothers between 17 to 19 years is similar to that for mothers between 20 to 24 years of age. In Bangladesh, the highest neonatal mortality rate was found among mothers below the age of 20 years. Neonatal mortality rate gradually decreased among mothers between 20 to 24 and 25 to 29 years of age (Islam *et al.*, 1982). The changing philosophy of marriage whether to go in for a baby or to avoid having a baby exerts an influence on rearing or destroying babies, particularly when a child is conceived by accident and is born against the desire and decision of the parents. A new norm of a 'no baby life' has been seen in western society; parents perceive that children affect their comfort and increase their social and economic problems. Such a philosophy in married life may increase the incidence of abortion and mortality. Marital problems and dissolution of marriage increases child abuse and this is further discussed under the norms about child care patterns. In the USA and in several European countries divorce is on the increase on account of several factors. It remains to be assessed to what extent marital problems and divorce lead to child abuse, injury and death of children, when marital problems or the dissolution of marriage cause sufferings to children. It may even lead to mortality not instantly but steadily through the influence of several other social and economic factors that emerge with the after-effect of divorce. For example, a husband may throw an object at his wife during the course of a fight which may accidentally hit the child leading to injury and even death of the child. This is common among several problem families in the USA and also elsewhere.

Marital dissolution, an outcome of divorce, affects the mortality level. The divorce rate in Japan was as low as 0.75 per 1,000 population during the early sixties (Kawashima and Steiner, 1961), it increased to 1.16 in 1972 because of changes in the age structure (UN, 1979). A low incidence of divorce adds to the stability and happiness of the family leading to the absence or minimal frequency of neglect of babies. On the other hand, an increase in the divorce rate in the USA and in the European countries leads to the abuse of children and ultimately leads to morbidity and mortality. The rate of divorce per 1,000 population in 1978 was only 1.16 in Japan in contrast to 5.17 in the USA, 2.93 in England and Wales and 1.39 in France (UN, 1979). The harmful effects of divorce on mortality in several other countries remains to be seen.

Parental Variables

Parental factors constitute a heterogenous cluster of variables beginning from biological, through social, cultural (life style) and psychological dimensions. It begins with the biological state of age, changes in the age at conception; conception at younger and older ages affect morbidity differentially. The age at first conception depends on the age at marriage and the socio-economic status and modernisation of the parents. Even the height or weight of the mother, which has a strong influence on mortality has its roots in the existing socio-economic status of the family. The biological state of the parent influences the mechanisms of absorption, metabolism, immunity level, certain genetic factors, and ultimately health itself. Universally, the mortality rate of infants and children born to mothers below 19 years of age and above 30 years is higher. Mortality among mothers below 19 years of age is caused by several social factors as well as poor physiological growth of the mother but mortality during the later stages of the reproductive period is perhaps due to the deterioration of the mother's health and other factors that are yet to be identified (Adlakha, 1970; Wray, 1971; Islam, 1978; Wyon and Gordon, 1971; Omran and Standley, 1976; Mahadevan, 1983).

Among the social dimensions, the status of the mother stemming from education, occupation and cultural influences of permitting a wife to emancipate herself or not in life and the consequences of the status of women affecting personal and sexual hygiene, health service utilisation are very important determinants of mortality. The status of women manifest different life styles and their changes in society. While many aspects of the life style particularly alcoholism, smoking and drug addiction have an adverse influence on mortality (UN, 1984: 27), certain other aspects of life style, namely, media exposure, use of soap and detergents and personal hygiene have a beneficial influence on the survival chances of babies. It differently affects mortality (Adlakha, 1970). An analysis of the leading causes of death among the US population over a period of one year (1975) revealed the average proportional allocation of the various factors leading to mortality: life style, 48.5 per cent; heredity or biological factors, 26 per cent; environmental factors, 16 per cent; and the medical system, 11 per cent (Nightingale cited in UN, 1984).

Psychological problems faced by parents, particularly emotional states of mind such as happiness or unhappiness, namely, adjustability, conflict, tension, frustration, suicide, develop among parents because of familial and non-familial problems. These psychic disturbances may cause accidents, injury and even homicide of infants. No doubt they may at the first instance lead to child abuse. Since the evidence of the influence of these factors on mortality is not perceptible because of their slow and indirect nature of influence they are not considered as hitherto important causes of mortality. Today, accidents constitute the single largest cause of child mortality in the USA (Wineberger, 1982 and 1983). The root cause of these accidents is not only motor accidents but also parental problems.

Maternal Nutrition The nutritional status of the mother which is an outcome of the food habits, social and economic status of the mother and family has a profound impact on mortality. The nutritional status may also be influenced by preferences regarding sharing of food among members of the family. The quality and quantity of food, and the processing mechanism may also influence the nutritional status. Chronic malnutrition stunts a woman's own growth, which may both increase the frequency of birth injuries and reduce the over all efficiency of the reproductive process. Acute maternal malnutrition may increase the risk of death of both mother and foetus because lesser nutrients are available for the growth of the foetus and for the maintenance of the health of the mother. It is well known that poor nutritional status plays a major role in increasing both susceptibility to and the severity of infection (Latham, 1968).

Lechtig *et al.* (1977) reported that both long-term and short-term nutritional status of the mother are independently related to infant mortality. Nutritional deficiencies in women may begin right before marriage and vary in severity at different periods of life depending upon the fluctuations of fortunes, inflation, drought, good crop and other familial factors. Therefore, studying the nutritional status of the mother at one point of time while ignoring the past history of nutritional status of the mother may yield data which is inadequate to assess the total influence of nutrition on mortality of infants and children in general.

Empirical studies have confirmed the influence of maternal nutrition on infant mortality in several countries (Antonov, 1947; Smith, 1947; Latham 1968; Habicht *et al.*, 1976; Winikoff, 1978; Wray, 1977). Maternal height has been associated with higher proportions of low weight at birth of infants and higher neonatal mortality (Baird, 1977; Mahadevan *et al.*, 1981). Maternal weight also affects neonatal mortality because underweight mothers and their foetus may be exposed to greater risk because of nutritional deficiency or due to other physiological or morbidity conditions. It may affect labour and birth injury as well. Very often the nutritional status of infants and children is studied and emphasis is given to supplementary nutritious feeding for children but not for expectant mothers. The root cause of nutritional deficiency of infants and infant mortality in several contexts is due to the nutritional deficiency of the mother both before and during pregnancy which, in turn, can be traced to other social, cultural and environmental factors.

Status of Women The status of women is a major factor that grows out of several other socio-economic and modernisation variables; when the status of women is lower mortality is higher and with an increase in the status of women infant and childhood mortality declines. An increase in the status of women improves child rearing, health and nutrition, and other child care practices leading to a decrease in mortality. It is evident in certain developed countries and in developing countries like Japan, China, Sri Lanka, Korea, Taiwan, Singapore, and in Kerala in India.

A woman's exposure to non-traditional sources of information such as media consumption, her interaction pattern and membership in clubs or other organisations, have revealed an association with infant mortality among the Turkish population (Adlakha, 1970). The author has confirmed the hypothesis that greater the exposure a woman has to non-traditional sources of information the less the degree of infant mortality she experiences. He observed that the association of infant mortality measures with the mass media index that he developed, is as pronounced as other background and status characteristics. The extent of various mass media exposure differentially influenced infant mortality. The mass media channels included readership and movie-going. The presence of radio and television in a dwelling unit may have a favourable influence on the feeding and nutritional habits of both parents and children. In Sri Lanka infant mortality rate is closely and positively correlated with the proportion of the population of Indian origin. It is also positively correlated with the proportion of female employees (Fernando, 1981). The differential life style of these populations may have contributed to the prevailing infant mortality pattern in Sri Lanka.

Neglect and Abuse Though the neglect and abuse of children have certain common features, they have to be studied separately in view of their independent influence on infant and childhood mortality. The presence of too many children in a poverty-stricken home leads to the neglect of children in an involuntary manner. The neglect of girls may be an outcome of the greater value attached to sons or to the first or last child. It must be remembered that failure to provide equal care, comfort, attention, opportunities and facilities to girls as compared to boys, has far-reaching consequences for the health and mortality of girls. This happens either partially or fully, explicitly or latently, voluntarily or involuntarily and consciously or unconsciously. In several developing countries children are abused by using them as child labour for the economic well-being of the family. They are kept at home either to perform agricultural labour or to assist in the household tasks of the mother. Though studies on child abuse have been undertaken in several countries and cultures from time immemorial, its influence on the mortality level of children is almost a new area of study. Harris, in his recent book on *Culture, People and Nature*, provides a detailed account on infanticide, malign aggression, and conscious and unconscious neglect (1975: 258). He reported that boys are consistently breast-fed for a longer time than girls. During natural calamities like epidemics and famine girls are neglected more than boys. Lang (1946: 150) reported that during epidemics in China, the death rate of girls was higher than that of boys. This is because during epidemics and famine one can observe the latent cultural practices manifested more vividly to show their real influence at one stroke.

The abuse of girls and children is manifested in a different way in the USA and several other developed countries. Children have become victims of the rapidly dissolving institution of marriage in the West. Several factors that are seen with an increase in divorce affect children more than anybody else. Incidentally, sexual abuse of teenage girls by males including the father is on the increase which gives rise to several psychological problems and suicide. In 1983 in a suburb of New York City, a paediatrician-cum-psychiatrist reported a case of sexual abuse of a daughter by her father. In this case, when the mother died, the father forcibly exploited the first daughter to satisfy his sexual urges. She had just reached puberty. This continued for several months and the girl became abnormal. The grandmother was unaware of this tragedy. When the child was treated for the psychiatric problem, the story was revealed and the doctor reported the same to the grandmother. This incident led to the eviction of the father from the family by the grandmother. According to the doctor, the girl is unlikely to return to normalcy and there are chances of her premature death.

Neglect is also seen in the treatment of illness, provision of food and other comforts. While discussing the causes of death among men and women Preston (1976) pointed out that diarrhoea is the predominant medical cause of death among females from the developing countries. The neglect of female children, particularly of later parities and in large families, was very common in South Central India, and this has increased their mortality level (Mahadevan *et al.*, 1981). It was noted that the neglect of female children through child marriage is associated with the timing, periodicity, and quality of food given, the degree of affection shown, the time spent in fondling the child, callous treatment of the infant, and the indifference expressed in cleaning and dressing the infant.

Conception and Pregnancy

Conception leading to pregnancy and several other events that occur during this period may be broadly classified into bio-social and pregnancy care factors. There are many specific factors associated with the different stages of pregnancy. One or more of these factors in different degrees affect pregnancy either directly or indirectly. The age at conception is the beginning of this process. Physiological conditions, particularly nutritional status, changes in the weight of the mother and physical growth have a tremendous influence on the outcome of pregnancy and the later survival of children. Of course, most of these factors have antecedent causal factors which may be either socio-economic or cultural in origin. Higher education, occupation and income may lead to improved nutritional status and health. On the other hand poverty may multiply all these social factors and may adversely affect conception and pregnancy. Cultural factors can affect the choice of food which may even undermine the beneficial effects of better economic status. The Pan American study on mortality as well as several other studies have shown the importance of the differential influence of maternal nutrition and maternal weight (PAHO, 1971; Wray, 1983; Mosley, 1977) on mortality. Interactions between frequent conception and high fertility also affect mortality (Knodel, 1978; Preston, 1972; Mahadevan *et al.*, 1981).

The physiological condition of pregnant women is not only controlled by the familial variables but also by several interventions that are introduced by governmental policies. The availability of a doctor and other paramedical persons at close proximity and the utilisation of their services at a reasonable cost, the personal and familial characteristics of pregnant women may together determine the healthy or poor advancement of pregnancy and its outcome. Beyond all these factors is the influential mother-in-law and her traditional life style which may nullify the favourable effect of several other factors. In an ongoing all-India study (1984) the adverse effects of religious fasting by pregnant and lactating mothers on the health and life of unborn children and neonatal ones were observed for the first time. This cultural factor is very common among the Muslims.

Perinatal Variables

Delivery care together with the follow-up measures adopted soon after delivery are important and visible determinants of infant mortality. A number of major variables that relate to this stage of human development can be broadly grouped into growth, delivery care, post-delivery care, and demographic characteristics. While most of the births in the developing countries take place at home, in developed countries births invariably take place in hospitals (Omran and Standley, 1976; PAHO, 1971). The place of birth influences the personnel and materials used for facilitating or damaging childbirth. As mentioned earlier most of the deliveries in the developing countries take place at home—whoever is available and has some experience in conducting a delivery will perform the role of a qualified birth attendant. In the absence of a sterilised knife and medicines they are bound to use a kitchen knife or sickle and any powder that is available for cutting and dressing the umbilical cord. As a result, this domiciliary delivery produces tetanus which claims the lives of many infants in the developing countries (Simmons *et al.*, 1979; Mahadevan *et al.*, 1981). Wrong handling of delivery by quacks or non-experienced persons also causes birth injury and infections.

The outcome of the delivery and the health status of the baby are influenced by several preceding events notably the variables grouped under conception and pregnancy care, other interventions, economic, ecological and cultural variables. Immaturity and underweight among infants are two imminent factors which are important determinants of mortality. Several empirical findings have demonstrated their adverse/favourable effect on mortality/survival of infants (PAHO, 1971; Omran and Standley, 1976; Knodel and Hermalin, 1983).

Post-delivery care and care during delivery are interrelated factors. If the delivery takes place in a hospital immediate care for a couple of days will be provided by qualified persons which may reduce the chances of neonatal mortality but such care is not available when the delivery takes place at home. Denial of colostrum which is very valuable for developing immunity in the baby for the next several months or even years, may be avoided for deliveries taking place in the hospital. Similarly appropriate prelacteal feeds free from the risk of infection may be given or avoided altogether when baby care is concentrated at the clinic. Furthermore, hygienic conditions during birth as well as the personal hygiene of the attendant are ensured in the clinic. Most of these risk free facilities and follow-up care are absent when the delivery takes place at home especially when the houses are poorly built, sanitation and hygiene are below the optimum standard and the background characteristics of the family members and birth attendants are unfavourable for safe delivery and post-delivery care. We only have empirical evidence on a few aspects of these factors but much more is needed today.

Several demographic factors, namely, birth order, sib-ship size, birth interval, sex of the baby, and number of family members, that exist at the time of delivery and have a very great influence on mortality. As stated earlier these factors are not independent by themselves, and are influenced sequentially by several antecedent factors which are related to the pregnancy period, and the post-delivery period. Birth order, sib-ship size and birth interval are determined by the age of the mother, personal characteristics of parents, influence of family members and family planning. They are also dependent upon the family size norm, value of children and other cultural practices. Factors such as the outcome of the preceding conception whether it was an abortion or a still-birth, and breast-feeding, also affect the birth interval.

Current knowledge about the differential influence of these factors on mortality from cross-cultural perspectives is fairly adequate to assess their importance and influence (PAHO, 1971; Omran and Standley, 1976; Knodel and Hermalin, 1983; Mahadevan *et al.*, 1981).

In developing countries most of the deliveries take place at home rather than in the hospital. Domiciliary delivery care is provided by untrained persons such as relatives and indigenous midwives who are largely responsible for neonatal mortality. When they cut the umbilical cord with an unsterilised knife and dress it with any available powder it generally leads to tetanus. Tetanus accounts for the single largest cause of infant mortality in several parts of northern and central India (Simmons *et al.*, 1982). It is also a major cause of infant mortality in Bangladesh (Kabir, 1982). Moreover, it is likely to be a major cause in several other developing countries but at present we do not have adequate data to confirm it.

Morbidity may be caused immediately through infection from another organism or through a defect in the physical system. But several antecedent factors contribute to the occurrence of morbidity. It may also be the result of food deficiencies, inhalation of poisonous gases or exposure to radioactive materials. When the immunity level of the body becomes low there is a greater likelihood of developing morbid conditions. The immunity level in children who do not receive colostrum and mother's milk for at least a few months becomes low. Nutritional deficiency may also make the infant susceptible to infections. A debate is going on with regard to the relationship between nutritional status, morbidity and mortality. Some scholars believe that a poor intake of nutrients due to poverty and cultural influences make children vulnerable to infection so that they succumb to disease leading to mortality (Mosley, 1983). However, others have argued that long-term morbidity condition deteriorates absorption and metabolism leading to deterioration of the body system. Under such conditions further nutritional intake may not be sufficient to maintain growth even though food may be available (Wray, 1981). Further research is needed to bridge the gap in our understanding on this matter.

The pattern of diseases is also changing in the developed countries whereas the type of disease has more or less remained the same in the developing countries. Even among the developing countries the prevalence and periodic occurrence of several diseases are not on the same level. Respiratory and gastrointestinal diseases are more or less common in several developing countries (WHO, 1978; Mata, 1983). But tetanus is being increasingly observed in certain countries (Simmons *et al.*, 1982; Chowdhury, 1982) particularly in South Asian countries. In other countries similar morbidity patterns may prevail but data on morbidity patterns is either inadequate or is not available. Parasitic problems and diseases caused by them are more frequent in the developing countries. Physiological defects leading to morbidity and the problem of survival normally are due to genetical problems. Several congenital malformations are the result of genetic defects. In the developed countries certain newly emerging or recently identified diseases like cancer and AIDS are partially the result of new life styles, environmental pollution and so on Such problems are comparatively less in the developing countries.

Parity Birth order exerts a definite pattern of influence on infant and childhood mortality in several countries and regions. Very often the risk of high mortality exists for the first and second order of birth, declines for the third and again increases from the fourth or fifth onwards. Since maternal age progresses with parity they face a similar mortality risk for the first and also the second baby because of the inadequate physical growth of the mother and the lack of adequate experiences in child care. When experience is gained and mothers learn from their past mistakes they are possibly able to avert mortality for the second born and if not, then for the third at least. However, from the fourth or fifth parity onwards mortality increases again because of other reasons. An increase in mortality for the second time from the fourth or fifth parity is possible only in high fertility communities. Where high fertility prevails, higher order parity increases the mortality risk largely because of the deteriorating health of the mother under poverty conditions, neglect, abuse and homicide resulting from several factors. When the value attached to son(s) is greater, a daughter is viewed as a liability. A daughter becomes a casuality in the family either consciously or otherwise at the hands of parents and relatives. In several Asian countries birth order showed a U-shaped or J-shaped pattern with pregnancy wastage including still-births across different cultural groups (Omran and Standley, 1976). Though in several studies in the developed countries birth order showed a J-shaped relationship with neonatal deaths and a direct relationship with post-neonatal mortality rates, the WHO (1976) study in Asian region did not show any definite pattern of association. In Uttar Pradesh State of India, parity showed a J-shaped distribution on infant mortality (Sakesena and

Srivastava, 1980). A large sample of data on the British have fully confirmed the observation that parity has an independent relationship with mortality even when the factor of social class is controlled (Heady and Morris, 1959). Major reasons for child abuse, neglect and homicide have been discussed elsewhere in this paper.

Sib-Ship Size Sibling size may affect infant and possibly childhood mortality in several ways. The existing food and other resources become inadequate for meeting the optimum needs of all babies and they suffer from want of required quantity and quality of food. The mother has little time and facility to look after the young ones who may suffer while the older children may look after themselves at least partially. Differential affection, attachment to the first born or last born or to sons or daughters may also add to the problems when the sib-ship size is large. The relationship between family size including the sib-ship size and the risk involved in the health status have been more or less confirmed in a cross national study (Omran and Standley, 1976). In a state in South India family size showed a positive association with childhood mortality but not with infant mortality (Mahadevan *et al.*, 1981). This variable is partly a function of economic and nutrition variables and family structure. Even when the economic status of a family is high, the sib-ship size may interfere with the needs of children, and may decrease the opportunities for parental care and affection. With the nuclear family on the increase help from other relatives as in the case of Chinese or Japanese stem-families is not available for children. All these factors together contribute to the existing direct relationship between sib-ship size and the mortality of infants and children. Knodel and Hermalin (1983) confirmed on the basis of data collected in Germany that sib-ship size is a more important correlate of infant mortality than birth order. This was partially supported in another historical analysis of data by Cohen. In the southern parts of India (Andhra Pradesh) parity and the number of members in the family showed a strong association with infant mortality while childhood mortality was associated with the latter variable alone (Mahadevan *et al.*, 1981).

Birth Interval Spacing is associated with the mortality of infants and children according to several studies (Knodel and Hermalin, 1983; Adlakha, 1970; Tylor, *et al.*, 1978; Frederick and Adelstein, 1973; Swenson, 1975). Short birth intervals are often associated with a markedly higher incidence of mortality than long intervals. How does birth interval act as a determinant of infant and childhood mortality? Short birth intervals reduce the probability of long durations of breast-feeding. Further, the mother is denied the opportunity to provide all forms of needed care to the infants. However, this may not be absolutely true in the case of affluent families where the availability of food and comfort can be compensated by adding/buying extra facilities. In several Asian countries pregnancy wastage including still-births showed a reversed J-shape relationship with pregnancy interval, the highest risks were associated with intervals of 3–5 years. Under one year mortality was high when birth intervals were short, that is, less than 1–2 years (Omran and Standley, 1976).

A study in the Narangwal area in North India revealed that neonatal, infant and second year mortality was highest when the birth interval was less than a year (Tylor *et al.*, 1978). The Khanna area in the same region also showed that the number as well as more closely spaced children faced higher risk of illness and death (Wyon and Gordon, 1971). Spacing of births normally has a favourable impact in promoting maternal and child health. No doubt all mothers space their pregnancies but spacing over two years is considered to be beneficial to both mother and child in terms of health. This hypothesis has been widely debated during the last half century in the developed countries and during the last two to three decades in the developing countries. Hobecraft *et al.* (undated) after analysing the WFS data on twenty-six countries for which information was available found that there was an increase in infant mortality when the preceding birth interval was less than two years. The strong association, the authors argued, observed between short birth interval and infant mortality was not due to any chance factor or to the influence of socio-economic status. They further confirmed that short birth interval affected not only infants but also children up to five years of age or less. It also affected the health of earlier parities in several ways. Elder children with short birth intervals between them had to compete with the younger children even for breast milk, parental care, and attention which summarily affect all the children under the existing poor economic set up in the absence of any supplementary food being given to such vulnerable group by voluntary agencies or the government. Maine and Wray (1983) analysing the WFS data conclusively confirmed that a very high significant level of increase in IMR was noticed in countries like Syria where the birth interval was short. In Turkey a decline in infant mortality with an increase in birth interval was observed; a negative relation was evident even when status and demographic variables were controlled (Adlaka, 1970). Birth interval and parity may together influence mortality, as childhood mortality in general is positively associated with the parity of mothers and negatively associated with the length of inter-birth intervals, because parity and sib-ship size also determine the birth interval. In Latin America, an increase in the birth interval and a reduction in the fertility level of younger and older women undoubtedly resulted in a lower incidence of prematurity and foetal growth retardation (Mata, 1983: 10). In twelve villages of Rajasthan in northern India, of the total number of infants who died in 1977 half of the infants were born within a birth interval of twelve months only.

Birth Weight Birth weight of an infant is a major determinant of infant mortality in several developing countries. Birth weight of the mother during pregnancy, pregnancy care, life style of the mother, dietary habits during pregnancy, and socio-economic status of the family are some of the important factors associated with the birth weight of babies. Lechtig *et al.* (1975) pointed out that maternal food supplementation during pregnancy has been shown to reduce the proportion of low birth weight of infants and even neonatal mortality. Hytten and Leiteh (1963) expressed that 'birth weight depends more on the environment provided by the mother than on any other single variable'. However, they felt that weight gain during pregnancy should be below twenty pounds. Certain other characteristics of the mother such as age, parity, nutrition, smoking habit are reported to be associated with birth weight and ultimately mortality. Nevertheless, the normal birth weight desired for a baby at the time of delivery is 2,500 grams or more, and anything less than this is considered as underweight.

The weight of the mother during pregnancy has a relationship with an increase in the weight of the mother during pregnancy and the birth weight of the baby. Similarly, the height of the mother also has a relationship with the birth weight of the infant. In an American study, an inverse association between maternal weight gain and pre-pregnancy weight was observed; women who weighed more before pregnancy have, on an average, lesser maternal weight gain. This relationship was consistently seen for both White and Negro gravidas (NINDS, 1972). The study further stated that the two weight variables, pre-pregnant weight and weight gain during pregnancy, are strongly associated with birth weight and foetal outcome; however other investigations of the data suggest that maternal height per se bears little significance. There is also an indication that taller women can expect to give birth to heavier babies though their pre-pregnant weights are similar. A strange phenomenon observed in this study was that Negro women in each height group, on an average, were five pounds heavier than their White counterparts. But Negro infants were, on an average, some 200 grams less in birth weight, which implies that a certain life style, food habits or other attributes of Negro women may be responsible for the lower birth weight among them.

All over the world

> approximately 22 million babies are born each year weighing less than 2,500 grams. Of these 25 per cent are in developing countries. In most developed countries approximately two-thirds of the babies with low birth weight are premature. Low birth weight appears to be one of the major biological mechanisms through which physical and social stress operate. The social environment that creates a child at birth exerts a powerful influence on the probability that it will live through infancy (Newland, 1981).

In Bangladesh premature birth is a major cause of infant mortality. Of this between 50 to 60 per cent infant deaths are neonatal. Nearly 60 per cent of infants die because of low birth weight, which is one of the multiple causes of death (Kabir and Howlader, 1980). Again in Bangladesh another study (Islam *et al.*, 1982) reported that 22 per cent of infants die from prematurity which is the second major cause of death. Another 18 per cent die exclusively because of malnutrition. In Rajasthan area of northern India 13 per cent of infants died of prematurity and another 19 per cent by nutrition deficiency (Gupta *et al.*, 1981). In India, 30 per cent infants who died had low birth weight problem (Bhargava and Mittal 1982).

A cross national Asian study (Omran and Standley, 1976) revealed cross-cultural variations in weight of babies. In Gandhigram (India) area in the same study, Muslim babies weighed more than the Hindus; boys generally weighed more than girls with the exception of the Gounders (Vellalas) and particularly the low social status babies of the Harijans, weighed much less. In Beirut area, Maronites were heavier, as well as taller than Shites, boys were slightly but consistently heavier than girls and the children from the low socio-economic group also weighed less. In Manila and Ankara, boys weighed more than girls and children from the low income group weighed less as in India and Lebanon. In Teheran too boys weighed less as compared to girls in the middle class group. Further American children were heavier as well as taller than Muslim children of the same age group. In all these Asian countries universally boys weighed more than girls. Therefore, in the causal relationship, socio-economic status and value of children and life style within different cultures may independently influence the birth weight and ultimately mortality in different Asian contexts. In Singapore, a multi racial society, between 1965–78 higher perinatal mortality rates occurred among the Indians and Malays compared to Chinese babies (Boon, 1979).

In Latin America while examining the influence of birth weight on neonatal and postneonatal mortality, Mata (1983:6) concluded that the marked differences in neonatal mortality between modern and traditional societies today is due almost entirely to variations in incidence of prematurity and foetal retardation. This evidence closely indicates the need for prevention of intrauterine malnutrition and premature delivery, more than fostering sophisticated neonatology.

Norms on Child Care and Socialisation

Norms on child care and socialisation are numerous and diverse in different cultural groups and in the developed and developing countries. Several aspects such as love and affection, neglect and abuse of children, breast-feeding and supplementary feeding, regular feeding and a number of factors on socialisation that are associated with them differentially influence the survival chances of children. Normally these factors are not perceived as major determinants of mortality. However, breast-feeding has received tremendous importance during the last few years. Neglect and abuse of children may lead to accidental deaths and even suicide. Parental and familial problems lead to the neglect of children. In the developed countries the neglect and abuse of children has become a common and major cause of mortality during the later part of this century, and has replaced the earlier dominant causes like diarrhoeal and respiratory diseases. However, the type of neglect and abuse that is seen in the developed countries does not form a major cause in the developing countries in the light of present evidence but infanticide is an imminent cause of mortality. However, neglect resulting from ignorance is a major cause in the developing countries.

Parental love and affection towards children are universal in nature but differences are observed in the manifestation of these behavioural traits which parents express towards children of different birth order, sons and daughters, first and last born (Mahadevan *et al.*, 1981). It is a delicate and difficult issue of study and one that has not been perceived, so far, as a causal factor leading to the neglect of children. It is, of course, the extra care shown towards children that may avoid the risk of mortality, whereas neglect may increase the risk of mortality. Depending upon the kinship bond between parents and children, the degree of love and affection shown, attending to or not attending to the demands of children for food, care, and comfort may differentially affect mortality in different cultures. At present it is a neglected area of research as far as mortality is concerned.

Breast-feeding Though the practice of breast-feeding is near universal in the developing countries, it is changing with the adverse influence of modernisation in several developing and developed countries. Breast-feeding is not an independent factor, it is influenced by the nutritional and health status of the mother, birth interval, the life style of the mother, cultural practices and so on. In a historical review of the importance of breast-feeding in Europe and North America, Wray (1978: 197) noted that for every disease category studied, breast-fed babies had lower morbidity rates than bottle fed babies. Further he found that infants breast-fed during the first few months of life seemed to fare better during the second half year than babies artificially fed since birth. Cunningham (1983) also reported that of the infants hospitalised for treatment, the incidence of bottle fed babies was three times higher than breast-fed babies in New York city. Bio-chemical, immunological, nutritional and psychological benefits of breast-feeding are well known today (Chandra, 1978; Jelliffe and Jelliffe, 1978; Eastham, 1977; Chen *et al.*, 1974; Jain *et al.*, 1970; Knodel, 1978; and Mahadevan *et al.*, 1981). As a result of a long duration of breast-feeding, amenorrhoea period is prolonged thereby facilitating the automatic postponement of next conception which also improves the chances of continuing breast-feeding. However, giving colostrum, the quantity, quality, and timeliness of breast-feeding, the availability of risk free breast milk (breast milk may also contain harmful chemicals like DDT, etc., when such substances enter the mother's system through the food she consumes), and the style of breast-feeding have a differential influence on children's survival.

Though breast milk is adequate for a baby up to four to six months depending upon the mother's health and the availability of breast milk, supplementary nutritious food is needed later. Poor selection, faulty preparation, and wrong procedure of feeding also affect the health of an infant. Very often factors such as poor personal hygiene, improper cleaning and sanitation while preparing supplementary food, which are a result of the socio-economic background of the mother lead to frequent occurrence of diarrhoea or dysentery among children (Chandrasekhar, 1972; Wegman, 1982). Even though breast-feeding and supplementary feeding are regular, poor absorption of the food into the body system will nullify all other favourable factors of the survival chances of children.

Breast-feeding is a major determinant of infant mortality under the existing circumstances in the developing countries. Several studies have confirmed its importance. In 1922 while studying 22,422 live born infants spread over 15 cities in America, an interesting association between the various types of feeding—exclusive breast-feeding, exclusive artificial feeding, and mixed feeding—at different months during infancy and mortality was confirmed. During the first month of life, the artificially fed infants had a mortality rate which was three times higher than that among the breast-fed infants. Similarly in the second month the mortality rate for the artificially fed infants was four times higher, in the third month nearly four times higher, in the fourth, fifth, sixth, seventh and eighth months it was more than five times, and in the ninth month it was slightly over three times the rate among breast-fed infants (Woodbury, 1922). After the ninth month the relative advantage of breast-feeding was much less. At each age level up to the eighth month the mortality rates for partially breast-fed infants was higher than that for the exclusively breast-fed and lower than that for the artificially fed infants. Among the German population breast-feeding was varied within Bavaria. Breast-feeding also showed a clear negative association with infant mortality. According to a regression equation, with 100 per cent breast-feeding the infant mortality rate would have been 148 and in the total absence of breast-feeding, the IMR would have been 335 (Knodel, 1978). The significance of breast-feeding was observed in Brazil as well. In Brazil, infant mortality was 10 per cent more for the non breast-fed babies. While for the sample population as a whole breast-feeding was highly associated with infant mortality (Goldberg *et al.*, 1982). In Chile, the failure to breast-feed has led to marasmus. In Guatemala, non breast-fed children were at least five times more likely to die in infancy than their counterparts (Pinal, 1981). An indirect estimate of the data on breast-feeding and its influence on infant mortality in four areas in different countries of the Pan American Health Organisation study has shown that mortality risks, during the second six months of life, are six to fourteen times higher for children who are not breast-fed than for those who are breast-fed (Wray, 1970). In Asian countries like Bangladesh, India, Sri Lanka and Malaysia, the influence of breast-feeding has been fully confirmed in several studies. In Sri Lanka, of 519 babies admitted for diarrhoea in a Colombo hospital, 50 per cent were infants and none of these infants were being breast-fed completely. Hence it was found that the absence of breast-feeding and the dependence on other foods increases the incidence of diarrhoea (Priyani, 1981). Breast-feeding also had an independent effect on infant mortality while socio-economic variables and age cohort were controlled in the Malaysian population (Holland, undated). In India the mean duration of breast-feeding was twenty-one months and it did not vary much across different communities. The duration of breast-feeding is negatively associated with infant mortality in South Central India (Mahadevan *et al.*, 1981).

Socialisation During socialisation in many cultures, differential treatment in terms of affection and discrimination is shown to children of different sexes, ages, and parity for several reasons. Discrimination can be seen in terms of feeding, manifestation of love and warmth, giving a bath, dressing, attending to the call and basic needs of children, seeking appropriate and timely medical care, prompt and timely attention, delayed care, seeking medical care from important or unimportant physicians, self medication and so on. No doubt some of these child care practices are visible while others are latent; it is not easy to quantify such factors through surveys but anthropological methods may be relatively more suitable.

Wray reported that (personal communication) children belonging to large families were more likely to be undernourished than those from smaller families. A similar association with low weight was reported by Greiner (1977). Infants from large families frequently suffered from diarrhoea, a major cause of death. In Colombia, the probability of malnutrition increased with birth order, later children in large families faced higher risk as compared to the first child (Drake and Fajardo, 1976). A possible explanation could be that special care is generally provided for the first baby. Emotional deprivation, according to Gardner (1972: 76), is perhaps 'the underlying cause of the spectacularly high mortality rates in the 18th and 19th century homes in Europe.' In his paper on 'Deprivation Dwarfism', Gardner (1972) argues that both environmental and emotional deprivation affect children's growth rates. He concludes that dwarfism can be produced by a complex interplay of hormonal and brain impulses heavily affected by parent-child interaction. The disappointment at the birth of a daughter is clearly manifested by the way the birth of a baby is celebrated. For example, a lower middle class upper caste Hindu, a typist in an office in India, was very upset when his second child was also a female. He did not have a look at the baby after he learnt about her sex and left the house for some time and even grew a beard, a sign of his frustration. His neglect of the mother and the baby continued for a couple of months till he reconciled himself to the situation (personal observation, 1983). This parental attitude is bound to affect the girl's health and longevity because of the absence of affection and love, and the conscious neglect and abuse on part of the parent. A few researchers have reported that boys receive more love than girls (Madan, 1965; Marshall, 1972; Agarwal, 1971; Newell, 1952; Mahadevan *et al.*, 1981). It is interesting to note that in the hilly regions of North India (Newell, 1952), and the villages in Kerala preferential allocation of affection is shown towards younger children. Whereas in Mewar region in the North (Aggarwall, 1971) the first child is shown greater affection. A recent study in South Central India revealed poor concern and care for weak babies of high parity couples (Mahadevan *et al.*, 1981).

Hygienic practices followed by several communities benefit or affect the health and life of infants. Neglect and absence of personal care of infants as seen in the lower socio-economic group and among mothers who are employed outside the home, affects the health and survival chances of infants. The Honans of China, a community in the mainland, follow generationally several positive hygienic practices which have reduced the mortality of infants (*Frontier* 8(36)). Paucity of data does not permit us to know the details of their child care practices. Similarly, several beneficial practices followed by the Brahmins in India reduces the mortality of young ones in their community as compared to the other caste and religious groups in India (personal observation). They do not allow neighbours, friends and even relatives to touch the new born baby for several days, this is done to avoid birth pollution but it acts more or less as a quarantine situation to the advantage of the infant. Their feeding practices and personal hygiene associated with pollution and purification also minimise the chances of infection through other persons. Further, use of soap and effective indigenous detergents for washing hands and vessels minimises the cyclical transmission of diarrhoeal diseases from faces to hand and mouth.

Food and Nutrition It is now well documented (NIH) that malnourished children are frequently more passive, demand less attention, and interact less with their environment, than their better nourished peers. This seems to produce a reinforcing phenomenon in which such children receive less stimulation and attention from their parents, less good parenting, more neglect and perhaps more malnutrition, disease, and impaired performance than children who are well nourished since birth.

In several African states between 20 to 70 per cent of the children under age 6 show mild to severe calorie malnutrition (PCM). The United Nations (Newman, 1979) report adds that due to the fact that the effects of malnutrition and infection are synergistic, it is difficult to isolate the effects of each. In Africa poor nutrition not only acts as a major determinant of infant and childhood mortality, but poor nutrition is also caused by poverty and other conditions particularly in the tropical regions (Gaisie, 1980).

In Asian countries, particularly in Bangladesh, malnutrition is a common cause of infant and childhood mortality. Due to malnutrition alone 18 per cent died in Bangladesh (Kabir and Howlader, 1980; Islam *et al.*, 1982). In Indonesia, 37 per cent of all children under age 7 were malnourished, 17 per cent of them severely so; while an astounding 80 per cent of the families consumed grossly inadequate diets providing not even the minimal energy supplies needed for daily activity. Chen *et al.* (1980) also examined death rates in relation to the nutritional status. Malnourished children had uniformly higher death rates as a result of each cause than better nourished children in Bangladesh. In Singapore severe nutritional deficiency was a major cause of death of infants (Boon, 1979). In Rajasthan area of India 19 per cent babies died indirectly due to nutritional deficiency and another 13 per cent died because of prematurity which may be caused by nutritional deficiency or other related factors (Gupta *et al.*, 1981). Several South East Asian countries showed a close correlation between infant mortality rates and calorie intake and protein intake. Calorie intake is again influenced by the entire food chain. The interrelationship between food, processing food, consumption pattern and their link with population growth has been very well conceptualised by Mosley (1979). He observed that the processing of food in different forms affect the nutritional status at different stages. It starts with: socio-economic development and subsequent food production and population growth. There are several other processes arising from food itself: (*a*) food availability per capita, (*b*) food distribution in society, (*c*) food selection and preparation, (*d*) food distribution in the family, (*e*) absorption and metabolism, finally leading to the nutritional status which, of course, is influenced by antecedent factors like food production on the one hand and population growth on the other.

In Bangladesh Chen *et al.* (1980) found that all malnourished children had uniformly higher death rates due to each cause than better nourished children. The discrimination shown to girls in providing food has contributed to their poor nutritional status and mortality. A UN report (PAG Bulletin, 1977) described the general pattern of discrimination shown to girls in Africa, where female children of all ages were given substantially less food as compared to the boys, and in terms of the priority regarding their food intake they ranked just above the pet dogs. On the island of St. Lucia, a significant correlation between low weight and age was simply explained by the fact that the child was a girl (Greiner, 1977). In Colombia, a differential malnutrition rate of 47.5 per cent for boys as against a rate of 64.6 per cent for girls was reported (Drake and Fajardo, 1976). In the same country discrepancies in the death rates of male and female children are also reported.

In the southern region of India nutritional deficiency particularly maternal anaemia, low haemoglobin level and low weight showed a significant association with infant mortality. Maternal anaemia resulting from a deficiency of iron affects the health of the infant through its effects on the foetus during pregnancy. The incidence of infant and childhood mortality was very high when mothers had a low haemoglobin level as compared to those who were graded normal (above 11.1 per cent) in all the groups. When the mother's weight was less (below 45 kgs) and she was short (less than 150 cms.) similar mortality was seen. It was observed that poor nutritional status was not solely due to poverty but was also the result of poor food choice (Mahadevan, 1983). In Narangwal area nutritional status showed a strong association with perinatal, neonatal, post-neonatal, and 1 to 3 year old mortality rates (Kielmann *et al.*, 1978).

The World Health Organisation differentiates poverty and education as two important first level causal factors. It is desirable because education is relatively more important than income. An educated person may boil the water before giving it to the baby even if protected water supply is not available. This has been observed in the rural areas of Kerala where people depend largely on well water and have a lower incidence of diarrhoea as compared to several urban areas in other parts of India, because the literacy level of women in Kerala is higher as compared to that in other places including New Delhi in India. In a recent study, Koopman and others (1981) observed the multiple influence of nutritional status rather than emphasising SES alone. They noted that 'The amount of food offered to a child will be a function not only of food availability at the family level, but of maternal care patterns, family structure, and the health of the child which leads him to demand food.' They also attribute that 'Any variance in the amount of food offered to a child could be a function of his anorexia. However, the growth deficits may not be due to anorexia. They could result from increased catabolism or malabsorption associated with infection.'

According to Mata *et al.* (1982) the implications of diarrhoea and other infectious diseases are—reduced food consumption, nutrient losses, metabolic alterations, hormonal imbalance, and alterations in immune function; these diseases manifest themselves as wasting, stunting, reduced activity, impaired learning and creativity and acute malnutrition and death (Mata *et al.*, 1982). He emphasises that infection and infectious diseases are the main determinants of chronic malnutrition in societies which do not experience persistent food shortage or famine. Similarly, Chen (1983) observed a strong correlation between diarrhoeal infections and malnutrition and between diarrhoeal infections and infant mortality. Another source of loss of nutrition of the mother and the weight at birth of the infant is repeated pregnancy at close birth intervals under poor socio-economic conditions. This process of steadily deteriorating maternal nutrition and health was referred to by Jelliffe (WHO) as the 'maternal depletion syndrome' which leads to low birth weight in infants, poor performance in lactation and high rates of mortality and morbidity. What is the remedy and what interventions must be introduced to control infant mortality through maternal malnutrition and nutritional deficiency of the infants per se? Supplementary feeding of pregnant and lactating women is one intervention that would lead to direct specific benefits for both infants and their mothers. Government sponsored nutritious meal programme for children may also help to reduce childhood mortality.

Morbidity Patterns and Genetic Problems

The twentieth World Health Assembly defined the causes of death as 'all those diseases, morbid conditions or injuries which either result in or contributed to death and the circumstances of the accident or violence which produced any such injuries' (WHO, 1977). Though several causes of death are known in both the developing and developed countries, precise quantitative influence of each cause is not sufficiently known. The individual influence of each cause needs to be known because causes vary across continents, countries and cultural groups. Whereas mortality in the developing countries in general is caused by communicable diseases, parasitic infestation, diarrhoeal and respiratory diseases, in the developed countries, accidents and congenital malformations assume greater importance. Of course, data on a comparative basis are not available to state how continent-wise and culture group-wise causes differ. Whatever knowledge we have today fully illustrates the influence of several social, ecological, nutritional status and other variables initially responsible for various diseases and which ultimately act as the immediate cause of death.

A question arises how do infectious diseases generate from several antecedent variables and ultimately cause death? This is normally through exposure to a communicable pathogen, which may be transmitted through several sources, the most common being oral and respiratory. The former is generally through drinking water and food, while the latter is transmitted through person-to-person contact, and airborne exposure (Saunders and Warford, 1976). Exposure to disease is obviously determined by a variety of environmental and behavioural variables such as housing, crowding, water, sanitation, personal and household hygiene and child care practices. The immunological ability of infants which is the result of several factors including parental origin, and nutritional status is another parallel factor which determines susceptibility or otherwise to disease and death. The human host possesses a variety of specific and non-specific defence systems against infectious agents, and the severity or intensity of the infection is determined by an interaction between the invading pathogen and the host defence systems (R.K. Chandra, 1978). Host defence is in part determined by the nutritional status. The final outcome of an infection depends upon the virulence of the invading pathogen, the host defence system, and the health behaviour in response to illness. Of course, several infections could be prevented by using appropriate immunisation and preventive measures with the help of modern technology (Chen, 1983).

Neonatal mortality is often produced by perinatal causes that affect the mother or the foetus. The Pan American Health Organisation Study (1973) outlined (*a*) maternal conditions during pregnancy; (*b*) difficult labour and injury at birth; (*c*) conditions of the placenta and cord; (*d*) hemolytic diseases of newborn; and (*e*) prematurity as important causes. Diarrhoea was a single major cause of neonatal mortality besides infection through the respiratory system. For the same population diarrhoea was also the single most important cause for early childhood mortality. This finding has been supported by several other findings. Walsh and Warren (1979) reported that the most common infectious causes of childhood deaths are diarrhoeal diseases (due to Shigella, and E. Coli bacteria and Rota virus), respiratory diseases (such as pertussis, diphtheria, tuberculosis, and bacterical and viral pneumonias), and other infections such as measles. The Rota virus is a common virus that repeatedly attacks children and leads to diarrhoea. It is mostly transmitted through the faeces or faecal infected sources. It begins to attack infants even before they reach their sixth month. But normally the incidence of diarrhoea is less during the first six months because infants are mostly breast-fed, and given extra care, besides possessing greater immunity during this period which they have inherited form the mother and through breast-feeding. This is not true in the case of mothers who fail to maintain personal hygiene and sanitation of the family surroundings.

In most developing countries diarrhoea is a major cause of death among infants during the later months and during the early childhood period. In Cuba, where infant mortality has drastically reduced, diarrhoea is still a major cause accounting for 39.6 per cent of infant deaths (Celade, 1977). In several other Latin American countries diarrhoea continues to be a major cause of death (Mata, 1983; PAHO, 1973). In Bangladesh, diarrhoea accounts for 10 per cent of infant mortality (Islam *et al.*, 1982) and an equal percentage of childhood mortality (Uddin *et al.*, 1981). In Rajasthan, diarrhoea accounted for 14.5 per cent of infant mortality in 12 villages (Gupta *et al.*, 1981). In South Central India, diarrhoea accounted for 17–18 per cent of deaths among infants and between 45–51 per cent of deaths among children across different cultural groups (Mahadevan *et al.*, 1981).

In India, according to the Registrar General (1971), dysentery, broncho-pneumonia and pneumonia appear to be the major causes of post-neonatal deaths; while neonatal deaths are mainly due to endogenic causes such as congenital malformation, birth injuries, and prematurity. In Africa, diarrhoea continues to be a major cause of death (UN, 1984).

Respiratory diseases acted as the second major cause of death both among infants and children. In Cuba, respiratory diseases are the second most important cause of infant deaths (Celade, 1977). In other Latin American countries as well respiratory diseases constitute an important cause of death among babies (Mata, 1983; PAHO, 1973). In Bangladesh, pneumonia accounted for 19 per cent deaths among children and another 9 per cent died because of fever (Islam *et al.*, 1982). In Rajasthan, pneumonia claimed the lives of 26 per cent infants (Gupta *et al.*, 1981). In Thailand, four major causes of death have been outlined which are responsible for neonatal deaths (Boonpasarit, 1982). Of 605 cases, with the exception of 14 cases who suffered from different congenital malformations, the rest died mainly due to pneumonia and sepsis (33–48 per cent) and rarely due to meningitis, gastroenteritis and enterocolitis. But in Sudan, malaria is an important cause of childhood mortality (Farah and Preston, 1981). In tropical Africa, communicable diseases in general act as the major killer of babies (Gaisie, 1980). In Africa, in general, the major causes of mortality among young children including infants are: malnutrition, parasitic diseases including malaria; communicable and infectious diseases especially measles; and respiratory and gastrointestinal diseases (Newman, 1979). Tetanus is a major cause of death in several developing countries of Asia, Africa and Latin America. It is primarily due to the unscientific procedure adopted by untrained persons while cutting the umbilical cord using an unsterilised pair of scissors, blades and knives. In Uttar Pradesh state of North India, it is reported to be the single major cause and accounts for 50 per cent of neonatal deaths. In Bangladesh, tetanus was the largest single cause of death (27.05 per cent) of children in the age group 1–4 years during 1981 (Chowdhury, 1982). In another study in Bangladesh, tetanus was the major factor accounting for 31 per cent of infant deaths (Islam *et al.*, 1982). The author further adds that tetanus also accounted for one-third of neonatal deaths and one-fifth of other infant deaths in Bangladesh. Several other authors have also reported the very high prevalence of tetanus in Bangladesh (D'Souza and Chen, 1983; Edmonston and Bairagi, 1982). In Latin America changes in the causes of mortality declined between 1955 to 1973 and this has been documented by Palloni and Wyrick (1981). They noticed that infectious diseases, except respiratory diseases and diarrhoea, were replaced by cardiovascular diseases and neoplasms. However, they feel that a part of the increase in cardiovascular diseases and neoplasms could be merely an artefact of improved diagnoses, and also attributed to environmental, economic and social changes. A major specific determinant observed in this study was the association between the decline in malaria and the concomitant decline in other infectious diseases. The major factors leading to the rapid process of mortality decline in Latin America were listed by Adamchak (1979), these are infectious diseases—influenza, pneumonia, bronchitis, and diarrhoea. He further stated that respiratory tuberculosis and other diseases of early infancy are responsible for about 12 per cent of the total deaths. However, later an apparent increase in deaths due to cardiovascular diseases and neoplasms has been observed. In the past cardiovascular diseases (probably of infectious origin) may have contributed positively to mortality decline, and may have accounted for as much as 28 per cent of the total decline (Adamchak, 1979).

Several other factors, particularly prematurity, congenital abnormalities, asphyxia, birth injuries and accidents caused neonatal deaths. Other infectious diseases like measles and chicken pox are important determinants of infant and childhood mortality. These are only a few specific causes of infant and childhood mortality but there are several other determinants which are yet to be identified and quantified.

Genetic Factors

At present empirical data on genetic factors which influence morbidity and lead to mortality is very scant. However, a few variables such as the weight at birth of the baby, height and weight of the mother, and children suffering from congenital diseases and several other factors hitherto unidentified may be important in understanding the genetic dimension of the determinants of infant and childhood mortality. Of these factors a good deal of information is available about the weight at birth of babies, the height and weight of mothers that influence the particular infant mortality. In the preceding section, the importance and differentials of birth weight in several parts of the world were discussed. But other aspects of genetical factors have perhaps not yet been explored. No doubt certain birth defects such as chromosomal abnormalities that are found to occur more frequently among women in the older age groups than among women in the intermediate age groups are known now. Such genetic problems lead to a higher incidence of neonatal mortality among older women (Apgar, 1979).

Sex ratio at birth is significantly different from 1:1 in favour of males; certain genetical factors influence the sex of the child. Conterio *et al.* (1977) offered two explanations for the adverse female sex ratio in society. They are: (*a*) different production rates of 'X' and 'Y' carrying sperm (moiotic drive) or differences in the efficiency with which the two types of sperm affect fertilisation (gametic selection); (*b*) differential mortality of male and female zygotes after fertilisation (p. 511). Conterio also expressed the view that the presence of sex linked lethal genes affects the sex ratio in still-births and live-births. However, the number of eliminated males should be very small. He concluded that:

> Undoubtedly the influence of the changes in differential mortality on the genetic structure, on its variability and on the evolution of human populations are profound and varied. Some of these influences can be directly detected and measured, while the others are much more difficult and often conflicting. The latter are the influences which involve the cultural dimension typical of man, and his ability to modify his environment.

However, certain specific genetic factors have shown their influence on infant and childhood mortality in very few isolated studies. Inbreeding is reported to affect the survival chance of the offspring. Li (1963) noted that 'The relative increase of homozygotes for recessive lethal or detrimental genes entails a decrease of biological fitness in children born to consanguineous parents.' Schull and Neel (1965) demonstrated with the help of data from Hiroshima that unrelated couples (before marriage) experienced least childhood mortality as against those couples who were either first, half or second cousins. Serna (1961), while analysing perinatal mortality and infant mortality on the basis of data collected by several authors, confirmed that the frequency of deaths is always higher though not always significant among offsprings of couples related before marriage. Consanguinity is also likely to affect abortions. Research in this direction is limited to a few isolated studies. Nevertheless, inbreeding seems to be a determinant of foetal wastage and mortality.

In Bangladesh, 12 per cent of the infants died as a result of a congenital illness which is genetically determined (Islam *et al.*, 1982), this was chiefly observed in the case of neonatal deaths. In Thailand, Boonpasant (1982) observed that of 605 neonatal deaths, 114 cases suffered from congenital malformations such as anencephaly, congenital heart disease, and multiple anomalies in that order of importance (p: 268). Out of the four major causes of death, namely, tetanus, prematurity, diarrhoeal diseases and congenital malformations, the congenital malformations were the fourth category in that order. In India, Agarwal *et al.* (1982) observed that perinatal loss was significantly higher in twin deliveries; mothers with a poor obstetric history and extreme in maternal age/parity. In Bangladesh, Islam *et al.* (1982) found that the age of the mother is a major influence on neonatal mortality. In Lagos in Nigeria, Abudu and Akinkugbe (1982) reported that trauma (23.2 per cent), congenital malformation and antepartum haemorrhage led to the loss of babies. An ecological analysis of infant mortality across different ethnic groups—Mexicans, Anglo Negroes—in Texas counties near the Mexican border revealed a negative and statistically significant correlation between percentage of Mexicans and neonatal mortality. Here neonatal mortality is influenced less by environmental factors and more by biological factors (Stockwell, 1962). This phenomenon is attributed to the characteristics of this ethnic group. But this group is highly migratory in nature and other babies have relatively more favourable birth weight. An explanation for this has been given by the author supporting the Baird's (1977: 6) hypothesis that 'women who migrate from depressed to more affluent areas in the UK tend to be taller, healthier and more able to and do have lower perinatal mortality rates than those in the same social classes who do not migrate'. Another explanation has been given for this phenomenon by Teller and Clyburn (1974). They believe that Mexican mothers cross the border and enter the US to give birth and subsequently return to Mexico thereby evading the number of births and deflating the number of infant deaths on the US side of the border. However, favourable birth weight cannot be explained by the place of delivery. Therefore, certain genetical characteristics as seen in the case of Chinese babies who face a lower mortality risk compared to Malay or Indian infants (Boon, 1979) may also account for this phenomenon among the Mexicans. Nevertheless, further work is needed to confirm the more plausible hypothesis.

Natural Calamity, Accidents and War

Several kinds of natural calamities, namely, floods, winds (cyclones), drought, earthquake, volcanic eruption, war, pestilence, and famine lead to mortality. The extent of mortality is determined by the severity of the calamity. The resulting damage in the wake of a calamity is greater in the developing countries as compared to the developed countries. Famine, for instance, claimed the lives of thousands of children during the early seventies in Bangladesh, before independence in Bengal in general, and even in Europe (Chowdhury and Chen, 1977; Susser, personal discussion). Every year many young lives are lost because of floods during the monsoons, and failure of crops due to drought. Similarly, war has a devastating effect on health and life both during and after the war depending upon the weapons used in the war. Atomic or chemical warfare not only kills children during the war period but even after the war. For example, children who are already born and those who are yet to be born may suffer from the fallout of atomic or chemical materials used in the war. It was seen in Japan and to a lesser degree in Vietnam. The effect of other natural calamities is not that devastating.

Chowdhury and Chen (1977) commented on the complexity and importance of famine in Bangladesh:

> Famine is a complex syndrome involving multiple interacting variables appears to be well substantiated by this empirical study of the three demographic variables. Mortality is the most sensitive and consistent indicator of the severity and duration of a crisis. But mortality is influenced by fertility and migration and mutual interactions between the three are possible.

They have also conceptualised the multidimensional nature of famine and its effect on morbidity and mortality. The Bangladesh famine dislocated agriculture, service and supplies, development and labourers, transport and communications. It also affected fertility, migration and mortality. In Bangladesh, during the famine which was a result of the 1971 and 1974 wars and subsequently an outcome of floods, the death rate was significantly higher (CDR 21 and 20 respectively) as compared to the two preceding normal years (CDR 14.6 each, Chowdhury and Chen, 1977: 411), and this was also seen after the famine. The mortality level was conspicuously high among young and old. According to Chen and Chowdhury they were affected more by the long-term biological and maternal care variables, malnutrition, increased chances of infection and spread of diseases and the crisis following the breakdown of infrastructure. Scarcity, poor sanitation, insufficient intake of nutritious food, epidemics, overcrowding of migrants affected the poor and the disadvantaged more. The disaster also affected more female children than male children. In 1971–72 period, the mortality level of female children in the 1–4 years age group was 54 per cent higher than that of male children whereas the differential was 40 per cent in the preceding five base-line years. D'Souza and Chen (1983) also reported that during the famine in Bangladesh the death rate of female children up to the age of 14 years was higher than that of boys. Similar adverse effects of famine in Europe have been reported by Susser (personal discussion) which further confirms the devastating and imminent effect of famine on mortality. Davis (1951: 38) identified three chief variables responsible for the early decline in mortality in India prior to independence: (*a*) elimination of wars and banditry; (*b*) control of famines; and (*c*) control of infectious diseases. Since independence during 1960s through 1970s there were several wars, partial famine due to drought and crop failure and, of course, infectious diseases which were controlled to a large extent.

Accidents which are a result of a fall or conflict may cause instant death or may lead to mutilation and even delayed death. These factors are independent of all the factors mentioned under the other categories and they are distinct by themselves. According to Wegman (1983, personal discussion), in recent times the single largest cause of death of children in the USA is due to accidents. Accidents are caused by motor vehicles, burning, falls and drowning. Accidents have thus replaced gastrointestinal and respiratory diseases as the major cause of death in the developed countries, while these diseases continue to be the major killers in the developing countries.

Interventions

Depending upon the economic conditions, developmental status, political set up and prevalence of diseases, several types of intervention programmes have been introduced in the developed and developing countries to control mortality. Most of the developing countries literally follow the intervention programmes which have been designed for the developed countries and which require massive investments but the existing resources of the less developed countries do not permit this investment. Do we really need such uniform intervention programmes at all? For instance, China controlled diarrhoea through extensive education of mothers to provide boiled water and food to children and implemented it successfully without even laying countrywide pipe-lines to provide protected water supply. Its resources did not permit a massive investment in protected water supply. But several other countries still continue to lay pipe-lines and perpetuate the risk of child mortality in spite of the very heavy financial investments already incurred in medical services. Therefore, the strategy of implementation of interventions is very important. Besides, several aspects of service programmes, preventive measures, curative and promotive aspects of health and social welfare programmes, affect mortality in different ways.

The present health policy of several developing countries on intervention programmes, is based on a blind adaptation of the system prevailing in the developed countries. While China and Costa Rica were able to simplify and selectively employ effective and suitable interventions with minimum cost and reduce childhood mortality, many other developing countries suffer from a lack of proper policy and directions and perpetuate high incidence of child mortality despite the higher investments already made in the field of health. Therefore, mere investment of resources for the success of any intervention programme may not be helpful. In this context, if all the birth attendants and pregnant women are provided with a small package of first aid materials and a sterilised knife, at least half of the neonatal mortality can be controlled in India, Bangladesh and in several similar countries. Therefore, interventions should be designed on the basis of research findings, and the urgency of the problem and the resources of the country. Inputs and interventions should be selective as the same intervention may not be effective in different countries and situations. It can either reduce mortality or it can be ineffective. These interventions should be designed in accordance with the culture, ecology and recurrence of natural calamities in addition to the needs from familial variables through child care practices. Risk factors resulting from nutritional status, morbidity patterns and infrastructural facilities may also have to be taken into consideration while deciding on interventions. These interventions reach people and can create an appropriate effect among the clients depending upon the potential of the interventions and the permissive influence of other life-affecting variables, particularly the acceptability of medical interventions, educational and socio-economic status of the clients and their relatives. In this context, a traditional and powerful mother-in-law may nullify the benevolent decision and influence of a mother in seeking timely medical aid and this may lead to child mortality.

Costa Rica in Latin America has made a breakthrough in the control of mortality resulting from gastrointestinal diseases through successful implementation of oral rehydration and rapid provision of intravenous fluid therapy. Massive supply of intravenous fluid solutions and implementation of oral rehydration therapy helped nearly 90 per cent of dehydrated children suffering from diarrhoeal diseases and averted their mortality (Mata, 1983). Mata recommends that if all women are taught oral rehydration procedure, rapid results can be achieved in controlling the inimical effect of water-borne diseases in the developing countries.

Mortality is an aggregation of the effect of separate causes with very different incidences. A decline in mortality usually occurs through medical or environmental advances that affect a specific cause (Keyfitz, 1977). Today we have several new drugs, better equipment and greater knowledge about controlling both environment and disease, and these are being used in different degrees in different countries. In spite of this the decline in mortality of infants and children is not rapid and adequate in several developing countries. Why is it so? Of course, developing countries such as Sri Lanka, Singapore, India (Kerala), China, Grenada, Cuba and Costa Rica in Latin America have achieved spectacular successes in rapidly reducing the mortality level during the last two to three decades. The success of Sri Lanka, India (Kerala), Grenada, Cuba and China is not exclusively through the inputs and interventions of modern innovations in medicine and equipment but largely through suitable political decisions, proper distribution of available resources particularly food, modernisation and effective and appropriate implementation of interventions and developmental programmes. No doubt improvement in sanitation in all these countries has affected a qualitative change in the ecology. But a rapid decline in the mortality level in Japan, Singapore and Costa Rica, is largely due to the adoption of modern medicine, improvement in environmental sanitation and socio-economic development. Hence differential policies, inputs and intervention patterns have contributed to the decline in the mortality level during the last three decades in different countries, except Japan. Therefore, a slow decline in mortality in most of the developing countries and regions may be due to the inappropriate implementation of the overall programmes in the health and development sectors. No doubt the developed countries controlled infant and childhood mortality over the centuries through inventions and greater adoption of modern medicines, massive investments, improvement in environmental and socio-economic conditions in general. However, in the developed countries, particularly in the USA, though mortality resulting from diarrhoea has declined, child abuse and accidents have either increased the mortality level or have prevented the overall mortality level to fall further.

In view of the various models of inputs and interventions that have been adopted to control infant and childhood mortality successfully in different countries with different political ideologies, administrative set up, developmental levels and at different periods of time, it is time to rethink on the future inputs in this direction. Should we go in for a massive investment in laying pipe-lines, digging wells, universalising latrines or should we educate women to boil water, be hygienic and clean, wear sandals, ensure the cleanliness of their village surroundings and roads, and emphasise more on the safety of perinatal, natal and post-natal care, and introduce all preventive cares? Of course, extending hospital facilities will not be an arbitrary decision but underutilisation of the existing hospital services is a question that needs to be considered again. What long-term and short-term measures are needed and what aspect should be given priority are some of the questions that arise now in view of the preceding discussion of knowledge, analytical framework and conceptual model. Therefore, future inputs may be differentially considered for different countries based on the existing socio-cultural situation, mortality levels and their severity, changes in determinants of mortality, cost and time needed to ensure mortality decline in a given situation and the nature of the overall developmental strategies to be followed in a country.

While historically examining the experiences of mortality control in England, Benjamin (1965: 52) observed that:

> The history of public health in the developed countries shows, however, that these services are of little avail against the major factor of poverty and it is only in recent decades, when the grosser forms of poverty have been eliminated in the developed countries that the real effect of freely accessible medical care of a high technical standard can be observed.

Benjamin (1965: 50) further added:

> It is worth recalling the experience of England and Wales at the beginning of this century when infant mortality was high. The factors involved, which were emphasised by public health workers at that time, were poverty, uncleanliness, over crowding, alcoholic indulgence and disease, poor water supply, unsatisfactory food storage conservancy, disposal as distinct from water-carriage of excreta, inefficient scavenging, but most of all abandonment of breast-feeding without adequate cause. It was known that breast-feeding was associated with lower incidence of infectious disease, especially enteritis. Vast campaigns of immunisation now protect communities against disease—diphtheria, smallpox, tuberculosis—which formerly killed great numbers.

However, according to a recent study in India (Mahadevan *et al.*, 1981), different inputs and strategies are needed for the control of infant and childhood mortality, since the former is largely caused by 'bio-familial factors' and the latter by 'familio-environmental factors'.

In conclusion, though scientific and technological progress, inventions of new drugs, creation of infrastructural facilities are very important to control mortality at all age levels, it may not be possible without the concurrent cultural changes, living conditions and life styles of the people. While diarrhoea which was a major cause of infant and childhood mortality has been successfully controlled in the developed countries, accidents and child abuse have emerged (Wegman, personal discussion) as major causes in the developed countries, factors which are not borne out of poverty or poor environmental sanitation or lack of scientific advancement. Thus causes of mortality in the developed countries are also changing which require not only technological advancement but also behavioural change. A similar view has been expressed by Vallin (1980) when he discussed mortality in the developed countries. According to him, the elimination of the most discriminating causes of death (alcoholism, accidents, suicide, tuberculosis) is less dependent on further scientific progress and more on a total change in the living conditions and behavioural patterns. He further stated that a decline in the high mortality level of the poorer classes (due to other causes such as cancer and cardiovascular diseases) also necessitates a thorough cultural transformation of the least privileged social classes rather than the discovery of new therapies or increased expenditure on hospitals. Further elimination of economic inequality is a major remedy for the existing differential mortality in the developed and developing countries or regions.

Measurement and Analysis

For measuring these variables through empirical research different approaches or a combination of more than one approach may have to be adopted based on the background of research(es) and the characteristics of the population considered for study. For instance, certain factors like infanticide/homicide can be properly and effectively studied using anthropological methods because the large-scale survey approach cannot be employed for the measurement of the same factor on account of the sensitive nature and difficulty involved in it. Further, this factor itself may be skipped or avoided in the discussion by the respondents for fear of the consequences. Barbara (1981) used anthropological methods for the study of infanticide in India. Similarly, for the study of certain other qualitative and sensitive variables different anthropological methods can be successfully employed. However, for the study of many ordinal and interval type of variables, the survey approach preceded by an in-depth study of a few cases may be ideal, valuable and feasible.

A combination of anthropological methods with other social research methods may be considered suitable for effectively studying several variables. Collection of data and different analysis techniques can be employed based on the strategy adopted in the study, the sample size, and type of variables. Again infanticide can be meaningfully discussed using the case study or descriptive approaches. For analysing many other variables different analytical and statistical procedures can be adopted. In a recent paper, Jayachandran, Yeung and Krishnan (1984) successfully adopted the path analysis technique for analysing mortality data using the present model. Other multivariate analysis can also be similarly used besides other tabulation. This model is being tested by the author in an all-India study. For want of space here further details of measurement problems are reserved for the author's forthcoming publications.

REFERENCES

ADAMCHAK, DONALD J. (1979). Emerging trends in the relationship between infant mortality and socio-economic status. (Mimeo).

ADLAKHA, A.L. (1970). *A study of infant mortality in Turkey*. Ph.D. thesis, University of Michigan.

AFGHANISTAN MINISTRY OF PUBLIC HEALTH. (Undated). *Infant and early childhood mortality in relation to fertility patterns*.

AGARWAL, V.K. *et al*. (1982). Some observations on perinatal mortality. *Indian Paediatrics*, 19(3): 233–38.

AGGARWALL, P.C. (1971). *Caste, religion and power: an Indian case study*. New Delhi: Sri Ram Centre for Industrial Relations.

AJPH, 1979. Infant mortality and morbidity in the International Year of the Child, *AJPH*, 69(9).

AKINKUGBE, A.O. (1982). Clinical causes and classification of perinatal mortality in Lagos. *International Journal of Gynaecology and Obstetrics*, December 20(6): 443–47.

ALFRED NIZARD & JACQUES VALLIN. (1977). *The effects of development on mortality differentials*. IUSSP Mexico Conference, pp. 441–480.

ANGELE PETROS-BARVAZIAN and M. BEHAR. (1978). Low birth weight: A major global problem. *Ambio*, 7(4).

ANKER, R. and KNOWLES. (1977). An empirical analysis of mortality differentials in Kenya at the macro and micro levels. *Population and Employment*, Working Paper No. 60, Geneva: International Labour Organization.

ANTONOV, A. (1947). Children born during the siege in Leningard in 1942. *Journal of Pediatrics*, 30: 250.

APGAR, V. (Ed.). (1979). Annals of the New York Academy of Sciences, 1970. *American Social Review*, 44: 286.

ARAKI, M. (1983). A 20th century miracle in a 19th century village-infant mortality . . . zero! *Joicfp Review*, (5): 6–12.

ARNOLD. *et al.* (1975). *The value of children: a cross-national study*. Vol. I. Honolulu: East-West Centre.

BAIRD, SIR DUGULD. (1977). Epidemiologic patterns over time. *In* D.M. Reed and F.J. Stanley (Eds). *The epidemiology of prematurity*. Baltimore: Schwarzenberg, pp. 5–15.

BAKER, HUGH D.R. (1979). *Chinese family and kinship*. New York: Columbia University Press.

BALIKOI. (1967). Female infanticide on the Arctic coast. *Man*, 2: 615–25.

BARBARA, D.M. (1981). *The endangered sex: neglect of female children in rural North India*. Ithaca: Cornell University Press.

BEAVER, M.W. (1973). Population, infant mortality and milk. *Population Studies*, 27: 243–254.

BEHM, H. (1980). Socio-economic determinants of mortality. *Latin America Population Bulletin*, (13): 1–15.

———. (1980). Economic and social determinants of mortality in Latin America. *Revistacubana de Administration de salud*, 6(1): 1–30.

BEHM, HUGO and DOMINGO A. PRIMATE. (1978). Mortalidad en los Primeros anos de vida en la America Latina LAP Center. Santiago, Chile, Year 6, No. 16.

BENEDICT, BURTON. (1972). *Social Regulation of Fertility*. *In* G.A. Harrison and A.J. Boyce (Eds.). *The structure of human populations*. Oxford: Clarendon Press, pp. 73–89.

BENJAMIN, B. (1965). *Social and economic factors affecting mortality*. Paris: Mouton and Co.

BHARGAVA, S.K., S.K. MITTAL. (1982). The challenge of infant mortality—need for a national commitment. *Indian Practitioner*, Editorial, 19 April (4): 287–289.

BHATTACHARYA, B.N. *et al.* (1980). Some sex differentials of infant mortality in rural areas. *The Journal of Family Welfare*, 27(2).

BONGAARTS, J. (1978). A framework for analysing the proximate determinants of fertility. *Population and Development Review*, Vol. 4, No. 1.

BOON, W.H. (1979). Perinatal morbidity and mortality in South East Asia. *Journal of the Singapore Paediatric Society*, 21 (3–4): 244–52.

BOONPASANT, Y. (1982). Causes of perinatal death at Ramathibodi hospital, Bangkok. *Thailand Journal of Tropical Paediatrics*, 28: 267–69.

BRAUM, P., Y.H. LEE, J.O. KLEIN, S.M. MARSEY, T.A. KLUM, D. CHARLES, P. LEVY, and E.H. KASS. (1971). Birth weight and genital mycoplasmas in pregnancy. *New England Journal of Medicine*, 284: 167–71.

BROOKS, C.H. (1975). The changing relationship between socio-economic status and infant mortality: an analysis of state characteristics. *Journal of Health and Social Behaviour*: 16: 291–303.

———. (1973). An application of path analysis to socio-economic, birth weight and medical resource, correlates of infant mortality in the United States: a study of countries and standard metropolitan statistical areas. Doctoral thesis. Michigan University.

BUTZ, WILLIAM, P. (1982). *Biological and behavioural influence on the mortality of Malaysian infants*. Santa Monica: Rand Corporation.

CALDWELL, J.C. (1979). Education as a factor in mortality decline: an examination of Nigerian data. *Population Studies*, 33(3): 395–413.

———. (1978). *In* Preston, *Mortality, morbidity and development*. Paper presented to the seminar on population and development in the ECWA region. (Mimeo).

CARTRELLE, P. and H. LERIDON. (1971). Breast feeding mortality in childhood and fertility in a rural zone of Senegal, *Population Studies*, 25: 505–33.

CARR SAUNDERS, A.M. (1955). *The population problem: a study in human evolution*. Oxford: The Clarendon Press. Pp. 201–02.

CASSEN, R.H. (1978). *India: population economy, society*. New York: Holmes and Meier Publishers Inc.

CELADE. (1977). *In* Preston, *Mortality, morbidity and development*. Paper presented to the seminar on population and development in the ECWA region. (Mimeo).

CEYHUN, ATUF KANSU. (1961). Infant mortality in Turkish villages. *The Turkish Journal of Paediatrics*, XXXI (2): 130.

CHANDRA, R.K. and P.M. NEWBERNE. (1977). *Nutrition, Immunity and Infection*. New York: Plenum Press.

CHANDRA R.K. (1978). Immunological aspects of human milk. *Nutrition Reviews*, 36: 265–72.

CHANDRASEKHAR, S. (1972). *Infant mortality, population growth and family planning in India*, London: George Allen and Unwin Limited.

CHASE, H.C. (1972). *A comparison of neonatal mortality from two cohort studies*. US Health Services and Mental Health Administration, US PHS 1000 Series 20 No. 13, Washington, D.C.

CHEN, C. LINCOLN. (1983). Child survival: levels, trends and determinants. *In Determinants of fertility in developing countries: a summary of knowledge*. Part A, Committee on Population and Demography. National Academy Press. Pp. 163–90.

————. (1981). Child survival: levels, trends and determinants. *In Determinants of fertility in developing countries: a summary of knowledge. NAS.*

CHEN L.C., A.K.M. ALAUDDIN-CHOWDHURY, and S.L. HUFFMAN, (1980). Anthropometric assessment of energy-protein malnutrition and subsequent risk in mortality among pre-school children. *American Journal of Clinical Nutrition*. 33: 1836–45.

CHEN, L.C., S. AHMED, M.C. GESCHE, and W.H. MOSLEY, (1974). A prospective study of birth interval dynamics in rural Bangladesh. *Population Studies*, 28: 277.

CHOWDHURY, A.K.M. (1982). The concept of intermediate level variables in mortality studies. *In* B. Edmonston and R. Bairagi (Eds.), *Infant and child mortality in Bangladesh*. Dacca: University of Dacca, 29–37.

CHOWDHURY, A.K.M. and C. CHEN. (1977). The dynamics of contemporary famine. Proceedings of the International Population Conference, Mexico City, 8–13 August. Liege: *IUSSP*, Vol. 1, 409–26.

CICRED and WHO. (1983). *Infant and child mortality in the third world*. Inter-Centre Cooperative Research Programme, Project No. 1: Final Report.

CLARK, A. (1976). Female infanticide as a means of maintaining caste dominance. Paper presented at the fifth Wisconsin Conference on South Asia. Madison, Wisconsin. (Mimeo).

COMMITTEE ON MATERNAL NUTRITION/FOOD AND NUTRITION BOARD, NATIONAL RESEARCH COUNCIL. (1970). Maternal Nutrition and the Course of Pregnancy, NAS, Washington, D.C.

CONTERIO, FRANCO, E. LUCEHETTI and L. SOLIANO. (1977). Effects of development on the interrelationships between genetic structure, the evolution of man, and mortality differentials. *IUSSP*, Mexico Conference, Vol. I, pp. 505–22.

CUNNINGHAM, A.S. (1983). Personal discussion.

————. (1981). Breast vs bottle: scientific evidence. Breast-feeding and morbidity in industrial countries: an up date. *In* Jelliffe and Jelliffe, *Advances in international maternal and child health*: Personal conversation.

————. (1977). Mortality in breast-fed and artificially-fed infants. *Journal of Paediatrics*, 90: 727–977.

DAVIS, K. (1951). *The population of India and Pakistan*. Princeton: Princeton University Press.

————. (1973). *Cities and mortality*. International Population and Urban Research Institute of International Studies, University of California, Berkeley, California, Reprint No. 433.

DAVIS, KINGSLEY and JUDITH BLAKE. (1956). Social structure and fertility: an analytical framework. *In* Charles B. Nam (Ed.), *Population and society*, 1968: 196–215.

DAVIS, CHRISTOPHER and MURRAY FESHBACH. (1980). *Rising Soviet infant mortality in the 1970s*. Washington, D.C.: US Department of Commerce, Bureau of the Census.

D'SOUZA, STAN and LINCOLN C. CHEN. (1983). Sex differentials in mortality in rural Bangladesh. *PDR* 6(2): 257–70.

DIVALE, WILLIAM and MARVIN HARRIS. (1976). Population warfare and social organization. *California Anthropologist*, 78: 521–38.

DIVALE, WILLIAM. (1971). Ibo population control: the ecology of warfare and social organization. *California Anthropologist*, 1: 10–24.

————. (1970). Explanation of primitive warfare: population control significance of primitive sex ratios. *New Scholar*, 2(2): 173–92.

DRAKE, W.D. and FAJARDO. (1976). The promotora program in Candelaria: A Columbian attempt to control malnutrition and disease: 1968–74, *Community systems foundation report.*

DUMONT, LOUIS. (1966). Marriage in India, the present state of the Question III: North India in relation to South India. *Contributions to Indian Sociology*, 9: 90–114.

———. (1964). Marriage in India, the present state of the question post script to part I–II Nayar and Newar. *Contributions to Indian Sociology*, 7: 77–98.

———. (1961). Marriage in India, the present state of the Question I: marriage alliance in South East Asia and Ceylon. *Contributions to Indian Sociology*, 5: 74–95.

EASTHAM, J.C.E. (1977). Family medicine. *In* Beverly (Eds.) *Nutrition, population and health: some implications for policy*. Population Council Working paper No. 3. p. 17.

EASTMAN, N.J. (1944). The effect of the interval between births on maternal and foetal outlook. *American Journal of Obstetrics and Gynaecology*: 48: 445–66.

EDMONSTON, BARRY and RADHESHYAM BAIRAGI. (1982). *Infant and child mortality in Bangladesh*. Bangladesh: Institute of Statistics Research and Training, University of Dacca. Pp. 1–223.

EVANS, A.S. (Ed.). (1976). *Viral infections of humans: epidemiology and control*. New York: Plenum Medical.

FARAH, A.M. (1981). Child mortality and its correlates in Sudan. Doctoral thesis.

FARAH, A.M. and S.H. PRESTON. (1982). Child mortality differentials in Sudan 1982. *PDR*, 8(2): 365–83.

———. (1981). *Child mortality differentials in Sudan*. Working paper No. 7. University of Pennsylvania. P. 23.

FEDERICI. *et al.* (1976). *In Mortality and health policy* UN, 1983, p. 21, mimeo.

FERNANDO, F.S. DALLAS. (1981). Factors influencing the infant mortality rate in Sri Lanka. *Journal of Biosocial Sciences*, 13, 281–86.

FERRY BENOIT and DAVID P. SMITH. (1983). *Breast-feeding comparative studies*. World Fertility Survey No. 23.

FIELD, G. MARK. (1983). The growth of infant mortality in the USSR: Reality or Statistics. *Obozrenie*, Paris, No. 51: 13–16.

FIELDING, E. JONATHAN. (1979). Smoking and pregnancy. *New England Journal of Medicine*, 9 February 1978; Birth disorders tied to smoking before pregnancy. *Washington Post*.

FREDERICK J. and P. ADELSTEIN. (1973). Influence of pregnancy spacing on outcome of pregnancy. *British Medical Journal*, 4: 753–56.

FREEDMAN, RONALD. (1961). The sociology of human fertility: a trend report and bibliography. *Current Sociology*, 10/11.

FREEDMAN, MILTON R. (1971). Social and ecological analysis of systematic female infanticide among the Netsilik Eskimos. *American Anthropologist*, 73(5): 1011–18.

Frontier, 8(36) January 1976, p. 3. *In* Ashok Mitra, *India's population*, Vol. I. New Delhi: Abhinav Publications. P. 265.

GAISIE, S.K. (1981). Mortality, socio-economic differentials and modernization in Africa. *In* UNECA *Population dynamics: fertility and mortality in Africa*. Pp. 441–63.

———. (1980). Some aspects of socio-economic determinants of mortality in tropical Africa (infants and children). *Population Bulletin*, (13): 16–25.

GARDNER, LYTT, I. (1972). Deprivation dwarfism. *Scientific American*, 227(1): 76–82.

GEORGE, F. BROWN, JAUN LAING and JASEN. (1982). *India's population policies and programs*. Bangkok: The Population Council Regional Office.

GOLDBERG, DAVID and ARJUN ADLAKHA. (1969). Infant mortality: estimates based on small surveys in the Ankara area. *In* Shorter and Guvenie (Eds.), *Turkish Demography*. Ankara: Hacettepe University. P. 141.

GOLDBERG, H.I. *et al.* (1982). Infant mortality and breast-feeding in north eastern Brazil. *PAA*, San Jose, Costa Rica: California Centro Latinoamericano de Demografia (*CELADE*). P. 93. Mimeo.

GOPALAN, C. (1958). Effect of protein supplementation and some so called 'Galactagogues' on lactation of poor Indian women. *Indian Journal of Medical Research*, Vol. 46, No. 2, pp. 317–25.

————. (1968). Nutrition and Family Planning. *Population Review*, 12 (1–2), 33–35.

Gordon, John E., Sohan Singh and John B. Wyon. (1965). Causes of death at different ages by sex and by seasons in a rural population of Punjab 1957–59. *Indian Journal of Medical Research*. September, 1906–1917.

Gordon, J.E., J.B. Wyon and W. Ascori. (1971). *In* Joe D. Wray, *Population pressure on families: Family size and child spacing in rapid population growth*, Vol. 2, Johns Hopkins University Press.

Government of India. 1982. Report of the Task force on minimum perinatal care. Pp. 1–49.

Granzberg, Gary. (1973). Twin infanticide: a cross-cultural test of a materialist explanation. *Ethos*, 4: 405–12.

Gray, R. (1981). *Birth intervals, post-partum sexual abstinence and child health. In* Hilary J. Page and Ron Lasthaege (Eds.) *Child-spacing in Tropical Africa: Traditions and change*, New York Academic Press.

Greiner, T.H. (1977). Infant food advertising and malnutrition in St. Vincent. Thesis, Cornell University.

Grimson, R.C. (1983). A standardized ratio for measuring the 'suddeness' of events with applications to the four leading causes of violent death and to certain birth defects. *Social Biology*, Vol. 27, No. 4, pp. 286–93.

Gupta, S.D., T.P. Jain, S. Joshi, D.K. Mangal. (1981). Infant mortality in Rajasthan villages, *Indian Paediatrics*, 18(2): 101–5.

Gurra, F. (1981). Determinants of infant mortality in Panama (1940–1974). *Celade*, p. 87.

Gwatkin, D.K. (1978). *Nutrition planning and physical well-being in Kerala and Sri Lanka*. Manuscript Overseas Development Council.

Gwatkin, D.R. and Sarah K. Brandel. (1980). *Infant and child mortality in the developing world, 1980–2000*. Washington, D.C., Overseas Development Council.

Habicht, J.P., A. Lechtig, C. Yarbrough, and R.E. Klen. (1976). Maternal nutrition, birth weight and infant mortality. *In* K.E. Elliot and Knight (Eds.). *Size at birth*. Associated Scientific Publishers.

Haldar, A.K. and N. Bhattacharya. (1969). Fertility and sex-sequence of children of Indian couples. *Sankhya*, Series B, 31: 144

Hamley, Susan, B. (1974). Fertility, mortality and life expectancy in pre-modern Japan. *Population Studies*, Vol. 28, No. 1, pp. 127–42.

Harris, J.T. (1975). *Culture, people and nature*. New York: Thomas Crowell.

Hauser, P.M. (1959). Demographic indicators of economic development. *Economic Development and Cultural Change*, 7, 98.

Heady, J.A., C. Daley, and J.N. Morris. Social and biological factors in infant mortality, II variations of mortality with mother's age and parity. *Lancet*, 1: 395–97.

Heady, J.A. and J.N. Morris. (1959). Social and biological factors in infant mortality: variation of mortality with mother's age and parity. *Journal of Obstetrics and Gynaecology of the British Empire*, 66: 577–93.

Hill, R. Reuben, Mayone J. Stycos and Kurt W. Beck. (1959). *The family and population control: a Puerto Rican experiment in social change*. Chapel Hill: University of North Carolina Press.

Holland, Bart. (Undated). *Breast-feeding and infant mortality: a Hazars model analysis of the case of Malaysia*. Princeton University.

Holly, W.L., A.L. Rosenbaum, and J.A. Churchill. (1969). Effect of rapid succession of pregnancy in Pan American Health Organisation, perinatal factors affecting human development. *PAHO*, No. 185, pp. 41–44.

Ho Ping-Ti, (1959). *Studies on the population of China, 1368–1953*, Mass: Cambridge, p. 60

Hull, Terence H. (1978). *Levels and trends of infant and childhood mortality in Indonesia*. Population Institute.

Hytten, Frank E. and Isabella, Leiteh. (1963). *The physiology of human pregnancy*. Philadelphia: F.A. Davis Company.

Isenman, Paul. (1978). *The relationship of basic needs to growth, income distribution and employment: The case of Sri Lanka*. Manuscript, policy planning and programme review department. The World Bank.

ISLAM, M.S. *et al.* (1982). Infant mortality in rural Bangladesh an analysis of causes during neonatal and postneonatal periods. *Journal of Tropical Paediatrics*. Dec: 28(6) 294–98.

ISLAM, M.S., RAHAMAN, M.M. AZIZ, K.M. RAHMAN, M. MUNSHID, M.K. PATWARI. (1981). *Infant mortality in rural Bangladesh an analysis of causes during neonatal and postneonatal period*. ICDDR Report, 14.

ISKARANT ALBERT, P. and PAUL V.J. OLIET. (1968). *Accidents and Homicide*. Mass: Harvard University Press.

JACQUELINE, M. MONDOT-BERNARD. (1977). *Relationships between fertility, child mortality and nutrition in Africa*, Paris OECD. Pp. 1–106.

JAIN, S.P. (1974). *Some aspects of mortality*. *In* A. Bose, P.B. Desai, A. Mitra and J.N. Sharma (Eds.), *Population in India's development: 1947–2000*. New Delhi: Vikas Publishing House. Pp. 319–33.

JAIN, A.K., T.C. HSU, P. FREEDMAN, and M.C. CHANG. (1970). Demographic aspects of lactation and postpartum amenorrhoea. *Demography*, 7: 255–71.

JARRETT, ROBERT E. (1970). *Environmental factors and childhood mortality*. Pan American Health Organisation and WHO.

JAYACHANDRAN, J. WEI-JUN J. YEUNG and P. KRISHNAN. (1984). A comparative analysis of infant and child mortality in Sri Lanka and Bangladesh using the Mahadevan analytical framework. Canada: Alberta University. (Mimeo).

JELLIFFE, D.B. and E.F.P. JELLIFFE. (1978). *Human milk in the modern world*. Oxford: Oxford University Press.

JOHNSON, N.E. and M.R. NELSON. (1982). Housing quality and child mortality in the rural Philippines. (Mimeo).

JONES, G. (1978). *Mortality manuscript*. A.N. University.

KABIR, M.H. (1982). *Determinants and consequences of infant mortality: a short review*. ISRT: Dacca University.

KABIR, M. and A.A. HOWLADER. (1980). Socio-economic determinants of mortality in Bangladesh. *Rural Demography*. 7(1–2), 1–13.

KANE, P.V. (1973). *History of Dharmasastra* (Ancient and medieval religious and civil law). Government Oriental Series, Class B, No. 6. Poona: Bhandarkar Oriental Research Institute.

KARKAL, M. (1980). Infant and child mortality in Maharashtra, India, IPPF, p. 16.

————. (1979). Child morbidity and mortality in Greater Bombay. Bombay: IIPS. (Mimeo).

KATHLEEN, NEWLAND. (1981). *Infant mortality and the health of societies*. World Watch Paper, 47.

KAUR, SURJIT. (1978). *Wastage of children*, Sterling Publishers, New Delhi.

KAWASHIMA, T. and K. STEINER. (1961). Modernization and divorce rate trends in Japan. *Economic Development and Cultural Change*. Pp. 213–39.

KELLIY, NARINDER, O. (1975). Some socio-cultural correlates of Indian sex ratios: case studies of Punjab and Kerala. University of Pennsylvania: Unpublished doctoral dissertation.

KENT, MARY MEDERIOS. (1981). *Breast-feeding in the developing world*: current patterns and implications for future trends. Reports on WFS (2), Population Reference Bureau. Pp. 1–43.

KEYFITZ, NATHAN. (1977). *Causes of death in future mortality*. International Population Conference, Mexico. Vol. I, pp. 483–503.

KHAN, M.E. (1973). Factors affecting spacing in births. *Journal of Family Welfare*, 20: 54–67.

KIELMANN, A.A. *et al.* (1978). The Narangwal experiment on interactions of nutrition and infections: II morbidity and mortality effects. *Indian Journal of Medical Research*, 68: 21–41.

————. (1978). The Narangwal nutrition study: a summary review. *American Journal of Clinical Nutrition*, 31: 2040–52.

KNODEL, JOHN. (1982). *Seasonal variation in infant mortality: an approach with applications*. Research Report No. 82–89. Michigan: Michigan Population Studies Centre.

————. (1978). Breast-feeding and population growth. *Science*, Vol. 198: 1111–15.

KNODEL, J. and ALBERT HERMALIN. (1983). Biological factors affecting infant and child moratlity: an assessment on 18th and 19th century reproductive histories. (Mimeo).

KNODEL, J. and E. VANDE WALLE. (1967). Breast-feeding, fertility and infant mortality: an analysis of some early German data. *Population Studies*, 21: 109–31.

KOHLI, K.L. (1977). *Mortality in India: A statewise study*. New Delhi: Sterling Publishers Pvt. Ltd.

KOOPMAN, S.J. (1983). Personal discussion.

KOOPMAN, S. J. *et al.* (1981). Food, sanitation and the socio-economic determinants of child growth in Columbia. *AJPH*, 71(1): 31–37.

KRISHNAN, P. (1975). Mortality decline in India, 1951–1961, Development Vs Public Health Programme, Hypothesis. *Soc. Sci & Medicine*, Vol. 9: 475–79.

KUTZIK, ALFRED, J. (1981). Infant mortality in the Soviet Union, Anti Sovietism in the United States, *New World Review*.

LANG, OLGA. (1946). *Chinese family and society*. New Haven: Yale University Press.

LATHAM, M.C. (1968). Science. *In* N.S. Scrimshaw, C.E. Taylor, J.E. Gordon, *WHO Monograph*, Series No. 57: 188, 561.

LATHAM, C. MICHAEL (1982). *The decline of the breast: an examination of its impact on fertility and health and its relation to socio-economic status*. Cornell University, International Nutrition Programme, No. 10.

LAUKARAN V.H. *et al.* (1981). *Research on determinants of infant feeding practices: a conceptual framework*. Working paper No. 15, The Population Council, New York.

LECHTIG, ARON, *et al.* (1977). Effect of maternal nutrition on infant mortality, *In* Henry Mosley, *Nutrition and human reproduction*, Plenum Press, pp. 147–74.

————. (1976). Low birth weight babies: world wide incidence, economic cost and programme needs. *WHO special supplement*.

————. (1975). Effect of food supplementation during pregnancy on birth weight. *Paediatrics*, 56: 508–20.

LEDERMANN, SULLY and JEAN BREDS. (1981). The dimensions of mortality. *Population*, Selected papers.

LEVINSON, FRANKLIN J. (1972). An economic analysis of malnutrition among young children in rural India. Unpublished doctoral dissertation, Cornell University.

LEWIS, O. (1958). *Village life in northern India*, Urbana: University of Illinois Press.

LI, C.C. (1963). Decrease of population fitness upon inbreeding. *Proc. Nat. Acad. Sci. 49: 439–45.*

LIN, R.Z. (1981). A brief account of 30 years mortality of Chinese population. *World Health Statistics Quarterly*, 34(2): 127–34.

LINCOLN, C. CHEN and A.K.M. CHOWDHURY. (1977). *The dynamics of contemporary famine*, Mexico: IUSSP. Pp. 409–26.

LINDA, G. MARTIN and SUZANNE CULTER. (1983). Mortality decline and Japanese family structure. Paper presented at the Pop. Assoc. Ame. at Pennsylvania.

LOEB, J. (1965). Weight at birth and survival of newborn by age of mother and total birth order: US early 1950, *Vital and health statistics*, Series 21, No. 5.

LOPEZ, ALAN D. and LADO, RUZICKA. (1983). *Sex differentials in mortality: trends, determinants and consequences*, ANU.

MADAN, T.N. (1965). *Family and kinship: a study of the pandits of rural Kashmir*. New York: Asia Publishing House.

MAHADEVAN, K. (1984). Strategies for population control. S. V. University, India.

————. (1984). Decision-making and diffusion in family planning. S. V. University, India.

MAHADEVAN *et al.*, (1981). Culture, nutrition and infant and childhood mortality: a study in South Central India. (Mimeo).

MAHADEVAN, K. (1979). *Sociology of fertility: determinants of fertility differentials in South India*. New Delhi: Sterling Publishers.

————. (1983). Rethinking on policy and interventions to control infant and childhood mortality in developing countries. Seminar on Social and Health Policy and Mortality Prospects. Paris: INED and IUSSP. (Mimeo).

————. (1984). Mortality, biology and society: analytical framework and conceptual model. Paper presented at a Seminar in S. V. University, India.

————. (1985). *Mortality in India: bio-social determinants*. Delhi: Mittal Publishers.

MAINE, DEBORAH and JOE WARRAY. (1983). The health effects of family planning. Centre for P. and F.H. New York. (Mimeo).

MALTHUS, T.R. (1978). The essay on the principle of population. *In An essay on population*. Vol. I. London: J.M. Dent and Sons Limited, 1952.

MARKIDES, K.S. and P.H. HELEN. (1980). Ethnicity and infant mortality in Texas countries. *Social Biology*, Vol. 27, No. 4: 261–71.

MARSHALL, J.F. (1972). Culture and contraception: response determinants to a family planning programme in a north Indian village. Doctoral dissertation, University of Michigan.

MATA, LEONARDO. (1983). Evolution of diarrhoeal diseases and malnutrition in Costa Rica and role of interventions. INED and IUSSP.

MATA, L. *et al.* (1982). Promotion of breast-feeding, health and growth among hospital-born neonates and among infants of a rural area of Costa Rica. *In* L.C. Chen and N.S.S. Scrimshaw (Eds.), *Diarrhoea and malnutrition, interactions mechanisms and interventions*. New York: Plenum Publications. Pp. 177–202.

MEEGAMA, S.A. (1981). The decline in mortality in Sri Lanka in historical perspective. IUSSP Conference, Manila. Pp. 143–64.

MEHTA, M.H. (1966). A study of the practice of female infanticide among the kanbis of Gujarat. *Journal of the Gujarat Research Society*, 28: 57–66.

MITRA, ASHOKA. (1978). *India's population aspects of quality and control*. A FPF and SSR Book.

MOHS, E. (1982). Infectious diseases and health in Costa Rica: the development of a new paradigm. *Ped. Infect. Dis. I*: 212–16.

MOSLEY, W. HENRY. (1983). Will primary health care reduce infant and child mortality? A critique of some current strategies with special reference to Africa and Asia. INED and IUSSP Conference.

MOSLEY, W.H. (1980). *Social determinants of infant and child mortality: some considerations for an analytical framework. A discussion paper*. Cairo: The Population Council.

————. (1979). Health nutrition and mortality in Bangladesh. *In* Sirageldin (Ed.), *Research in human capital development*. Greenwich: Jai Press. Pp. 77–94.

————. (ed.) (1977). *Nutrition and human reproduction*. New York: Plenum Press.

NAG, MONI *et al.* (1981). An anthropological approach to the study of the economic value of children in Java and Nepal. *Current Anthropology*, Vol. 19, No. 2.

NAIR, P.S.G. (1980). Regional variations in infant and child mortality vis-a-vis, rural health services: a case study of Kerala. Paper presented at the Indian Association for the Study of Population Conference, Bombay.

NAM, C.B. and T.M. HARRINGTON. (1983). Factors shaping the morbidity-mortality expectations of youth: a socialization model. (Mimeo).

THE NATIONAL ACADEMY OF SCIENCES. (1970). *In* A.M. Farah, *Child mortality and its correlates in Sudan*. Doctoral thesis, p. 32.

NATIONAL CENTRE FOR HEALTH STATISTICS. (1973). *A study of infant mortality from linked records by age of mother, total birth order, and other variables*; US 1960, live birth cohort, Washington, D.C., US department of Health, Education and Welfare (Vital and Health Statistics, Series 20, No. 14).

NCHS, US Government. (1979). Trends in breast-feeding among American mothers, Series 23, No. 3: 1–39.

NATIONAL INSTITUTE OF NEUROLOGICAL DISEASES and STROKE. (1972). *The women and their pregnancies*. US Department of Health, Education and Welfare, National Institute of Health.

NEWMAN, J.S. (1979). Nutrition disease and mortality in young children. *In Population dynamics: fertility and mortality in Africa. UNECA*: 513–26.

NEWELL, WILLIAM, H. (1952). Gaddi Kinship and Affinity Terms. *Man in India*, 32(2): 86–110.

NEWLAND, K. (1981). *Infant mortality and the health of societies*. World Watch Institute, Paper 47.

NIH. (1978). *Malnutrition learning and behaviour*. Department of Health Education and Welfare, No. 76–1036.

NORA, FEDERICI *et al.* (1976). Urban/rural differences in mortality, 1950–1970. *World Health Statistics Report*, Vol. 29, Nos. 5–6. Pp. 249–378.

OMRAN, A.R. and C.C. STANDLEY. (1976). *Family formation patterns and health*. Geneva: World Health Organisation.

Oshima, Harry T. (1967). Food consumption, nutrition and economic development in Asian countries. *Economic development and cultural change*, Vol. 15, No. 4.

Padmanabha, P. (1983). *Census of India (1981)*: Series 1, Paper II.

Pakrasi, Kanti B. (1970). *Female infanticide in India*. Calcutta: ISI.

———. (Undated). The genesis of female infanticide. Pp. 255–81.

———. (1968). Infanticide in India – a century ago. *Bulletin of the Socio-economic Research Institute*, Calcutta, Vol. 2, pp. 21–30.

Palloni, Alberto and Randy Wyrick. (1981). Mortality decline in Latin America: changes in the structure of causes of deaths. *Social Biology*, Vol. 28. Nos. 3–4.

Pan American Health Organization. (1973). *Patterns of Mortality in childhood*. Report ot the Inter-American investigation of mortality in childhood. Washington.

———. (1971). *Inter-American investigation of mortality in childhood*. Washington, D.C.: PAHO.

Panikkar, P.G.K. (1975). Fall in mortality rates in Kerala: an explanatory hypothesis. *Economic and Political Weekly* November, 22: 1811–18.

Patel, M. (1980). Effects of the health service and environmental factors on infant mortality: the case of Sri Lanka. *Journal of Epidemiology and Community Health*. 34(2): 76–82.

Pethybridge *et al.* (1974). *In* A.M. Farah, *Child mortality and its correlates in Sudan*. Doctoral thesis, p. 33.

Pichat, Jean Bourgeois. (1981). Measuring infant mortality: I. principles and methods, II. causes of death. *Population*.

———. (1980). The causes of infant mortality in developed countries during the last five years. *In* P.M. Boularger, and D. Tabutin. *La Mortalitees enfants dans le monde et dans l'Historic*, Liege, Belgium: Ordina Editions.

———. *(1963). In* Preston (1978). Mortality, morbidity and development. Paper presented to a seminar on population and development in the ECWA region.

Pichat, J.B. and Chia-Lin Pan. (1956). Trends and determinants of mortality in under-developed areas in trends and differentials in mortality. *Milbank Memorial Fund*. Pp. 11–25

Pinal, Jorge H.D. (1981). Breast-feeding and infant mortality in Guatemala. PAA Paper, Washington, D.C. (Mimeo).

Poffenberger, Thomas. (1980). Child rearing and social structure in rural India: towards a 1
cross-cultural definition of child abuse and neglect. *In* J. Korbin (ed.), *Child abuse: a cross cultural perspective*. New York: Garland.

Pool, L. (1982). Is Newzealand a healthy country? *Newzealand Population Review*, 8: 2–27.

Popkin, Barry, M. Richard, E. Bilsborrow, John S. Akin. (1982). Breast feeding patterns in low-income countries. *Science*, 218: 1088–93.

Poulain, M., D. Tabutin. (1980). Infant and childhood mortality in Europe and North America from the 19th century to the present. *In* P.M. Boulanger, D.L. Tabutin. *Mortalite des enfants dans le monde et dans l'histoire*. Liege, Belgium: Ordina editions: 119–57.

Preston, Samuel H. (1979). Causes and consequences of mortality decline in less developed countries during the twentieth century. *In* Richard Easterlin (eds.), *Population and Economic Change in Less Developed Countries*, N.B. Eco. Research.

———. (1978). Mortality, morbidity and development. Presented to Seminar on Population and Development in the ECWA Region. (Mimeo).

———. (1976). *Mortality patterns in national populations: with special reference to recorded causes of death*. New York: Academic Press.

———. (1975). The changing relation between mortality and fertility trends. *Public Health Reports*, 831: 715–27.

———. (1975). Introduction in S.H. Preston (ed.) *The effects of infant and child mortality on fertility*. New York: Academic Press.

———. (1975). The changing relation between mortality and level of economic development. *Population Studies*, 29(2), 231–48.

———. (1972). Interrelations between death rates and birth rates. *Theoretical population biology*, 3(2): 162–85.

Preston, S.H. and Gardner. (1976). Factors influencing mortality levels in Asia: international comparisons and Japanese case study. Paper presented at the Seventh Summer Seminar on Population. Honolulu: East-West Population Institute.

PRESTON, S.H. and V.E. NELSON. (1974). Structure and change in causes of death: an international summary. *Population Studies*, 28(1): 19–51.

PRIYANI, E. SOYSAM. (1981). The advantages of breast-feeding. *In Assignment Children: A Journal concerned with Children, Women and Youth*. December, Geneva: UNICEF.

PROTHRO, EDWIN T. (1967). *Child rearing in the Lebanon, Harvard Middle Eastern Monograph*, VIII Cambridge, Mass.: Harvard University Press.

PUFFER, R.R. and C.V. SERRANO. (1975). *Birth weight, maternal age and birth order: three infant determinants of infant mortality*. Washinghton D.C.: Pan American Health Organisation Publication No. 294.

————. (1973). *Patterns of mortality in childhood*. Report of the Inter-American Investigation of Mortality in Childhood, Scientific Publication No. 262. Washington, D.C.: PAHO and WHO.

————. (Undated). Patterns of mortality: papers New Guinea survey in breast vs bottle: the scientific evidence for breast-feeding. (Mimeo).

Registrar General. (1980). *Survey on infant and child mortality: a preliminary report*. New Delhi: Office of the Registrar General, India.

Registrar General of India. (1971). *Causes of deaths: a survey report for 1971*.

RELE, J.R. and U.P. SINHA. (1973). Birth and death rates in India during 1961–1970. A census analysis. *Demography India*, 2(2), 267–77.

RISLEY, H.H. (1908). *The people of India*. Calcutta: Thacker, Spinck and Company, pp. 148–68.

RIVERS, W.H.R. (1906). *The Todas*. London: MacMillan and Company Limited. Pp. 478–79 and 518–21.

ROHDE, JOHN E., TERENCE H. HULL and LUKAS HENDRATA. (Undated). Who dies of what and why. (Mimeo).

ROLLET, CATHERINE, (1981). Infant feeding, fosterage and infant mortality in France at the end of 19th century. *Population*.

ROSS, JOHN A. (Ed.). (1982). *International Encyclopedia of Population*. Vols. I and II. The Free Press: Pp. 342 and 1912–79.

RUZICKA, L.T. (1982). *Mortality in India: past trends and future prospects*. Conference on the Population of India, Oxford.

RUZICKA, L.T. and TARA KANITKAR. (1972). Infant mortality in the urban setting: the case of Greater Bombay. *In* K.E. Vaidyanathan (ed.), *Studies on mortality in India*. GIRH & FP Monograph, p. 200.

SAKSENA, D.N. and J.N. SRIVASTAVA. (1980). Bio-social correlates of perinatal mortality: experiences of an Indian hospital. *Journal of Bio-social Sciences*, 12(1): 69–81.

SALIMANO, GEORGIO R. and VINE MARTY. (1980). Malnutrition, infection and infant mortality. *In* S.H. Preston (ed.), *Biological & social aspects of mortality and the length of life*. Liege: Ordina Editions, pp. 83–111.

SAUNDERS, J. ROBERT and J. JEREMY WARFORD. (1976). *Village water supply, economics and policy in the developing world*. World Bank Research Publication, The Johns Hopkins University Press.

————. (1976). *Village water supply economics and policy in the developing world*. Baltimore: The Johns Hopkins University Press.

SCRIMSHAW, S.C.M. (1978). *Infant mortality and behaviour in the regulation of family size*. PDR. Pp. 383–404.

SCRIMSHAW, N.S., C.F. TYLOR and J.E. GORDON. (1968). *Interaction of nutrition and infection*, Monograph Series No. 27, Geneva: WHO.

SCRIMSHAW, N.S., M. BEHAR, M.A. GUZMAN and J.E. GORDON. (1969). *Archives of Environmental Health*, 18, 51.

SERNA, A. (1961). Cited by Franco Conterio *et al.*, p. 516.

SHAH, F.K., and H. ABBEY. (1971). Effects of some factors on neonatal and post-neonatal mortality. *Milbank Memorial Fund Quarterly*, 49: 33–57.

SHAPIRO, SAM, EDWARD R. SCHLESINGER, ROBERT E.L. NESBITT JR. (1968). *Infant, perinatal, maternal and childhood mortality in the United States*. Massachusetts: Harvard University Press.

SHIGAMATSU, ITSUZO. (1983). Decline of mortality in Japan. US-Japan conference on aging. (Mimeo).

SIMMONS, GEORGE B., CELESTE SMUCKER, STAN BERN STEIN and ERIC JENSEN. (1982). Post-neonatal mortality in rural India: implications of an economic model. *Demography*. Vol. 19, No. 3: 371–89.

SIMMONS, GEORGE B., CELESTE SMUCKER and B.D. MISRA. (1979). Some aspects of infant and childhood mortality in rural North India. *Social Action*. Vol. 29: 249–69.

SMUCKER, C.M., G.B. SIMMONS, BERNSTEINS, B.D. MISHRA. (1980). Neonatal mortality in South Asia. The special role of tetanus. *Population Studies*, 34(2).

SOUSA, P. (1975). The decline of breast-feeding in Brazil. *J. Trop. Fed. Environment Child Health*, 2, 212.

SRINIVASAN, K., P.C. SAKSENA and TARA KANITKAR. (1979). *Demographic and socio-economic aspects of child in India*. Bombay: Himalaya Publishing House.

SRIVASTAVA and D.N. SAKSENA. (1981). Infant mortality differentials in an Indian context: follow-up of hospital deliveries. *Journal of Biosocial Sciences*, 13, 467–78.

STEIN, Z.S., M.W. SUSSER, G. SAENGER, and F. MAROLLA. (1973). *Intelligence test result of individuals during gestation in World War II*, Famine in the Netherlands, 198: 708–713.

STEIN, Z.S. and M.W. SUSSER. (1977). Famine and fertility *In* Henry W. Mosley. *Nutrition and human reproduction*. Plenum Press, pp. 123–45.

STOCKEL, J. and A.K.M.A. CHOWDHURY. (1972). Neonatal and post-neonatal mortality in a rural area of Bangladesh, *Population Studies*, 26(1): 112–20.

STOCKWELL, E. (1966). Use of socio-economic status as a demographic variable. *Public Health Reports*, November 1981, p. 966.

———. (1962). Infant mortality and socio-economic status. a changing relationship. *Milbank Memorial Fund Quarterly*, 40: 101–11.

———. (1961). Socio-economic status and mortality. *In US Public Health Reports*, 76(12): 1081–86.

———. (1960). The measurement of economic development. *Economic Development and Cultural Change*, 8, 419.

SWENSON, I. (1975). Expected reduction in foetal and infant mortality by prolonged pregnancy spacing in rural Bangladesh. *BDS*, 2: 1–16.

TAMBIAH, S.J. (1973). Dowry and Bridewealth and the property rights of women in South Asia. *In* Jack Goody and S.J. Tambiah, (eds.), *Bride wealth and dowry*. New York: Cambridge University Press, pp. 59–169.

TAWIAH, E.O. (1981). Some demographic social differentials in infant and early childhood mortality in Ghana. *In* UNECA, *Population dynamics: fertility and mortality in Africa*, pp. 464–87.

TELLER C.H. and S. CLYBURN. (1974). Trends in infant mortality. *Texas Bus. Rev.*, 48(10): 240–264.

THE POPULATION COUNCIL (WANAR). (1980). *The health and mortality in infancy and early childhood: report of a study group*. Cairo.

THURSTON, EDGAR. (1975). *Ethnographic notes in Southern India*. Delhi: Cosmo Publications. Reprint 1907 edition.

TYLOR, C.E. *et al.* (1978). The Narangwal experiment on interaction of nutrition and infections: project design and effects upon growth. *Indian Journal of Medical Research*, 68 (supplement): 1–20.

UDDIN, M.M., and MIA A. ALAUDDIN. (1981). Causes of death in children 1–4 years old in Bangladesh. ISW & R, Dacca University. (Mimeo).

UEDA, KOZO. (1983). *Recent trends of mortality in Asian countries*. Tokyo: SEAMIC.

UNITED NATIONS. (1984). Mortality and health policy. (Mimeo).

———. (1980–81). *Demographic yearbook*.

———. (1979). *Demographic yearbook*.

———., Department of Economic and Social Affairs. (1975). Poverty unemployment and development policy: a case study of selected issues with reference to Kerala state, ESA/29.

————. (1975). *The population debate: dimensions and perspectives*. Vol. I, New York.
————. (1973). *The determinants and consequences of pop. trends*, New York.
————. (1961). *International definitions and measurement of levels of living: an interim guide*, New York.

VAIDYANATHAN, K.E. (Ed.). (1972). *Studies on mortality in India*. Gandhigram, Tamil Nadu: The Gandhigram Institute of Rural Health and Family Planning.

VALLIN, J. (1980). Socio-economic determinants of mortality in industrialised countries. *UN Bulletin*, (13): 26–41.

VAN DE WALLE, FRANCINE. (1979). Infant mortality and the demographic transition. Paper presented to the Princeton-European Fertility Seminar, Princeton, N.J. 23–29 July.

VILQUIN, ERIC. (1981). Infant mortality by month of birth, The case of 19th century Belgium population.

VISARIA, P.M. (1969). Mortality and fertility in India: 1951–1961. *Milbank Memorial Fund Quarterly*, 47(1)/1, 91–116.

WALSH, J.A. and K.S. WARREN. (1979). Selective primary health care: an interim strategy for disease control in developing countries. *New England Journal of Medicine*, 301: 967–74.

WATERLOW, J.C. (1972). Classification and definition of protein calorie malnutrition. *British Medical Journal*, 3: 566.

WEGMEN. (1982). Personal discussion.
————. (1983). Personal discussion.

WESTERMARCK, E. (1921). *The history of human marriage*. Vols. 2 and 3 (fifth edition), London: MacMillan and Company Limited, pp. 164–65.

WOLFERS, D. and S. SCRIMSHAW. (1975). Child survival and intervals between pregnancies in Guayaguil, Eucador. *Population Studies*, 29: 479–96.

WORG, H.B. *et al.* (1971). *Breast feeding in Singapore*. Singapore: Choon Kee Press.

WOODBURY, R.M. (1922). The relation between breast and artificial feeding and infant mortality. *American Journal of Hygiene*, 2(6): 668–87.
————. (1925). *Causal factors in infant mortality, a statistical study based on investigations in eight cities*, Children's Bureau Publications, No. 142, pp. 60–67 & 125.

WOOLF, B. and J. WATERHOUS. (1945). Studies on infant mortality. Part I, *Journal of Hygiene*, XLIV, 2: 67–98.

World Bank. (1975). *Health: sector policy paper*, Washington, D.C.

World's Children Data Sheet of the Population Reference Bureau Inc. (1982). Washington, D.C.: UNICEF.

World Health Organisation. (1981). *Drinking water and sanitation—1981–1990: a way to health*. Geneva.

WHO. (1978). *Risk approach for maternal and child health care*. No. 39.
————. (1978). *Report on social and biological effects on perinatal mortality*. Vols. I and II, Geneva.
————. (1977). *Manual of the international statistical classification of diseases*. Injuries and Causes of Death. Vol. I, Geneva, p. 699.
————. (1976). Causes of infant mortality. *Chronicle*, Vol. 30, 198–201.

WHO and SUDAN MINISTRY. (1981). *Infant and early childhood mortality in relation to fertility patterns*. Report of an ad-hoc survey 1974–1976.

WHO and UNICEF Alma Ata. (1978). *Primary Health Care*. Geneva.

WINEBERGER. (1982). Personal discussion.
————. (1983). Personal discussion.

WINIKOFF, BEVERLY. (1978). *Nutrition, population and health: some implications for policy*. The Population Council: Working Paper No. 3, pp. 1–32.

WRAY, D. JOE. (1983). Supplementary feeding of pregnant and lactating women. (Mimeo).
————. (1981). Feeding and survival: historical and contemporary studies of infant mortality and morbidity. *In* Jelliffe and E.F.P. Jalliffee (eds.), *Advances in international maternal and child health*. Oxford: Oxford University Press.
————. (1978). *In* W.H. Mosley (ed.), Nutrition, and human reproduction. New York. Plenum Press, p. 197.

———. (1977). Maternal nutrition, breast-feeding and infant survival. *In Nutrition and human reproduction*. New York: Plenum Press.

———. (1971). *Population pressure on families: family size and child spacing in rapid population growth*. The Johns Hopkins Press, pp. 403–61.

———. (1970). Paper presented at the Conference on Nutrition and Reproduction. Bethseda: National Institute of Health.

WRAY, D. JOE. and A. AGUIRRE. (1969). Protein calorie malnutrition in Candelaria, Columbia, I prevalence: social and demographic factors. *Journal of Tropical Paediatrics*, 15: 76–98.

WYON, J.B. and J.E. GORDON. (1971). *The Khanna study: population problems in the rural Punjab*. Harvard University Press, p. 193.

YAMAGUCHI, K. (1975). *An outline of the demographic trends in Japan*. Population Research Series Foundation, Japan: Institute for Research on Population Problems, No. 4.

YANG, SHU OW and BRIAN F. PENDLETON. (1983). Socio-economic development and mortality levels in less developed countries. *Journal of Biosocial Sciences*, Vol. 27, No. 3, pp. 220–29.

YERUSHALMY, J. (1945). On the interval between successive births and its effect on survival of infants-1: an indirect method of study. *Human Biology*, 17: 65–106.

YERUSHALMY, J., C.E. PALMER, and M. KRAMER. (1940). Studies in child-birth mortality. Age and parity as factors in puerperal fatality. *Public Health Reports*, 55: 1195–1220.

YERUSHALMY, J. (1938). Neonatal mortality by order of birth and age of parents. *American Journal of Hygiene*, 28: 244–70.

17

FACTORS SHAPING THE MORBIDITY-MORTALITY EXPECTATIONS OF YOUTH: A SOCIALISATION MODEL

CHARLES B. NAM AND THOMAS M. HARRINGTON

Although demographic research on fertility and geographic mobility has been approached from both macroanalytical and microanalytical perspectives, demographic studies of morbidity and mortality have been largely based on macro perspectives with little emphasis on micro orientations.

Some have argued that significant breakthroughs in understandings of fertility and geographic mobility patterns occurred when micro studies were blended with macro studies. In that way, the heterogeneity of aggregate relationships could be discovered and more specific behavioural patterns that account for variations in fertility and mobility outcomes could be detected.

Micro Approaches in Morbidity-Mortality Research

We are now beginning to see the micro perspective emerge as a forceful approach in morbidity and mortality research, although population analysts have been slower to adopt it than those in other research disciplines.

The survey mechanism has been used to gather data on the relationship between a number of common daily habits or health practices, and health and mortality. One example of this type of study is the Alameda County Health and Ways of Living Study in California (Belloc, 1982). Based on an area probability sample of the non-institutional adult residents of an urban community, the self-administered questionnaire included items on disabilities, impairments, chronic conditions, symptoms, personal health practices, family life, social activities, employment, childhood background, census-type demographic variables, and psychological questions. Mortality follow-ups were done at five-and-a-half year and nine-and-a-half year intervals. Among 6,928 respondents in 1965, there were 717 deaths reported by 1974. A relationship was observed between good health practices and lower mortality rates that was independent of social class and physical status. Good health practices included not smoking, drinking moderately if at all, maintaining body weight within desirable limits, exercising in leisure activities, sleeping seven or eight hours a day, eating breakfast, and not eating between meals.

A recent analysis of the impact of day-to-day behaviour on morbidity and mortality concludes that 'the individual's role, and the "choice" people make for health-compromising or health-promoting behaviours are already of great importance for half of all diseases and causes of premature mortality in developed nations' (Nightingale, 1981).

A study by the Center for Disease Control indicated that among the ten leading causes of death in the US, factors associated with the life style of persons accounted for 52 per cent of all deaths. This percentage varied by causes of death, but ranged from at least 23 per cent to as much as 70 per cent (Center for Disease Control, 1980).

The National Survey of Personal Health Practices and Consequences, conducted among adults in 1979, revealed that 17 per cent admitted to doing no more than a fair to poor job of taking care of their health, two-fifths considered themselves overweight, more than half agreed that they exercised less than needed, 15 per cent (but more men than women) had no regular source of medical care, more than half had not had a general physical examination during the previous twelve months, one-third were current cigarette smokers, one-third of the smokers smoked twenty-five or more cigarettes a day, one-fourth rarely or never ate breakfast, two-thirds seldom or never used seat belts when travelling in a car, and one-fourth felt a great deal of emotional stress on the job (Danchik, 1981).

These analyses of the personal behaviours of adults that are life-threatening point to a vastly neglected area of social science research on morbidity and mortality. Even more neglected is research on the bases for these behaviours.

Micro Studies of Youth Concerning Morbidity-Mortality

It has been strongly argued that the beliefs and behaviour patterns of adults have their roots in the developmental years of childhood and in youth. According to some authorities, comprehending the health and mortality-avoidance learning process in these earlier years will permit us to devise educational programmes 'not only to protect children from immediate health dangers, but also to develop a foundation that will later influence adult patterns of behaviour' (Gochman and Parcel, 1982: 102).

Specialists in the area of health beliefs and behaviours of children and youth have focused on the perceptions, meanings, and importance that youngsters attach to the general concepts of health and illness, actions such as their nutrition and exercise patterns, their acceptance of health-focused educational programmes, and their response to being ill and to therapeutic interventions (Gochman and Parcel, 1982: 101). These have been viewed largely from the perspectives of Piaget's stages of cognitive development, psychological expectancy theory, and social learning theory (Kalnins and Love, 1982; Lewis and Lewis, 1982).

Researchers using the social learning approach have stressed the modelling behaviour of parents, peers, and television, and the impact of school-based programmes (Bruhn and Parcel, 1982; T.E. Dielman *et al.*, 1982). While the findings have not been entirely consistent, evidence indicating the importance of these learning sources for the health practices of children and youth appears to be substantial.

The Population Socialisation Model

Demographers interested in early life-cycle determinants of demographic behaviours have adopted what is now called a population socialisation approach. In the general social science literature, socialisation refers to the processes through which persons learn and internalise norms of behaviour in their family, community, and other membership groups. Population socialisation deals with these processes as they relate to the population phenomena. Consideration is given to particular population content areas, socialisation agents, the nature of the socialisation process, and demographic outcomes (Philliber, 1980). Studies which had employed this perspective have largely focused on fertility orientations (Gustavus, 1975), but also include those related to geographic mobility (Lyons, Nam, and Ockay, 1980) and mortality (Nelson and Honnold, 1980). What differentiates research using the population socialisation model from other health-related research on children is the primary interest in factors which have observed or predictable demographic outcomes.

Conceptual Scheme for Morbidity-Mortality Socialisation

The conceptual scheme shown in Figure 1 depicts the sets of variables involved in the morbidity-mortality socialisation model. Demographic and socio-economic background factors define the variety and nature of socialisation effects. Genetic and health backgrounds may, however, lead directly to morbid states. Through socialisation effects involving various socialisation agents, both knowledge and awareness about health and survival, and values and attitudes related to those phenomena, are formed and interact with each other. The individual generates expectations about future health and survival that derive from values, attitudes, knowledge, and awareness and which influence later practices and behaviours. These practices and behaviours have morbidity and mortality consequences and are based on values and attitudes, and interact with knowledge, awareness, and expectations. Morbid states result either from genetic and health backgrounds or risky practices and behaviours. Mortality, in turn, flows from morbid states or practices and behaviours with high mortal risks (such as those that increase the chances of fatal accidents). The key explanatory component of this diagram is the category representing socialisation effects, but socialisation consequences, such as knowledge, are uniquely important as well.

The Expectations Component of the Model The category of 'expectations' is seen as a critical intervening condition in this model because individuals often

FIGURE 1 Conceptual Scheme of Morbidity-Mortality Socialisation Model

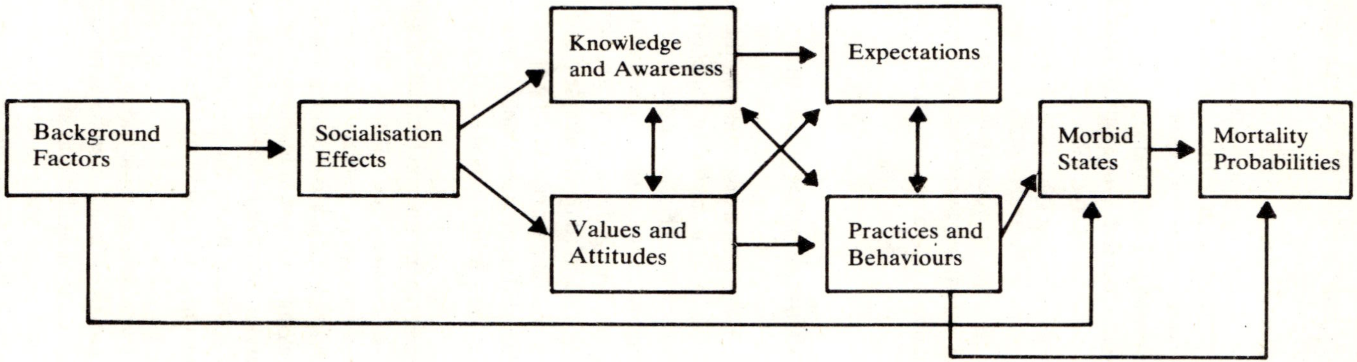

Background Factors
Demographic
Socio-economic
Genetic & health

Socialisation Effects
Models for health & risk-taking behaviour
Parents' influences
Peers' influences
Other agents' influences
Perceptions of others' behaviours

Knowledge & Awareness
Health conditions in society
Own health conditions
Others' health conditions
Risks of illness & deaths
Causes of illness & death
Deaths of known persons

Values & Attitudes
About health-related behaviour
About life itself
About good health itself

Expectations
About future illnesses
About ability to control own health
About own age at death
About own cause of death

Practices & Behaviours
Regular medical check-ups
Sanitary practices
Health maintenance behaviours
Risk-taking behaviours

Morbid States
Acute illnesses
Chronic illnesses

regulate their behaviour according to their anticipations of what the future will hold for them. In the case of the morbidity-mortality model, expectations about future illness and mortal risks may alter practices and behaviours that can either hasten or retard morbidity and mortality.

One such expectation, a person's expected age at death or subjective life expectancy (referred to as SLE) has been studied by a number of researchers. SLE of adults has been found to vary inversely with manifest death anxiety, preferences regarding mode of body disposal, desire for children, job success, and personality variables (Nelson and Honnold, 1980: 10–11).

The determinants of SLE have also been analysed. One study of adult white males showed that respondents based their subjective life expectancies disproportionately on their relatives' longevity, and that poor health practices (for example, smoking and low levels of exercise) were related to lower SLE (Hamermesh, 1982). Another study of students between 18 to 29 years of age revealed that those exposed to early death of near relatives expected significantly shorter lives themselves, more so for females than males (Nelson and Honnold, 1980).

Pilot Study of Morbidity-Mortality Socialisation

The authors conducted a pilot study in 1981, based essentially on the Socialisation model and emphasising the morbidity and mortality expectations of youth. The study was conducted in three phases. First, questionnaires were administered in classrooms to 290 students in the sixth, eighth, tenth and twelfth grades of a public school in Florida which has a fairly representative cross-section of community youth. Second, one-third of the sample (or approximately 100 youth) was personally interviewed in a private setting of the school about a week or two later. Third, a slightly different questionnaire was mailed to the mothers of the one-third sample, and follow-ups were conducted to maximise the returns.

The *student questionnaire* included items on personal and familial background factors; awareness of health conditions of oneself, one's family, and society; perceived health-related attitudes and behaviours of parents and other socialising agents; one's own values, attitudes, and behaviours concerning health care practices and morbid and mortal risk-taking; interest in learning more about health maintenance; and expectations about own future health and survival.

The *student interview* encompassed selected items on the original questionnaire. Our interest was in both probing for more specific details and verifying accuracy of initial responses.

The *questionnaire sent to mothers* dealt with demographic and socio-economic background factors, assessment of the health of family members, perceptions about health dangers, values and attitudes about health and life, expectations about own future health and survival, own health and risk-taking behaviour, and perceived socialisation of the surveyed child regarding health and risk-taking.

Pilot Study Findings

According to the findings about expectations, two-thirds of the youth mentioned cancer as the most threatening disease, with only 9 per cent indicating heart disease and an equal percentage reporting high blood pressure (Nam, Freeman, Harrington, and LeNoir, 1982). The youth's estimates of their own life expectancy averaged only slightly above what actuarial estimates would be. However, there was considerable variability in the response pattern, with the distribution skewed somewhat to the high side and three-fourths of the sample suggesting ages at death several years above or below the mean. Over half of the sample gave 'old age' as their eventual cause of death, with only 12 per cent reporting cancer and 8 per cent heart disease (Table 1).

These response patterns were highly stable across categories of population characteristics of the youth. Factors like age, sex, race, religion, or psychological characteristics of the youth did not account for these patterns.

The findings surprised us in a number of respects. First, why did such a high percentage of the sample view cancer as the most threatening disease in their lives? Moreover, why did such a low percentage cite heart ailments? Second, if cancer is seen as most threatening, why was old age regarded so prominently as one's probable cause of death? Why was heart disease as well as cancer mentioned so infrequently? Third, although the average respondent gave a reasonable answer to the question about life expectancy, a high proportion responded in a manner which departed significantly from the factual base.

We speculated that the youth correctly perceived the relative risks of the various diseases but their awareness of the suffering associated with most forms of cancer helped to define that as a more threatening disease. Moreover, the large investments made in cancer research and the publicity given to it in the public media have created an aura of dominance of that illness. Also, heart disease is seen as more preventable than cancer. The substantial reports of old age as cause of own death suggest that a cause which is not associated with suffering in the minds of young people is preferable. Our follow-up interview revealed that the young people had a difficult time articulating what old age meant in terms of bodily processes. Variance in estimates of own life expectancy might be attributed to different perceptions of future gains in longevity. It seemed that additional explanations may be obtained from an analysis of socialisation effects.

We first examined the effect of a socialisation consequence, the knowledge dimension. The questionnaire included items on the major cause of death for men and women in the US today, the life expectancy for men and women in the US today as well as in 1900, and the comparative life expectancy of different population categories.

The youth demonstrated a fairly good level of knowledge on the average, about life expectancies and causes of death today, but were less accurate in their perceptions of life expectancies at the turn of the century and comparative life expectancies (Table 2). Moreover, there was substantial variability around the average response on those items (Nam and Harrington, 1982).

When estimates of own life expectancy were cross-tabulated by items on mortality knowledge, it was discovered that SLE was related to knowledge about mortality conditions in the society (Table 3). Youth who approximated current life expectancies in the general population more often gave the most actuarially correct SLEs. Where life expectancies were underestimated, SLEs were generally lower; where they were overestimated, SLEs were generally higher. Also, youth who correctly identified college graduates as having longer life expectancy than high school graduates more often gave actuarially correct SLEs. On the other hand, knowledge levels seemed to be unrelated to perceived threatening diseases or expected causes of one's own death.

There were, therefore, some indications of a learning factor related to expectations about mortality, but neither the extent of it nor the nature of the learning process was clear.

Although mothers and fathers are only two of many socialisation agents, there is reason to assume that they are among the most important agents for children and youth in their acquisition of knowledge about health maintenance and mortality avoidance. Questions asked of both the youth and their mothers (about themselves and their husbands) about morbidity and mortality expectations and risk behaviours provide a basis for examining whether there is an association between the several sets of data that would suggest socialisation effects. This analysis is limited, however, because the pilot study dealt with the response records of only 83 mother-youth pairs and for some items the reporting was less. The relationships have been summarised in Table 4. Although they are tentative, they provide some presumptive evidence of the nature of socialisation effects.

Table 4 shows the responses of the youth and their mothers which are cross-classified for three *expectations* items. Regarding the most threatening disease, the relationship is not statistically significant but there is some loading on cancer reports by both mothers and young people. There is, likewise, no significant relationship between the two groups, reports of subjective life expectancy, although a substantial number of both groups gave SLEs that were greater than expected. The relationship for expected cause of own death was also lacking, even though the percentage of mothers reporting 'old age' was nearly as high as that for youth. These findings seem to indicate a lack of parental modelling effects.

Morbidity and mortality-related *behaviours* were more likely to show associations. The use of seat belts by mothers and the young people was substantially related. Present drinking behaviours of both groups were significantly linked when the full subsample was included; when the sample was restricted to tenth and twelfth graders and the younger groups' reports on expected drinking at the age of 25 (not shown), the relationship was even stronger. However, there was no significant relationship between the drinking and smoking behaviour of the younger group and that of their fathers' (as reported by mothers).

Reliance on the mothers' reports of the fathers' behaviour patterns was necessary because of methodological limitations in the pilot study. It could not be arranged to obtain information from the fathers directly. We did collect data on the youth's *perceptions* of their parents' drinking and smoking behaviours. This provided an alternative source of information about the fathers. Moreover, it could be that perceptions of parental behaviours by the youth is more relevant as a socialisation experience than the actual (or mother-reported) behaviour of parents.

The data do, in fact, show different relationships. The younger group's self-reported drinking was not only significantly related to perceptions of their mothers' drinking (just as it was to the mothers' reports of their drinking), but it was also significantly related to their perceptions of their fathers' drinking behaviour (whereas that relationship was not established when the mothers' reports of the fathers were used). Similarly, the younger group's self-reported smoking was significantly associated with their perception of their fathers' smoking behaviour. This finding is interesting in light of research which shows that peer effects are greater than parental effects with regard to smoking behaviour (Eckert, 1983).

The overall effects of parents on the younger group's expectations and behaviours in this pilot study are mixed. However, the mothers' responses do reveal some association, especially with regard to risk-taking behaviours, and the knowledge effect found suggests some form of socialisation process. It may be that the effects of other agents, especially peers, schools, and mass media, are stronger effects, and may even counteract those of the parents for some types of behaviour. Also, the stronger effects of the younger group's perceptions than the mothers' reports of behaviours indicate that how the youth perceive their parents (and perhaps other socialisation agents) may be more important than how the parents actually behave.

Further Research on the Socialisation Model

The evidence from a variety of studies, including our own, regarding the utility of the socialisation model for understanding the elements of the morbidity-mortality process is substantial enough that additional studies are needed which are more comprehensive in coverage of theoretically important variables and different populations. Such studies should incorporate longitudinal features that enable follow-up of the youth sampled over time to assess subsequent health-related practices and behaviours and later morbid and mortal outcomes. They should also include a more sizable sample than that included in the pilot study, greater emphasis on the variables which the pilot research and other studies suggest are most important, information about a variety of socialisation agents, and more intensive data collection efforts. Such studies will help us determine what lessons can be learned about the morbidity-mortality process in the early life-cycle that can inform educational programmes oriented to health maintenance and mortality avoidance as well as enrich microanalytical theories of morbidity and mortality changes.

REFERENCES

Belloc, Nedra B. (1982). Personal behaviour affecting mortality. *In* Samuel H. Preston (Ed.), *Biological and social aspects of mortality and the length of life*. Liege: Ordina Editions, pp. 449–77.

Bruhn, John G. and Guy S. Parcel. (1982). Current knowledge about the health behaviour of young children: a conference summary. *Health Education Quarterly*, 9:2 and 3, 142–66.

Center for Disease Control. (1980). *Ten leading causes of death in the United States, 1977.* Atlanta: Center for Disease Control.

Danchik, Kathleen M. (1981). Highlights from Wave I of the national survey of personal health practices and consequences: United States, 1979. *Data from the National Health Survey*, Series 15, No. 1. Washington: US Government Printing Office.

Dielman, T.E. *et al.* (1982). Parental and child health beliefs and behaviours. *Health Education Quarterly*, 9:2 and 3, 60–77.

Eckert, Penelope. (1983). Beyond the statistics of adolescent smoking. *American Journal of Public Health*, 73:4, 439–41.

Gochman, David S. and Guy S. Parcel. (1982). 'Introduction' to Children's health beliefs and health behaviours. *Health Education Quarterly*, 9:2 and 3, 5–7.

Gustavus, S.O. (1975). Fertility socialization research in the United States: a progress report. Papers of the East-West Population Institute, No. 35. Honolulu: East-West Center.

Hamermesh, Daniel, S. (1982). Expectations, life expectancy, and economic behaviour. Revised version of National Bureau of Economic Research Working Paper, No. 835.

Kalnins, Ilze and Rhonda Love. (1982). Children's concepts of health and illness—implications for health education: an overview. *Health Education Quarterly*, 9:2 and 3, 8–19.

Lewis, Charles E. and Mary Ann Lewis. (1982). Children's health-related decision making. *Health Education Quarterly*, 9:2 and 3, 129–41.

Lyons, Morgan, Charles B. Nam and Kathleen A. Ockay. (1980). Residential morbidity orientations of children and youth. *Population and Environment*, 3:1, 23–34.

Nam, Charles, B. and Thomas M. Harrington. (1982). Factors related to youth's expectations about health and longevity. Paper presented at the annual meeting of the Southern Regional Demographic Group, Greensboro, North Carolina.

Nam, Charles, B., Garry M. Freeman, Thomas M. Harrington and Rebecca G. Lenoir. (1982). Youth's expectations about their health and longevity. Paper presented at the annual meeting of the Population Association of America, San Diego, California.

Nelson, Lynn D. and Julie A. Honnold. (1980). Socialization and demographic determinants of mortality expectations. *Population and Environment*, 3:1, 10–22.

Nightingale, Elena. (1981). Prospects for reducing mortality in developed countries in day-to-day behaviour. International Population Conference, Manila 1981. Vol. 2, Liege: IUSSP, pp. 207–29.

Philliber, Susan G. (1980). A conceptual framework for population socialization. *Population and Environment*, 3:1, 3–9.

APPENDIX

TABLE 1 Youth's Reports of Most Threatening Disease to Self, Expected Eventual Cause of Death, and Subjective Life Expectancy

	Most Threatening Disease to Self	
	No.	*Per cent*
Total Reporting	289	100
None	8	3
Cancer	192	66
Heart disease	26	9
High blood pressure	25	9
Stroke	12	4
Diabetes	8	3
Kidney disease	6	2
Old age	6	2
Other	6	2
	Expected Eventual Cause of Death	
	No.	*Per cent*
Total Reporting	278	100
None	21	8
Cancer	32	12
Heart disease	21	8
High blood pressure	4	1
Stroke	5	2
Diabetes	2	1
Kidney disease	1	—
Old age	154	55
Other	38	14
	Subjective Life Expectancy	
	No.	*Per cent*
Total Reporting	290	100
Below 43 years	14	5
43–47	1	—
48–52	4	1
53–57	3	1
58–62	14	5
63–67	11	4
68–72	36	12
73–77	24	8
78–82	61	21
83–87	35	12
88–92	40	14
93–97	16	6
98 and above	31	11

TABLE 2 Youth's Beliefs About Life Expectancies in the Society

(a) Life Expectancy for US Males Today?

	No.	*Per cent*
Total Reporting	290	100
Below 43 years	3	1
43–47	4	1
48–52	4	1
53–57	4	1
58–62	29	10
63–67	33	11
*68–72	100	35
73–77	46	16
78–82	34	12
83–87	19	7
88–92	9	3
93–97	2	1
98 and above	3	1

(b) Major Cause of Death for Men Today?

	No.	*Per cent*
Total Reporting	289	100
*Heart disease	129	44
Accidents	58	20
Cancer	50	17
Stroke	13	5
High blood pressure	31	11
Other	8	3

(c) Life Expectancy of US Males in 1900?

	No.	*Per cent*
Total Reporting	290	100
Below 43 years	31	11
*43–47	13	4
48–52	51	18
53–57	19	7
58–62	59	20
63–67	20	7
68–72	34	12
73–77	12	4
78–82	22	8
83–87	12	4
88–92	6	2
93–97	5	2
98 and above	6	2

(d) Who Lives Longer?

	No.	*Per cent*
Total Reporting	285	100
Males	32	11
*Females	217	76
Same	36	13
*Whites	127	45
Blacks	40	14
Same	118	41
*Whites	105	37
American Indians	100	35
Same	79	28
High school graduates	27	9
*College graduates	112	40
Same	144	51

* Asterisk is placed by correct response, as indicated by official statistics

TABLE 3 Youth's Subjective Life Expectancy Cross-classified by Beliefs About Life Expectancy at Birth of US Males Today and Educational Differences in Life Expectancy

	Life Expectancy at Birth of US Males Today		
SLE	*Per cent under Correct Figure*	*Per cent near Correct Figure**	*Per cent over Correct Figure*
Total	27	35	39
Less than 58	50	32	18
58–62	79	14	7
63–67	46	45	9
68–72	39	39	22

(Table 3 *contd.*)

73–77 †	17	58	25
78–82	16	36	47
83–87	20	34	46
88–92	15	25	60
93–97	13	44	44
98 or more	23	23	55

	*Belief in Who Lives Longer***		
SLE	*H.S. Graduate*	*College Graduate*	*Same*
Total	10	40	51
Less than 58	9	41	50
58–62	8	31	62
63–67	0	36	64
68–72	11	43	46
73–77 †	0	57	44
78–82	12	35	53
83–87	12	38	50
88–92	8	44	49
93–97	13	38	50
98 or more	13	33	53

† Range including life expectancy for sample (both males and females)
* 68–72 years for males
** College graduates have longer life expectancy

TABLE 4 Relationships of Variables for Youth and Parents

Youth	*Mother*	*Father*	*Chi Square*
Most threatening disease	Most threatening disease		NS
Subjective life expectancy	Subjective life expectancy		NS
Expected cause of death	Expected cause of death		NS
Use of car seat belts	Use of car seat belts		P<0.001
Self-reported drinking	Self-reported drinking		P<.05
Self-reported drinking		Mother's report of drinking	NS
Self-reported smoking		Mother's report of smoking	NS
Self-reported drinking	Youth's perception of drinking		P<.04
Self-reported drinking		Youth's perception of drinking	P<.04
Self-reported smoking		Youth's perception of smoking	P<.01

NS is not statistically significant

18

Culture, Nutrition and Infant and Childhood Mortality

K. MAHADEVAN, P.R. REDDY, M.S.R. MURTHY, P.J. REDDY, V. GOWRI and S. SIVA RAJU

Introduction

The present paper is an outcome of a study on the determinants of infant and early childhood mortality, namely, socio-cultural, economic, demographic factors, community health, and nutrition. However, this paper focuses only on the cultural, nutritional and certain health aspects related to mortality. Factors related to social and demographic correlates of mortality are discussed elsewhere. The study was carried out among the three major cultural groups: the Muslims, the Harijans and the Caste Hindus belonging to Andhra Pradesh in South Central region of India. This trichotomous division of the population was considered on the basis of the major cultural and socio-economic differences found among them from the primary and secondary sources of data. It was also presumed that mortality may differ across these groups in view of their inter-cultural variations. Differences in mortality have also been observed among the different states in India. The infant mortality rate was as high as 200 per thousand live births in UP (Simmons *et al.*, 1978) as against only one-fourth of this figure in Kerala. In the caste hierarchy of India's population, the Caste Hindus are economically advanced with a complex culture as compared to the Harijans who belong to the lowest strata of Indian society. The Muslims have a distinct culture of their own. Evidences of demographic differences among these three cultural groups—the Muslims, the Harijans and the Caste Hindus—have also been reported in three major studies in the past (Davis, 1951; Omran and Standley, 1976; Mahadevan, 1979). Thus, the present study focused on these three major and diverse groups of population which constitute the major segment of the population in Indian society.

Mortality studies hitherto carried out in India were largely based on data generated through census, sample registration scheme, hospital data and National Sample Surveys (Chandrasekhar, 1972; Vaidyanathan, 1972; Jain,

Acknowledgements The authors acknowledge the sincere and efficient assistance rendered by all the investigators for data collection and analysis: Mrs. J. Ashalatha, Mrs. P. Seeta, Miss. A. Padmini, Mrs. A. Swarnalatha, Miss A. Surekha and Mr. S. Usman Shariff. They are also grateful to the UGC for the financial assistance given for this project and to S.V. University administration for their support in processing the funding assistance.

1974; Agarwala, 1975; Kulkarni, 1975; Kohli, 1977). Most of these studies were carried out using secondary data and therefore had several limitations and also could not explain much of the variance in the mortality behaviour of Indian population. However, a few major studies have been carried out in different parts of India generating primary data on mortality (Wyon and Gordon, 1971; Ruzicka and Kanitkar, 1973; Upadhyay, 1974; Omran and Standley, 1976; Simmons *et al.*, 1978; Smucker *et al.*, 1980).

These studies followed different methodologies and focused on different variables depending upon the background of the researcher and the aim of the study. While most of the studies followed the survey approach, the Khanna Study (Wyon and Gordon, 1971) was more in-depth in nature. The variables covered, so far, focused on service delivery systems, health and demographic aspects, and socio-demographic factors. As these studies did not cover several cultural and nutritional variables which have great relevance in understanding mortality behaviour in a traditional society like ours, the present paper aims to bridge this gap in knowledge in the field of cultural and nutritional determinants of mortality.

Methodology

For the present study a multistage proportionate random sampling technique was followed. As a first stage, Chittoor District in Andhra Pradesh in South Central India was purposively selected to minimise the cost and distance involved in the survey and to facilitate maximum utilisation of the time of the researchers. Out of the twenty community development blocks in Chittoor District, Chandragiri and Pulicherla blocks were selected at random. These blocks lie west and south of Tirupati town.

From these two community development blocks, proportionately 3,000 households from the three major cultural groups—the Caste Hindus, the Harijans and the Muslims—were selected at random since each constitutes almost one-third of the universe. An equal number of 1,000 respondents each were selected uniformly from these three cultural groups. Respondents were eligible couples with a minimum of one living child. From every family only one eligible couple was considered.

Perspectives studies on the relation of maternal nutrition to infant and child mortality are scarce even though area development plans and programmes of the VI Plan (Ministry of Social Welfare, 1980) focused on the poor health conditions of women and children (Mitra, 1980). It has been observed that maternal nutrition affects the birth weights of infants (Chandra, 1971). Against this background, the present investigation was planned to study the existing nutritional status of rural women in the context of the food choices in the family and clinical observations and to relate the incidence of infant and child mortality among the selected families to the existing nutritional status of the mother. Though the ideal method of relating maternal nutrition to infant and child mortality would be longitudinal studies covering a period of at least two or three years from the time of birth of the baby, due to the retrospective nature of the present study in its totality, the existing nutritional status of the mother was viewed retrospectively to the incidence of infant and child mortality.

For an examination of the nutritional status, information on various aspects such as dietary habits of the family, health status and child care practices was gathered. The child care practices considered were prelacteal feeds, duration of breastfeeding, frequency of feeding, care and affection, methods of cleaning the vessels, etc.

The sample for this study with regard to food choices, general nutritional status of mothers and infant/child mortality, comprised 3,000 eligible women belonging to the Muslim, the Harijan and the Caste Hindu categories. These findings are not reported here and can be seen in the main report.

However, the study of a sub-sample, wherein the nutritional status of the mother was assessed in greater detail and correlated to infant and child mortality, consisted of 235 eligible women fairly well distributed among the three groups. The results of this study are discussed here.

The sub-sample of 235 respondents was divided into control and experimental groups on the basis of no deaths and incidence of deaths respectively, in the infant and child categories. For the nutritional study, 139 families who have experienced infant and childhood mortality during the last five years were taken as the experimental group and a matching group of 96 families from similar socio-economic background where mortality did not occur during the same period were taken as the control group. This formed the sub-sample of the entire sample. The physical health status of the mother was assessed through anthropometric measurements such as height and weight, and on the basis of the physical observation of the nutritional deficiency for both mother and child. The physiological health status of the mother was assessed through blood tests for determining the levels of haemoglobin (Crossby, Munn and Furth, 1954).

With regard to the expression of haemoglobin per cent, four levels were indicated as follows: less than 8.4 per cent as very severely deficient; 8.4 to 10.0 per cent as severely deficient, 10.1 to 10.9 per cent as moderately deficient and 11 per cent as normal. For each category—the Muslims, the Harijans and the Caste Hindus—the data was classified according to control and experimental observations, for anthropometric as well as haemoglobin measurements. A comparison was made between experimental and control observations in each category apart from the comparison between the Muslims, the Harijans and the Caste Hindus. Though laboratory examination of stool was planned for detection of worm infestation, it could not be carried out due to lack of cooperation on part of the subjects, this is a limitation of this study. A schedule was designed for the survey and data was collected by trained post-graduates in population studies and nutrition. Data was analysed using a computer.

Age at Infant Death

Considering the level of infant mortality at different culturally important periods of infancy, that is, the same day after birth, two to seven days, eight to twenty-eight days, one to three months, and fourth month to less than a year, significant variations in mortality level during these periods of child rearing were observed (see Table 1). Critical periods of infant mortality were observed during the first week, and twenty-eight days after delivery. It was also higher at the time when supplementary weaning foods were introduced. During the first week after delivery 40 per cent of infants died, and 55 per cent died within twenty-eight days. However, Simmons *et al.*, (1978) reported a slightly higher proportion of infant deaths (60 per cent) during the neonatal period on the basis of a sample of population in UP—a state in North India. However, infant mortality varied across different cultural groups. During two to seven days period, a slightly higher percentage of infants from the Harijan families died followed by the Muslims and lastly by the Caste Hindus. But these differences were not very great. However, significant differences in the incidence of mortality were noticed during the period extending from the fourth month to less than one year. While 36 per cent of the Muslim infants died during this period, only 32 per cent and 29 per cent of the Harijan and the Caste Hindu infants respectively, died during the corresponding period. The higher incidence of mortality during two to seven days period may be due to poor maternal health, tetanus neonatorum—a result of the unscientific way of cutting the umbilical cord with a household knife or a unsterilised blade, asphyxia and partially starving the baby for the first three days. Incidentally, such practices were more among the Harijans as also the incidence of infant deaths. The phenomenal increase in infant mortality during the fourth month to less than one year period may be on account of the various problems and practices observed for infant weaning because most children were given supplementary food during this period. Inadequate and improper weaning foods given to the baby may lead to mortality during the post-neonatal period. For the first three months after birth, intensive care is provided to the mother and child, the child is exposed freely to the outside world from the fourth month onwards. This

TABLE 1 Per Cent Distribution of Infant Deaths by Age at Death, Religion and Caste

Age at Death	*Three Cultural Groups*			*Total*
	Muslims	*Harijans*	*Caste Hindus*	
Same day	18.0	14.8	16.2	16.3
	(57)	(47)	(44)	(148)
2–7 days	23.7	24.9	22.9	23.9
	(75)	(79)	(62)	(216)
8–28 days	11.4	11.7	14.8	12.5
	(36)	(37)	(40)	(113)
1–3 months	10.7	17.0	17.3	14.9
	(34)	(54)	(47)	(135)
4–12 months	36.2	31.5	28.8	32.4
	(115)	(100)	(78)	(293)
Total	100.0	100.0	100.0	100.0
	(317)	(317)	(271)	(905)

sudden exposure subjects the child to the risk of contact with sick and unhealthy persons and environment. Over and above this, the differential and wrong methods of infant weaning and child care practices and the various immunisation measures differentially provided against communicable diseases account for the existing differential incidence of mortality during this period across these three cultural groups. Another differential mortality specifically noticed between the Muslims and the non-Muslims was the lower incidence of mortality seen among the Muslims during the period extending from the eighth day to the third month as against the other two groups. This may be due to existing differences in child care practices between these two groups.

Contrary to this, the Muslims have relatively high incidence (18 per cent) of mortality on the day of birth followed by the Caste Hindus (16 per cent) and the Harijans (15 per cent). This is a reflection of the health status of the mother during pregnancy and the intranatal care provided besides the delivery attended to by suitable persons and the problems occurring at the time of delivery (Table 2).

Male mortality was greater as against female mortality during neonatal and post-neonatal periods across the three cultural groups (table not shown). This has confirmed the findings of the studies conducted in Albania and the United Arab Republic (UN, 1968). A similar trend of greater proportion of male mortality was noticed among the children from the Harijan and the Caste Hindu families but an opposite trend, that is, higher female mortality as against male mortality, was noticed among the Muslims. The higher incidence of female deaths among the Muslims may be due to the neglect of girls. On the other hand, higher male mortality cannot be due to neglect but may be due to other reasons. One possible reason may be that females have two chromosomes and males have one, which probably confers a survival advantage on females (Naeye *et al.*, 1971), and another reason may be that the boys are more exposed to outside environment and subjected to various types of risks.

Socialisation and Breast-Feeding

During the first four years of childhood, children grow up around the mother, the primary kinship group, traditional birth attendant and the immediate neighbourhood. Therefore, the growth/survival chance of children (1–4 years) depends largely on the nature of child rearing practices. However, the interaction of children with other individuals and the neighbourhood in the overall socialisation process extends beyond the environment of the family. The concept of socialisation consists of a number of variables beginning from the perinatal period through infancy and the first four years of early childhood period. They consist of: (*a*) types of medicines and prelacteal feeds given during the first week after delivery; (*b*) timing of commencement of breast-feeding; (*c*) commencement of supplementary food; (*d*) periodicity of feeding; (*e*) duration of breast-feeding; (*f*) nature of bath given; and (*g*) fondling of children including care and affection. An attempt was made to understand the prevalence, patterns and the influence of these socialisation factors on infant and childhood mortality which were, hitherto, not fully explored in many of the earlier mortality studies.

Prelacteal Feeds

Prelacteal feeds of different kinds were given in various patterns to the infants in these cultural groups during the first one or two days before regular breast-feeding. Such feeds included sweetened water, honey, goat's or donkey's milk, musk, and Gorojanam in addition to the milk of other lactating mothers. Neem oil was also given to the infant against epilepsy. Sweetened water was found to be the major food for infants from the Muslim households (94 per cent) and the Caste Hindu households (92 per cent) followed by the Harijan households (76 per cent). The remaining Harijans (20 per cent) gave the milk of other lactating women as an infant food while others gave the infants other types of prelacteal feeds. It is a widely held belief that the newly delivered mother's milk is harmful to infants and may affect their health. This belief is contrary to the many advantages of colostrum or first milk, which is rich in immunoglobulins and other nutritional factors. When infants are not fed colostrum and breast milk, they are exposed to a number of morbid conditions (Priyani, 1981; Chandra, 1978). All the families (90 per cent) belonging to these groups usually resumed breastfeeding on the third day of birth, with the exception of a small proportion of women resuming it on the second or the fourth day. In addition to giving breast milk, 50 per cent of families in all the groups gave Kasthuri and Gorojanam (musk) on the third and the fifth days. This practice was more widely observed among the Harijans (64 per cent) followed by the Muslims (51 per cent) and the Caste Hindus (48 per cent). From the ninth day to two years, a paste prepared out of Capsicum annum (karam), Acorus calamus (vaza), Zingiber officinale (sonti), Myristica fragrance (fruit kernel = jaji kaya = and Arill seed = Japatri), Quercus infectoria (Masikaya), Smitax glabra (Paringi chekka), Piper nigrum (nalla miriyalu), Piper cubeba (vayu miriyalu), Terminalia chebulla (karkkaya), Cucurmaaromatica (kasthuri pasupu) and castor oil was give separately on alternative days. The medicinal value of these preparations and their pharmacological effects on the new born infants is an important subject for research (table not given).

Breast-Feeding

Breast-feeding is a common practice among the rural women. Prolonged breast-feeding for about twenty-three months is in fact common among the lower socio-economic groups in India (Belawady, 1980). Duration of breast-feeding bears a relationship to the nutritional status of the child and even to the incidence of infant and child mortality (Woodbury, 1922; Knodel, 1978; Wray, 1977). Introduction of supplementary foods is a complementary practice to breast-feeding and weaning in various stages. Among the rural population it has been observed that only about 2 per cent of the cases are solely breast-fed at the age of twenty-three months (Belawady, 1980). Supplementary foods are introduced at various stages, the highest frequency distribution of age at supplementation being between twelve and seventeen months for the rural group (Belawady, 1980)..

Breast-feeding continued unabated for a period of one year or more among these groups. However, the mean duration of breast-feeding for the entire population was twenty-one months. The duration of breast-feeding, therefore, varied among the Muslims (eighteen months), the Harijans (twenty-four months) and the Caste Hindus (twenty months). In the Khanna area of Punjab, the mean duration of breast-feeding extended up to twenty-six months (Wyon and Gordon, 1971). Breast-feeding has influenced infant and childhood mortality as a third most important factor in the absence of supplementary foods given for a long time, that is, one year among these groups. Though the duration of breast-feeding was longer among the Harijans in comparison with the other groups, it could not minimise the degree of infant and childhood mortality among them because of the mother's poor nutritional status and the inadequate feeding of supplementary or weaning foods. But a negligible proportion of mothers—the Harijans (2 per cent), the Muslims (5 per cent) and the

TABLE 2 Per Cent Distribution of Infant and Child Deaths among the Religious and Caste Groups by Duration of Breast-Feeding

Duration of Breast-Feeding	*Three Cultural Groups*			
	Muslims	*Harijans*	*Caste Hindus*	*Total*
1 *Infant Deaths*				
No lactation	44.79 (142)	36.28 (115)	35.42 (96)	39.00 (353)
Up to 28 days (most of the cases from 8 to 28 days)	14.20 (45)	17.67 (56)	22.51 (61)	17.90 (162)
1–6 months	27.13 (86)	32.80 (104)	25.46 (69)	28.62 (259)
7 to less than 12 months	13.88 (44)	13.25 (42)	16.61 (45)	14.48 (131)
Total	100.00 (317)	100.00 (317)	100.00 (271)	100.00 (905)
2 *Child Deaths (1–4 years)*				
No lactation	5.71 (8)	2.80 (4)	3.36 (4)	3.98 (16)
Less than 12 months	39.29 (55)	32.86 (47)	34.45 (41)	35.57 (143)
12 months to 24 months	50.00 (70)	41.26 (59)	52.10 (62)	47.51 (191)
25+ months	5.00 (7)	23.08 (33)	10.08 (12)	12.94 (52)
Total	100.00 (140)	100.00 (143)	100.00 (119)	100.00 (402)

Caste Hindus (8 per cent), did not breast-feed for a period of six months or less due to lack of milk in them. Only 0.3 per cent women did not breast-feed the infants for fear of losing physical fitness and beauty.

Breast-feeding significantly depresses infant mortality across the three cultural groups (see Table 2). Similar findings have been reported from several other studies particularly the Pan American Health Organisation study conducted in Central and South America (1973), in Germany (Knodel, 1978), Bangladesh (Kent, 1981), Bolivia (Koopman, 1983). It has been well documented that breast milk has anti-infective factors which protect the infant against infections such as diarrhoeal diseases, Otitis media particularly with low standards of environmental hygiene (Jelliffe, 1976; Nutrition Reviews, 1983). A similar relationship which childhood mortality, however, was not observed. The largest proportion of infant mortality among the Muslims (45 per cent), the Harijans (36 per cent) and the Caste Hindus (35 per cent) occurred in the absence of any breast-feeding. A high degree of infant mortality occurred between one to four days after birth. These babies were not breast-fed for the first three days and instead a variety of prelacteal feeds referred to earlier were given. Therefore, neonatal mortality may have been due to the absence of breast-feeding for the first three days, poor health and physiological condition of the mother, and due to the complications resulting from the removal of the umbilical cord. The magnitude of the incidence of neonatal mortality suggests that urgent and imaginative health education coupled with hospital delivery or delivery attended to by qualified persons should be considered. The relationship between infant mortality and the duration of breast-feeding among the three cultural groups showed a somewhat inconsistent trend—a drastic drop in mortality with an increase in the duration of breast-feeding among the groups. During one to six months of breast-feeding infant mortality is higher than that recorded for the period up to twenty-eight days and for the period—seven to less than twelve months of breast-feeding. This may be related to the greater nutritional needs of the infant at this stage combined with a lack of adequate breast milk, limited knowledge of the mother regarding the nutritional needs of the infant and a greater handling of the baby by the family members with a chance for infections.

Contrary to the findings seen above, breast-feeding did not show such a significant relationship with childhood mortality. This was because the largest proportion of children from the Muslim households (50 per cent), the Harijan families (41 per cent) and the Caste Hindu households (52 per cent) died even after being breast-fed for more than one year. These children may certainly have died due to the influence of other environmental factors. By the time they are one year of age rural children are exposed to a wider environment beyond their home.

Initiation of Supplementary Foods

Mothers tend to terminate breast-feeding largely because of: lack of milk, conception and also refusal of the child to feed. Bitter applications of herbs on the breast by mothers were also reported to terminate breast-feeding (14.2 per cent). A few mothers (8.8 per cent) adopted fear techniques while the others (8.4. per cent) continued to breast-feed their infants for prolonged periods. All these practices revealed the need for educating mothers to encourage breast-feeding for a longer period of time. Supplementary food was very often given after cooking. Most of the infants below the age of six months were given mother's milk, animal milk and/or both and tinned milk. Other supplementary foods like iddli* were given from the sixth month only. Between six to twelve months, a very small proportion of infants were fed supplementary foods: the Muslims (13 per cent), the Harijans (7 per cent) and the Caste Hindus (9 per cent). Actual supplementary feeding began only after one year of age for most of the Muslim infants (62 per cent), the Harijan infants (75 per cent) and the Caste Hindu infants (65 per cent). In many cases there was no regular, fixed timing for feeding though it often coincided with the meal-times of adults or with the time when the babies began to cry. The supplementary food given to children up to 4 years included animal milk, iddli and rice in addition to mother's milk. For the intermittent feeds, biscuits, fruits, sweets, etc., were also given. Fruit as a supplementary food was given by 12 per cent of the Muslims, 5 per cent of the Harijans and 17 per cent of the Caste Hindus. The larger intake of fruits by the Caste Hindus may be due to their high socio-economic status and/or better understanding of the nutritional value of fruits. The supplementary foods given during early childhood consisted chiefly of cereals with the exception of pulses used in the preparation of iddli. Therefore, a qualitative change in the food habits of this age group of children with a balanced supply of supplementary nutrition may promote the growth of the children. In many of these families in the three groups children between one to four years of age experienced a high incidence of morbidity due to imbalanced dietary habits.

Breast-feeding, weaning and supplementary food practices among all the rural groups of women studied, follow similar patterns observed in other studies. Mothers breast-feed the babies from the third day onwards, if there is lactation. Prolonged breast-feeding was a common practice among these mothers and in 2 per cent of the cases the children were solely breast-fed up to twenty-three months. Supplementary foods were initiated anywhere between six to thirteen or more months; the highest frequency distribution of age at supplementation being between seven and twelve months. Weaning was followed using a variety of techniques and the majority of the mothers weaned their children between thirteen and twenty-four months. Supplementary foods were mostly cereal/pulse combinations and commercial biscuits.

Care and Affection

A definite pattern of initiation of the first bath for the infants from these three groups was observed on odd days starting from the first day to the eleventh day. A greater proportion of the Muslims (36 per cent) and the Caste

* A batter made out of soaked rice and blackgram and steam cooked.

Hindus (35 per cent) bathed the infants on the first day itself. While the Harijans bathed them only on the third day, slightly more than a quarter of the Muslims (28 per cent) and the Caste Hindus (26 per cent) were found to bathe the new born on the same day. Others without much variation, among these groups, bathed them on the fifth, seventh, ninth and eleventh days.

The degree of care and affection shown to infants and children among these three groups revealed remarkable differences. These differences were noticed specifically with reference to (*a*) affection in terms of fondling of children; (*b*) parental attention in terms of time spent; (*c*) comforts given; (*d*) food provided; and (*e*) education planned. The affection shown to children was found to be uniform without any special preference for a particular child—the Muslims (38 per cent), the Harijans (37 per cent) and the Caste Hindus (47 per cent). The differences in the proportion of women showing affection to the first born (11 to 18 per cent) and the last born (9 to 11 per cent) were more or less same in all these groups. However, a slightly higher proportion of Muslims (18 per cent) showed equal affection to the first child and to sons. Apart from the varied affection shown, there was also differential affection shown to a combination of children or to a particular child which, of course, was negligible. These patterns of affection would have been much more pronounced if all the mothers had responded without much inhibition (table not added here). In spite of the best efforts made to gather this information from mothers, in certain cases mothers tried to conceal their real response in the presence of their children or neighbours. Further, data were not collected separately for infants and children because of the delicacy involved in an enquiry of this nature. It is clear that affection does not seem to influence infant mortality much in spite of the diverse nature of affection shown to different children of specific order and sex. The incidence of childhood mortality was extensively observed when the degree of affection was uniform: the Muslims (51 per cent), the Harijans (49 per cent) and the Caste Hindus (61 per cent). On the other hand childhood mortality was found to be high when differential affection was shown more for the first child—the Muslims (25 per cent), the Harijans (15 per cent) and the Caste Hindus (9 per cent). Nevertheless, the magnitude of mortality did not increase much among the first born but mortality increased substantially among the subsequent children. Similarly, when more affection was shown to sons and to the last child, there was a tendency to manifest a low level of mortality among them, though this was not statistically significant. This variable is important because the nature of affection shown to different order and sex, at least latently, culminates in differential care and attention to respective children which may lead to a diverse incidence of morbidity and mortality. But more in-depth studies are needed to isolate the importance of this factor and to confirm the exact nature of its influence on infant and childhood mortality.

However, the nature of comforts, attention, food and the provision for education for the children have not shown any appreciable variation. Nevertheless, the last child (19 to 23 per cent) seems to receive equal attention in all the groups. Further research and in-depth case studies may yield conclusive findings about child care patterns and their influence on both infant and childhood mortality.

Nutritional Status of Mothers

Clinical Observations

All the mothers of the large sample underwent a clinical examination for the symptoms of nutritional deficiencies of iron, riboflavin and vitamin C. This method of examination is only a crude estimate of the nutritional status as compared to the more refined and precise methods such as biochemical assessments. While clinical examination facilitates detection of nutritional deficiencies, it complicates our interpretation as the observed physical changes may result from other than nutritional factors such as exposure to wind, sun or the lack of teeth. Many of the individual lesions cannot be associated with a single nutrient deficiency. Angular lesions and Cheilosis, for example, can be due to deficits in both riboflavin and niacin (McGanity, 1973). With this understanding an attempt has been made to examine the results for the incidence of deficiency states among the three groups in the total sample. The possibility of relating infant and childhood mortality to the general nutritional status of the mother has been explored.

Among the three deficiencies observed, anaemia was found to be prevalent to the maximum extent (Table 3). However, only 19.4 to 26.4 per cent cases observed externally had anaemia among the three groups. The rest of the cases may possibly be at various stages of iron deficiency. Since biochemical assessments were not carried out on the total sample it was not possible to clearly

TABLE 3 Per Cent Distribution of the Muslims, the Harijans and the Caste Hindu Mothers as related to their Nutritional Status

Nutritional Status (Clinical Symptoms)	*Three Cultural Groups*			
	Muslims	*Harijans*	*Caste Hindus*	*Total*
Iron Deficiency				
Observed	24.5 (245)	26.4 (264)	19.4 (194)	23.4 (703)
Not observed	75.5 (755)	73.6 (736)	80.6 (806)	76.6 (2297)
Riboflavin Deficiency				
Observed	5.8 (58)	8.0 (80)	4.2 (42)	6.0 (180)
Not observed	94.2 (942)	92.0 (920)	95.8 (958)	94.0 (2820)
Vitamin C Deficiency				
Observed	3.2 (32)	3.0 (30)	1.7 (17)	2.6 (79)
Not observed	96.8 (968)	97.0 (970)	98.3 (983)	97.4 (2921)

define the degree of deficiency state with regard to iron. Riboflavin deficiency was found to be the next highest in the total population. Here, again, biochemical parameters may have yielded a better classification of the deficiency state. Ascorbic acid deficiency was the least reported deficiency not only in the present study but also in other nutritional surveys (Mohanram, 1965; Srikantia *et al.*, 1970). For all the three deficiency states reported at present, the Muslims and the Harijans showed a greater incidence of the deficiencies than that observed in the case of the Caste Hindus. These results are in concordance with the results of food choice patterns of the large sample survey where a greater proportion of the Caste Hindus were observed to have a better choice of foods than the Muslims and the Harijans (for details see the original report).

Considering the total population, anaemia was seen to be prevalent to the extent of 23.4 per cent, riboflavin deficiency 6 per cent, and vitamin C deficiency 2.6 per cent. In the present context, though an attempt was made, infant and child mortality data could not be correlated with the existing nutritional status of the mother, probably because the clinicial method of assessment of the nutritional status of the mother has its limitations and it does not give a precise and true estimation of the status of the mother.

Anthropometric Measurements and Haemoglobin Estimations

With regard to infant and childhood mortality in the experimental group, a relationship between the height of the mother and mortality, between the weight of the mother and mortality, and between the haemoglobin level of the mother and mortality was determined on the basis that infant and childhood mortality may be a reflection of the nutritional status of the mother.

Data regarding the height and weight of the mothers were arranged at intervals of 5 centimetres and 5 kilograms respectively, and the per cent distribution of the cases for the three groups—the Muslims, the Harijans and the Caste Hindus—was studied at each level. With regard to height, the number of cases in the lower range of 140 to 150 centimetres was generally higher for the experimental groups than for the control groups. The number of cases in the 150 to 160 centimetres range was higher for the control group of women than for the experimental group. When weight was considered, no such trend was observed. Weight is a criterion that could vary within short periods of time depending on various other factors, apart from the nutritional status (see Table 4).

Height/Weight Indices of Mothers

The infant and childhood mortality distribution in the experimental group of the sub-sample as related to the height/weight of the mothers indicated that the maximum number of deaths occurred at the lower levels of both height and weight of the mothers. The least number of deaths were observed when either

TABLE 4 Per Cent Distribution of Respondents According to Height and Weight among the Religious and Caste Groups

Height and Weight	*Three Cultural Groups*							
	Muslims		*Harijans*		*Caste Hindus*		*Total*	
	Exp.	*Cont.*	*Exp.*	*Cont.*	*Exp.*	*Cont.*	*Exp.*	*Cont.*
Height (cms.)								
140–145	18.4 (9)	12.1 (4)	20.0 (10)	15.8 (6)	15.0 (6)	32.0 (8)	18.0 (25)	18.8 (18)
145–150	40.8 (20)	45.5 (15)	36.0 (18)	26.3 (10)	42.5 (17)	16.0 (4)	39.6 (55)	30.2 (29)
150–155	26.5 (13)	24.2 (8)	28.0 (14)	50.0 (19)	32.5 (13)	44.0 (11)	28.8 (40)	39.6 (38)
155–160	14.3 (7)	18.2 (6)	16.0 (8)	7.9 (3)	10.0 (4)	8.0 (2)	13.7 (19)	11.5 (11)
Total	100.0 (49)	100.0 (33)	100.0 (50)	100.0 (38)	100.0 (40)	100.0 (25)	100.0 (139)	100.0 (96)
Weight (kgs.)								
≤40	26.5 (13)	36.4 (12)	46.0 (23)	39.5 (15)	40.0 (16)	36.0 (9)	37.4 (52)	37.5 (36)
41–45	44.9 (22)	36.4 (12)	34.0 (17)	39.5 (15)	32.5 (13)	32.0 (8)	37.4 (52)	36.5 (35)
46–50	12.2 (6)	21.2 (7)	14.0 (7)	18.4 (7)	12.5 (5)	24.0 (6)	12.9 (18)	20.8 (20)
51–55	16.3 (8)	6.0 (2)	4.0 (2)	2.6 (1)	7.5 (3)	8.0 (2)	9.4 (13)	5.2 (5)
56+	–	–	2.0 (1)	–	7.5 (3)	–	2.9 (4)	–
Total	100.0 (49)	100.0 (33)	100.0 (50)	100.0 (38)	100.0 (40)	100.0 (25)	100.0 (139)	100.0 (96)

Standard weight of the Indian adult woman is 45 kgs

the height or the weight of the mother was above 150 centimetres or 50 kilograms respectively (Table 5).

Haemoglobin Per cent of Mothers

At all levels of anaemia expressed as a percentage of haemoglobin, the incidence is higher in the experimental group than that observed in the control group across the three cultural groups. When the total number of respondents were classified into three categories on the basis of their haemoglobin status—very severely deficient, severely deficient and moderately deficient, the experimental group showed 7.9 per cent, 51.1 per cent and 12.9 per cent respectively as against 3.1 per cent, 23.9 per cent and 4.2 per cent respectively for the control group. For cases with the haemoglobin level of 11 per cent or more, the percentage distribution of respondents in the experimental and control groups was 28.06 and 68.75 respectively. When a comparison was made between the

TABLE 5 Per Cent Distribution of Infant and Childhood Deaths by Height and Weight of the Respondents among the Religious and Caste Groups

Height and Weight	*Three Cultural Groups*			
	Muslims	*Harijans*	*Caste Hindus*	*Total*
Infant Mortality: Height (cms.)				
140–145	13.6 (6)	16.7 (9)	20.5 (9)	16.9 (24)
145–150	36.4 (16)	37.0 (20)	43.2 (19)	38.7 (55)
150–155	36.4 (16)	31.5 (17)	31.8 (14)	33.1 (47)
155–160	13.6 (6)	14.8 (8)	4.5 (2)	11.3 (16)
Infant Mortality: Weight (kgs.)				
≦40	38.6 (17)	53.7 (29)	45.5 (20)	46.5 (66)
41–45	38.6 (17)	29.6 (16)	40.9 (18)	35.9 (51)
46–50	16.0 (7)	13.0 (7)	2.3 (1)	10.6 (15)
51–55	6.8 (3)	1.9 (1)	6.8 (3)	4.9 (7)
56+	–	1.9 (1)	4.5 (2)	2.1 (3)
Childhood Mortality: Height (cms.)				
≦150	78.4 (29)	41.7 (5)	55.6 (10)	65.7 (44)
151–155	5.4 (2)	25.0 (3)	22.2 (4)	13.4 (9)
156+	16.2 (6)	33.3 (4)	22.2 (4)	20.9 (14)
Childhood Mortality: Weight (kgs.)				
≦40	16.2 (6)	58.3 (7)	16.7 (3)	23.9 (16)
41–45	46.0 (17)	25.0 (3)	22.2 (4)	35.8 (24)
46–50	32.4 (12)	8.3 (1)	22.2 (4)	25.4 (17)
51+	5.4 (2)	8.3 (1)	38.9 (7)	14.9 (10)

Muslims, the Harijans, and the Caste Hindus with regard to the incidence of anaemia of different degrees, it was observed that at the normal level of 11 per cent and above, the Muslims in the experimental group have the lowest figure of 20.41 per cent as against 32.0 and 32.5 per cent for the Harijans and the Caste Hindus respectively. For the deficient cases, the highest incidence was observed for all the three groups at the severely deficient state of anaemia. However, the

Caste Hindus, had a lower incidence of anaemia—37.5 per cent as against 57.2 per cent and 56.0 per cent for the Muslims and the Harijans respectively. Among the three cultural groups when all the three levels of deficiency are considered together, the Muslims had the highest incidence of anaemia—79.6 per cent as against 68.0 per cent and 67.5 per cent for the Harijans and the Caste Hindus respectively (Table 6). The mothers in the control group, however, had a better nutritional status as compared to those in the experimental group. The per cent incidence was 24.2, 34.2 and 36.0 for Muslims, Harijans and Caste Hindus respectively.

TABLE 6 Per Cent Distribution of the Muslim, the Harijan and the Caste Hindu Respondents by their Haemoglobin Assessment

Degree of Anaemia (Haemoglobin: gms./100 ml. of blood)	*Three Cultural Groups*							
	Muslims		*Harijans*		*Caste Hindus*		*Total*	
	Exp.	*Cont.*	*Exp.*	*Cont.*	*Exp.*	*Cont.*	*Exp.*	*Cont.*
Very severely deficient (≲8.4)	12.2 (6)	6.0 (2)	4.0 (2)	2.6 (1)	7.5 (3)	0.0 –	7.9 (11)	3.1 (3)
Severely deficient (8.4 to 10.0)	57.1 (28)	18.2 (6)	56.0 (28)	23.7 (9)	37.5 (15)	32.0 (8)	51.1 (71)	23.9 (23)
Moderately deficient (10.1 to 10.9)	10.2 (5)	0.0 –	8.0 (4)	7.9 (3)	22.5 (9)	4.0 (1)	12.9 (18)	4.2 (4)
Normal (11.0+)	20.4 (10)	75.8 (25)	32.0 (16)	65.8 (25)	32.5 (13)	64.0 (16)	28.1 (39)	68.8 (66)
Total	100.0 (49)	100.0 (33)	100.0 (50)	100.0 (38)	100.0 (40)	100.0 (25)	100.0 (139)	100.0 (96)

The incidence of infant and childhood mortality was related to the nutritional status of the mother expressed as severity of anaemia (Table 7). The greatest incidence of infant and childhood mortality occurred when the mothers' anaemic state was reported as severe. Considering the total sample, the percentage of deaths was 51.3 and 53.7 for infants and children respectively, when the mothers had 8.4 to 10.0 per cent haemoglobin level. This trend was noted in all three categories (the Muslims, the Harijans and the Caste Hindus), when respondents with different levels of deficiency of haemoglobin were considered together and compared to the normal cases, it was found that the former experienced a higher incidence of infant and childhood deaths than the latter. With regard to infant mortality, the percentage of incidence was 71 among the deficient mothers while it was only 29 per cent among the normal mothers. Considering childhood mortality, the incidence was 80.6 per cent among the deficient mothers as compared to 19.4 per cent among the normal mothers. Comparing the three cultural groups. the highest per cent of infant and child

TABLE 7 **Per Cent Distribution of Infant and Childhood Deaths as per Haemoglobin Per Cent Among the Religious and Caste Groups**

Level of Anaemia (Haemoglobin: gms./100 ml. of blood)	*Infant*				*Child*			
	Muslims	*Harijans*	*Caste Hindus*	*Total*	*Muslims*	*Harijans*	*Caste Hindus*	*Total*
Very severely deficient (≤8.4)	13.6 (6)	3.7 (2)	11.4 (5)	9.2 (13)	18.9 (7)	16.7 (2)	11.1 (2)	16.4 (11)
Severely deficient (8.4 to 10.0)	68.2 (30)	50.0 (27)	36.4 (16)	51.4 (73)	59.5 (22)	58.3 (7)	38.9 (7)	53.7 (36)
Moderately deficient (10.1–10.9)	4.5 (2)	9.3 (5)	18.2 (8)	10.6 (15)	10.8 (4)	0.0 –	16.7 (3)	10.4 (7)
Normal (11.0+)	13.6 (6)	37.0 (20)	34.0 (15)	28.9 (41)	10.8 (4)	25.0 (3)	33.3 (6)	19.4 (13)
Total	100.0 (44)	100.0 (54)	100.0 (44)	100.0 (142)	100.0 (37)	100.0 (12)	100.0 (18)	100.0 (67)

mortality among the deficient mothers was observed among the Muslims; for the deficient mothers the per cent of infant mortality was 86.4, 63.0 and 66.0 for the Muslims, the Harijans, and the Caste Hindus respectively. The percentage of child mortality for the deficient mothers from the Muslim, the Harijan, and the Caste Hindu households, was 89.2, 75.0 and 66.7 respectively (Table 7). These observations confirm the fact that the Muslim mothers in this study were at a lower level of nutritional status than either the Harijan mothers or the Caste Hindu mothers. This, in turn, appears to account for the greater incidence of infant and child deaths observed among the Muslim families. Similar findings were reported from Narangwal area of Northern India which showed a strong association of perinatal, neonatal, post-neonatal, and one to three years mortality rates with the nutritional status (Kielmann *et al.*, 1978).

Morbidity and Medical Aid

Causes of Death

The major factors influencing infant mortality were found to be mostly maternal: prematurity, malnutrition, asphyxia, neonatorum, and delivery problems (the Muslims, 31 per cent; the Harijans, 28 per cent; the Caste Hindus, 25 per cent). As expected, the role of these factors was significantly less for the Caste Hindus compared to the other groups. It could be due to the better health status and greater utilisation of health facilities. Water-borne diseases (diarrhoea, dysentery, cholera, etc.) occupy the second place among the causes of infant mortality (the Muslims, 18 per cent; the Harijans, 19 per cent; the Caste Hindus, 18 per cent). Air-borne diseases constitute the third major cause of death which ranged from 11 to 17 per cent. Other causes of death included fever, fits, stomach ailments, respiratory problems, animal bites and so on. One-fifth of the respondents were unaware about the causes of death and 4 to 8 per cent attributed it to unscientific causes like wind, ghost, and evil eye. In other words, one-fourth of the respondents either did not know the causes or knew only the wrong ones (Table 8).

The major causes of childhood mortality in all the groups were water-borne and air-borne diseases and fever (the Muslims, 82 per cent; the Harijans, 80 per cent; the Caste Hindus, 79 per cent). The pattern of disease reported here more or less fully concurs with the findings obtained from several other developing countries (Mata, 1983; Mosley, 1983; Preston, 1979; Wyon and Gordon, 1971; Chen *et al.*, 1980; Omran and Standley, 1976). The environmental factors play an important role in childhood mortality, as children move away from parents after a year or even earlier and are exposed to environmental hazards. During this period, children play near ponds and ditches, and eat and drink whatever comes their way. Though the incidence of childhood mortality due to water-borne diseases was high in general, the incidence was more significant among the Harijans (Table 8). Under-reporting regarding the specific causation of death of infants was much less compared to the reporting of the causes of childhood mortality. Other causes of death of children included malnutrition, fits, stomach problems, respiratory and liver troubles, animal bites/accidents, tumours, and internal haemorrhage. While maternal causes and water-borne diseases constituted together the major causes of infant mortality, environmental factors played a major rôle for the higher incidence of childhood mortality. This observation is further confirmed in the succeeding section where the individual contribution of a number of variables is discussed. Thus, it may be confirmed and generalised that major factors of infant mortality were largely rooted in maternal and familial sources while the causes of childhood mortality stem from familial and environmental sources. As many of these factors are controllable, appropriate interventions should be planned for manipulating these causes of death.

Medical Aid for Babies and Mothers

Whenever infants and children fell ill, the larger proportion of the Muslims, the Harijans and the Caste Hindus, no doubt, sought the help of medical and paramedical persons. However, a fair proportion of the infants and children suffered in the absence of any medical facilities. The Harijan children suffered a lot without any medical aid. An encouraging factor noticed here was that the medical help sought from non-medical persons, particularly the indigenous quacks, has declined to a large extent. The Muslims account for only 2 per cent of this category as against a greater number of Harijans (5 per cent) and the Caste Hindus (4 per cent) (table not given).

The nature and types of complications reported during pregnancy and the type of treatment received varied significantly. A considerable proportion of the Muslims (57 per cent), the Harijans (50 per cent) and the Caste Hindus (66

TABLE 8 Per Cent Distribution of Causes of Infant/Early Childhood (1–4 years) Deaths Among the Religious and Caste Groups

S. No.	Causes of Death	Infants				Children			
		Muslims	*Harijans*	*Caste Hindus*	*Total*	*Muslims*	*Harijans*	*Caste Hindus*	*Total*
1.	Don't know	20.18 (64)	20.19 (64)	17.34 (47)	19.34 (175)	3.57 (5)	3.50 (5)	6.72 (8)	4.48 (18)
2.	Maternal factors (prematurity, malnutrition, asphyxia, neonatorum, delivery)	30.91 (98)	27.76 (88)	25.09 (68)	28.07 (254)	2.86 (4)	2.80 (4)	3.36 (4)	2.99 (12)
3.	Water-borne diseases (diarrhoea/dysentery, jaundice, cholera, etc.)	18.30 (58)	19.24 (61)	17.71 (48)	18.45 (167)	47.14 (66)	51.05 (73)	44.54 (53)	47.76 (192)
4.	Air-borne diseases (tetanus, chicken-pox, measles, etc.)	11.67 (37)	11.36 (36)	16.61 (45)	13.04 (118)	25.71 (36)	24.48 (35)	26.05 (31)	25.37 (102)
5.	Superstitious reasons (wind, ghost, evil eye, etc.)	5.36 (17)	7.89 (25)	3.69 (10)	5.75 (52)	2.14 (3)	2.80 (4)	1.68 (2)	2.24 (9)
6.	Stomach problems	4.42 (14)	5.99 (19)	2.58 (7)	4.42 (40)	0.71 (1)	0.70 (1)	0.84 (1)	0.75 (3)
7.	Fever	2.84 (9)	3.47 (11)	6.64 (18)	4.20 (38)	9.29 (13)	4.90 (7)	9.24 (11)	7.71 (31)
8.	Fits	1.26 (4)	1.89 (6)	3.69 (10)	2.21 (20)	1.43 (2)	2.80 (4)	1.68 (2)	1.99 (8)
9.	Animal bites/accidents	0.63 (2)	1.26 (4)	0.74 (2)	0.88 (8)	2.86 (4)	5.59 (8)	3.36 (4)	3.98 (16)
10.	Respiratory troubles	1.58 (5)	0.32 (1)	2.21 (6)	1.33 (12)	2.86 (4)	0.70 (1)	–	1.24 (5)
11.	Others (liver, tetanus, internal haemorrhage, congenital malformation)	2.84 (9)	0.63 (2)	3.69 (10)	2.32 (21)	1.43 (2)	0.70 (1)	2.52 (3)	1.49 (6)

per cent) sought the services of qualified doctors or paramedical persons during the antenatal period. Extensive utilisation of medical facilities was largely by the Muslims and the Caste Hindus and to a certain extent by the Harijans also. This was an exceptional finding for this area because of the availability of a Medical College Hospital, the government and private hospitals in Tirupati within a distance of four to twenty kilometres. In spite of the availability of such facilities, nearly one-third of the respondents did not seek any prenatal care. However, the antenatal service rendered by neighbours, friends and relatives drastically reduced to 5 to 6 per cent among the three cultural groups. Nevertheless, 4 to 10 per cent of the Caste Hindus, through the Muslims to the Harijans did initiate the treatment first from non-medical persons and ultimately went to qualified persons when they were disillusioned with the service of the former (table not given). In this context, the incidence of infant mortality and post-abortion complications was higher among people who sought the antenatal services rendered by non-qualified persons. If this population is educated to accept modern medical facilities, morbidity, infant and maternal mortality could be reduced to a minimum.

Delivery Attendants

The general pattern of delivery care was provided largely by the indigenous *dai* (who was untrained). It varied significantly across the three cultural groups accounting for 47 per cent, 66 per cent and 53 per cent of deliveries attended to by these *dais*. It revealed that a higher proportion of the Harijans and the Caste Hindus utilised their services in contrast to the Muslims. The services rendered by the trained *dai* were very negligible (1 to 2 per cent). The age old practices of having neighbours/relatives/friends conduct the delivery were still in vogue among one-tenth of the respondents in each of the two groups—the Muslims and the Caste Hindus—compared to one-fifth of the Harijans. ANMs, and nurses constitute the second largest category of birth attendants; they conduct a substantial number of the total deliveries among the Muslims (26 per cent) as against 13 per cent and 5 per cent among the Caste Hindus and the Harijans respectively. It was seen that only a small fraction of the respondents (11 per cent Muslims, 7 per cent Harijans and 19 per cent of the Caste Hindus) sought the services of allopathic doctors. The findings suggest that a concerted effort may be made to provide extensive maternal and child health services to the Harijans and the Caste Hindus. An in-depth study is also necessary to find out why an overwhelming proportion of the Harijans do not utilise the services of qualified birth attendants (Table 9).

Cutting the Umbilical Cord

The types of instruments used for severing the umbilical cord; the type of personnel who sever the cord and the material used for dressing the cord are given in Table 10. The instruments used for severing the cord include (*a*) unsterilised household knife; (*b*) new blade; and (*c*) scissors; which were used

TABLE 9 Per Cent Distribution of Respondents by Category of Birth Attendants at the Last Delivery

Birth Attendants	*Three Cultural Groups*			
	Muslims	*Harijans*	*Caste Hindus*	*Total*
Neighbours/Relatives/ Friends	14.7	19.6	13.8	16.0
	(147)	(196)	(138)	(481)
Indigenous *dai*	47.3	65.7	52.7	55.2
	(473)	(657)	(527)	(1657)
Trained *dai*	1.3	2.2	1.6	1.7
	(13)	(22)	(16)	(51)
Nurse/ANM	25.6	5.2	12.6	14.5
	(256)	(52)	(126)	(434)
Doctor	11.1	7.3	19.3	12.6
	(111)	(73)	(193)	(377)
Total	100.0	100.0	100.0	100.0
	(1000)	(1000)	(1000)	(1000)

differentially across the three cultural groups. A significant proportion of the Harijans (73 per cent), half of the Caste Hindus and slightly less than half of the Muslims (46 per cent) used the household knife for severing the cord. This is a dangerous practice leading, very often, to tetanus neonatorum. This unhygienic practice accounted for a very high proportion of neonatal mortality. The other instruments particularly the new blade could also lead to complications but as scissors were invariably used by trained birth attendants there were usually no complications. It was encouraging to note that an increasing proportion of the Muslims (41 per cent) and the Caste Hindus (33 per cent) as against a negligible proportion of the Harijans (15 per cent) used a pair of scissors for severing the umbilical cord. It is likely that doctors and ANMs used a pair of scissors; while neighbours, relatives and untrained *dais* used the household knife and a new blade. The single largest group of individuals who severed the umbilical cord comprised unqualified persons like neighbours, and indigenous *dais*, accounting for 62 per cent among the Muslims, 88 per cent among the Harijans and 68 per cent among the Caste Hindus, while others sought the services of ANMs or doctors. Simmons *et al.*, (1978) also found that an overwhelming proportion of deliveries in Uttar Pradesh, a state in North India, were attended to by untrained *dais* which caused large-scale tetanus accounting for 60 per cent of perinatal mortality.

A variety of materials ranging from harmless to fatal were applied by the untrained persons for dressing the cord. These preparations consisted of ash, mud removed from an earthen stove, areca nut powder, pepper, myristica fragrance (jajikaya), zeera (vamu), sompu, latex from creepers, talcum power, and medicated powder. Most of the people, with the exception of 38 per cent of the Muslims, 16 per cent of the Harijans and 33 per cent of the Caste Hindus, used different types of unscientific materials for dressing the cord. These materials along with a non-sterilised knife and a blade used for severing the cord together accounted for infant mortality. Unless health education for the public in general and intensive training to all the indigenous *dais* coupled with supervision by trained health supervisors are not provided immediately, infant mortality resulting from unhygienic practices cannot be controlled. These practices significantly contributed to infant mortality largely among the Harijans followed by the Muslims and the Caste Hindus.

TABLE 10 Per Cent Distribution of Respondents in Relation to the Practices Adopted During Deliveries Among the Religious and Caste Hindus

Practices during Delivery	*Three Cultural Groups*			
	Muslims	*Harijans*	*Caste Hindus*	*Total*
Material Used for Cutting the Umbilical Cord				
Household knife	45.5 (455)	72.8 (728)	50.6 (506)	56.3 (1689)
New blade	13.2 (132)	11.9 (119)	16.3 (163)	13.8 (414)
Scissors	41.3 (413)	15.3 (153)	33.1 (331)	29.9 (897)
Personnel Who Cut the Cord				
Neighbours/relatives	15.1 (151)	19.9 (199)	13.6 (136)	16.2 (486)
Trained/untrained Indigenous *dai*	47.3 (473)	67.7 (677)	54.8 (548)	56.6 (1698)
Doctor/Nurse/ANM	37.6 (376)	12.4 (124)	31.6 (316)	27.2 (816)
Material Applied to the Cord				
Nothing applied	15.4 (154)	13.6 (136)	14.9 (149)	14.6 (439)
Ash/mud taken from earthen stove	22.2 (222)	50.2 (502)	31.4 (314)	34.6 (1038)
Areca nut powder/pepper/ jajikaya/vamu/sompu/ latex of creeper	1.8 (18)	6.9 (69)	2.3 (23)	3.7 (110)
Talcum powder	22.9 (229)	12.9 (129)	18.1 (181)	18.0 (539)
Medicated powder	37.7 (377)	16.4 (164)	33.3 (333)	29.1 (874)
Total	100.0 (1000)	100.0 (1000)	100.0 (1000)	100.0 (3000)

Taking into account the two major factors discussed earlier, namely, source of drinking water and material used to sever the umbilical cord did show differential influence in producing infant mortality among the three cultural groups (see Table 11). It was seen that the source of drinking water was uniformly negatively related to infant mortality among the three groups, but the material used for cutting the umbilical cord was negatively influencing both among the Harijans and the Caste Hindus but not among the Muslims. The analysis of variance revealed that the source of drinking water influenced infant mortality at a very high significant level (at 1 per cent) whereas the material used to sever the umbilical cord influenced infant mortality at 5 per cent level of significance only.

TABLE 11 Correlation Coefficient of Health Variables Influencing Infant Mortality

Variables	*Three Cultural Groups*			
	Muslims	*Harijans*	*Caste Hindus*	*Total*
Material used to cut the umbilical cord	−.3060	−.8789	−.7446	−.7340
Source of drinking water	−.8839	−.8740	−.8455	−.8633

Thus, nearly 53 per cent of infant mortality occurred within twenty-eight days after delivery. The critical periods after delivery are during the first three days and up to twenty-eight days besides at the time of commencement of supplementary feeding. Several socialisation variables such as prelacteal feeds, breast-feeding, initiation of supplementary foods, care and affection together affect the incidence of infant mortality. Nutritional deficiency stemming from poor food choice and poverty together affect the health of the mother which, in turn, leads to infant mortality. Anaemia of the mother is associated with infant mortality which is the result of nutritional deficiency, poverty and lack of proper and timely medication. While socialisation occurring within the context of the family environment, breast-feeding and nutritional factors (bio-familial) influence infant mortality, morbidity like diarrhoea, dysentery and communicable diseases resultant of familial and environmental factors (familio-environmental) predominantly influences childhood mortality. *Hence we can conclude that by and large bio-familial factors affect infant mortality and familio-environmental factors affect early childhood mortality.*

REFERENCES

AGARWALA, R.D. (1975). Influence of infant and child mortality on fertility in India. Paper presented at the seminar on Infant Mortality in Relation to Level of Fertility, Bangkok. Paris: CICRED Series.

BELAWADY, B. (1980). Breast feeding practices in Andhra Pradesh. Proceedings of the Nutrition Society of India, 26: 39–47.

CHANDRA, H. (1971). Birth weights of infants of different economic groups in Hyderabad. Proceedings of the Nutrition Society of India, 10: 99–103.

CHANDRA, R.K. (1978). Immunological aspects of human milk. *Nutrition Reviews*, 36: 265–272.

CHANDRASEKHAR, S. (1972). *Infant mortality, population growth and family planning in India.* London: George Allen & Unwin Ltd.

CHEN, L.C., A.K.H. ALLAUDDIN CHOWDHURY and S.L. HUFFMAN. (1980). Anthropometric assessment of energy—protein malnutrition and subsequent risk of mortality among pre-school aged children. *American Journal of Clinical Nutrition*, 33(8): 1836–45.

CROSBY, MUNN and FURTH. (1965). Estimation of haemoglobin. *Med. Journal*, 693, 1954. As described by B.L. Oser, *Hawk's physiological chemistry*, Chapter 29. Blood Analysis, p. 1096. Bombay: Tata McGraw Hill Company Ltd.

DAVIS, KINGSLEY. (1951). *Population of India and Pakistan*. Princeton: Princeton University Press.

JAIN, S.P. (1974). Some aspects of mortality. *In* Ashish Bose *et al.* (Eds.), *Population in India's development*. New Delhi: Vikas Publishing House.

JELLIFFE, D.B. (1976). World trends in infant feeding. Proceedings of the Nutrition Society of India, 20: 1–18.

KENT, MARRY MEDERIOS. (1981). Breast-feeding in the developing world. Current patterns and implications for future trends. Report on WFS (2), Population Reference Bureau, pp. 1–43.

KIELMANN, A.A. *et al.* (1978). The Narangwal experiment on interactions of nutrition and infections: II morbidity and mortality effects. *Indian Journal of Medical Research*, 68: 21–41.

KNODEL, J. (1978). Breast feeding and population growth. *Science*, Vol. 198, pp. 1111–15.

KOHLI, K.G. (1977). *Mortality in India: a state-wise study*. New Delhi: Sterling Publishers.

KOOPMAN, S.J. (1983). Personal discussion.

KULKARNI, G.A. (1975). Relation between infant mortality and fertility. *Sample Registration Bulletin*, 9 (1–2).

MAHADEVAN, K. (1979). *Sociology of fertility*. New Delhi: Sterling Publishers.

MATA, LEONARDO. (1983). Evolution of diarrhoeal diseases and malnutrition in Costa Rica and role of interventions. IUSSP.

MCGANITY, W.J. (1973). The clinical assessment of nutritional status. Proceedings of the Miles symposium 1973. The assessment of nutritional status. Presented by the Nutrition Society of Canada, University of Saskatchewan, pp. 47–64.

MINISTRY OF SOCIAL WELFARE, GOVERNMENT OF INDIA and UNICEF. (1980). Proceedings of Regional Consultation Meeting of Projects of Social Inputs under Area Development Programmes.

MITRA, ASOK. (1980). Presidential address. Fourth Gopalan Oration Green Power and Freedom from Hunger and Symposium on Nutrition and Fertility of Women. Proceedings of the Nutrition Society of India, 26: 14–19.

MOHAN RAM, M. (1965). Studies on ascorbic acid nutrition. *Indian Journal of Medical Research*, 53: 891.

MOSLEY, HENRY, W. (1983). Will primary health care reduce infant and child mortality? A critique of some current strategies, with special reference to Africa and Asia. INED & IUSSP Conference.

NAEYE, R.W., G.S. BURT, D.L. WRIGHT, W.M. BLANC and D. TATTER. (1977). Neonatal mortality, the male disadvantage. *Paediatrics*, 48(6): 902–6.

NUTRITION REVIEWS. (1983). Breast feeding prevents otitis. *Media*, 41: 241.

OMRAN, A.R. and STANDLEY. (1976). *Family formations patterns and health*. Geneva: World Health Organisation.

PAN AMERICAN HEALTH ORGANIZATION. (1973). Patterns of mortality in childhood. Report of the Inter-American Investigation of Mortality in Childhood, Washington.

PRESTON, SAMUEL, H. (1979). Causes and consequences of mortality decline in LDCs during the twentieth century. *In* Richard Easterlin (Eds.), *Population and economic change in LDCs*. N.B. Eco. research.

PRIYANI, E. SOYSA. (1981). The advantages of breast feeding in assignment children. A journal concerned with children, women and youth in December. UNICEF, Palais des Nations 1211, Geneva, 10 Suisse.

RUZICKA, G.T. and T. KANITKAR. (1973). Infant mortality in Greater Bombay. *Demography—India*, 2(1): 41–55.

SIMMONS, G.B., C.M. SMUCKER, B.D. MISRA and P. MAJUMDAR. (1978). Patterns and causes of infant mortality in rural Uttar Pradesh. *Journal of Tropical Paediatrics and Environmental Child Health*, 27(5): 207–16.

SMUCKER, C.M., G.B. SIMMONS, S. BERNSTEIN and B.D. MISRA. (1980). Neonatal mortality in South Asia: special role of tetanus. *Population Studies*, 34(2): 321–35.

SRIKANTIA, S.G., M. MOHAN RAM, KAMALA KRISHNASWAMY. (1970). Human requirements of ascorbic acid. *Journal of Clinical Nutrition*, 23: 59–62.

UNITED NATIONS. (1968). *Demographic Year Book*.

UPADHYAY, B.N. *et al.* (1974). A study of perinatal, and neonatal mortality in rural population, Lucknow, India. *Canadian Journal of Public Health*, LXV, p. 219.

VAIDYANATHAN, K.E. (Ed.). (1972). *Studies on mortality in India*. Monograph Series No. 5. Gandhigram Institute of Rural Health and Family Planning.

WOODBURY, R.M. (1922). The relation between breast and artificial feeding and infant mortality. *American Journal of Hygiene*, 2(6): 668–87.

WRAY, J. (1977). Maternal nutrition, breast-feeding and infant survival. *In* Henry Mosley (Ed.), *Nutrition and human reproduction*. New York: W. Plenum Press, pp. 197–230.

WYON, J.B. and J.E. GORDON. (1971). *The Khanna study: population problems in the rural Punjab*. Cambridge: Harvard University Press.

V

A Profile

19

C. Chandrasekaran: A Profile

ASHA A. BHENDE AND V.S. BADARI

Dr C. Chandrasekaran, a pioneering demographer, has many distinctions to his credit. A Chair for Statistics was created at the All-India Institute of Hygiene and Public Health and he was appointed the first Professor of Statistics at the Institute in 1945. He was designated as the Director of the Demographic Training and Research Centre (now the International Institute for Population Sciences), Bombay, by the Government of India and appointed to this position in 1959. He has served the United Nations system in several capacities. He was appointed to a senior position in the Population Division, United Nations, New York, as early as 1948 and has subsequently held the positions of Regional Advisor on Population Policies and Programmes at ECAFE; Inter-regional Advisor in Population at United Nations, New York; Senior Advisor on Policy and Planning of the United Nations Fund for Population Activities; and Population Specialist of the International Bank for Reconstruction and Development.

Dr. C. Chandrasekaran was the first person from the developing countries to have been nominated the President of the International Union for the Scientific Study of Population and was subsequently elected as Honorary President of the Union. He was nominated the first Chairman of the Technical Advisory Committee of the World Fertility Survey. He is a member of the International Statistical Institute and a life member of the Indian Academy of Sciences.

Dr. C. Chandrasekaran was a Rockfeller Fellow during 1947–48. He was awarded the Watumull prize for his studies on family planning in 1964, and was the recipient of the University of Michigan Sesquecentennial Award in 1967. He was nominated Associate Life Member of Institut d' Egypte, Cairo, in 1981.

The Chandrasekaran–Deming formula for the estimation of birth and death rates using dual-records and Dr. C. Chandrasekaran's work on the evaluation of family planning programmes are well known. A detailed list of his contributions in the field of Population Studies is presented here, it will be valuable as a reference for future research.

List of Publications of Dr. C. Chandrasekaran*

The efficiency of statistical tools and a criterion for the rejection of outlying observations (with E.S. Pearson). *Biometrika*, 1936, Vol. 28: 308–20.

The effect of differential response to soil variation on the validity of parabolic curves. Dissertation, London University, 1938.

Analysis of empty wagon mileage. *Traffic Circular*, 1940, No. 38, B.B. & C.I. Railway.

A method to get the significance limit of a type of test criteria (with Mary G. Francis). *Sankhya*, 1941, Vol. 5, Part 2: 165–68.

The measurement of the red cell diameter—standard Price-Jone's curve for an Indian population (with L. Everard Napier and P.C. Sen Gupta). *Indian Journal of Medical Research*, 1941, Vol. 29, Part 4: 903–16.

A study of registration of birth and its utilisation for public health administration in maternity and child welfare (with L. Ghosh). *Indian Medical Gazette*, 1942, Vol. 77, No. 1: 243–52.

The health status of a Bengal village (with B. Mukerjee). *Indian Medical Gazette*, 1942, Vol. 77, No. 6: 371–78.

A note on the inverse sine transformation. *Sankhya*, 1942, Vol. 6, Part 2.

Pleural effusion in artificial pneumothorax (with P.K. Sen and K.N. De). *Indian Medical Gazette*, 1943, Vol. 78, No. 10: 498–505.

Studies in haemolysis: statistical analysis. *Indian Journal of Medical Research*, 1943, Vol. 31. No. 1: 84–101.

Development of Indian infants (with L. Ghosh and M. Sen). *Indian Journal of Pediatrics*, 1944, Vol. 11, No. 1.

Distribution of Fisher's g_1 for samples of three from a continuous rectangular distribution. *Current Science*, 1944, No. 1.

Results of a survey undertaken in a rural area in Bengal in maternity and child welfare (with L. Ghosh), *Indian Medical Gazette*, 1944, Vol. 79, No. 3: 123–30.

Errors in the use of field testing outfits for disinfection of water (with T.R. Bhaskaran *et al.*). *Water and Waterworks Engineering*, 1945, Vol. 47, No. 582.

The planning of a tuberculosis survey to determine the morbidity rate with special reference to Indian conditions. *Indian Medical Gazette*, Special Tuberculosis Number, 1945, Vol. 80, No. 1: 42–51.

Groundwater pollution in West Bengal (with B.R. Dyer and T.R. Bhaskaran). *Indian Journal of Medical Research*, 1945, Vol. 33, No. 1: 23–62.

The possibility of using pigeons in toxicity determinations of organic compounds (with N.K. Dutta). *Quarterly Journal of Pharmacy and Pharmacology*, 1946.

* The assistance given by Miss Sarah Abraham and Mr. R.T. Randhera, Librarian and Assistant Librarian respectively of the International Institute for Population Studies, Bombay, and Mr. N.S.A. Rao, Research Librarian, Population Centre, Bangalore, in the preparation of this list of publications is gratefully acknowledged.

Statistical science and public health. *Indian Health Gazette*, 1946.
Statistical assessment of the efficiency of anti-cholera inoculation. *Indian Journal of Medical Research*, 1947, Vol. 35, No. 3: 153–76.
Inquiry into vital loss up to one year of life (with L. Ghosh). Calcutta: Government of India Press, 1948.
Some aspects of Parsi demography. *Human Biology*, 1948, Vol. 20, No. 2: 47–89.
A study on the standards of maturity for Indian infants (with L. Ghosh). Calcutta: Government of India Press, 1948.
The effect of the change in mortality conditions in an age group on the expectation of life at birth. *Human Biology*, 1949, Vol. 21, No. 2.
On a method of estimating birth and death rates and the extent of registration (with W. Edwards Deming). *Journal of the American Statistical Association*, March 1949, Vol. 44: 101–15. Also *International Programme of Laboratories for Population Statistics*. Chapel Hill: University of North Carolina, 1949, Reprint Series 1971, No. 1.
Reproductive patterns of Bengali women. Report of the Indian Council of Medical Research. New Delhi, 1949.
A theorem on the correlation coefficient for samples of three when the variables are independent. *Annals of Mathematical Statistics* (now *Annals of Statistics*), 1951, Vol. 22, No. 1: 132–33.
Cultural patterns in relation to family planning in India. Third International Conference on Planned Parenthood, Bombay, 1952, 73–79.
On the concept and use of orthogonal semi-polynomials (with P.N. Chakraborty). *Sankhya*, 1952, Vol. 12, Parts 1 & 2: 141–50.
Population dynamics and food resources. Report of a Symposium on Food and Population. Mysore: Central Food Technological Institute, 1952.
Fertility trends in India. Proceedings of the World Population Conference, Rome, 1954, 1: 827–40.
Pilot study on the rhythm method of family planning in India. United Nations: T.A.A., 1954.
Use of household sample in the United Nations/Government of India population study in the Mysore State. Proceedings of the World Population Conference, Rome, 1954, 6: 227–52.
Fertility survey in Mysore State. Current research in human fertility. *Milbank Memorial Fund Quarterly*, 1955, 11–23.
Government family planning programme and its implementation. *Alumni Association Bulletin*, Calcutta: All-India Institute of Hygiene and Public Health, 1955, Vol. 2, No. 3.
India's future population. *Indian Public Health Association Souvenir*, 1955.
On the power function of a test of significance for the difference between two proportions (with S.P. Agarwala and P.N. Chakraborty). *Sankhya*, 1955, Vol. 15, Part 4: 381–90.
Opinion research in human fertility. *The Journal of Family Welfare*, 1955, Vol. 1, No. 2.
Some aspects of field surveys of human population. *Alumni Association Bulletin*,

Calcutta: All-India Institute of Hygiene and Public Health, 1957, No. 6.

Need for surveys and for assessment of rural health work. *Alumni Bulletin*, Calcutta: All-India Institute of Hygiene and Public Health, 1957, No. 6.

Present day demographic trends. *Indian Journal of Public Health*, 1957, Vol. 1, No. 1: 1–6.

India's population problems. *Asian Population Studies*, 1958, Vol. 2, No. 1: 217–25.

Cultural factors and the propagation of family planning in the Indian setting. Report of the Sixth International Conference on Planned Parenthood, 1959: 67–72. Also *The Journal of Family Welfare*, 1959, Vol. 5, No. 3: 43–51.

Sample surveys for additional information. Paper submitted to the ECAFE Seminar on Sampling, Bangkok, 1959.

Survey of the status of demography in India. *In* P.M. Hauser and O.D. Duncan (Eds.), *The study of population* Chicago: University of Chicago Press, 1959.

Pilot studies on the rhythm method of family planning (with D.V.R. Murthy). *Studies in Family Planning*, 1960, 29–43.

Population policy in India. *In* S.N. Agarwala (Ed.), *India's population*. Asia Publishing House, 1960, 143–50.

Report on the demographic and economic survey of Thailand, United Nations, 1960.

The use of sampling to extend and supplement information obtained from population census. United Nations Seminar on Evaluation and Utilisation of Population Census Data in Asia and the Far East. Bombay, 1960.

Demography as an essential basis of overall planning. *In* R.N. Saksena (Ed.), *Sociology, social research and social problems in India*. Asia Publishing House, 1961, 178–88.

Family planning. *Problems of the Third Plan—a critical miscellany*. New Delhi: Government of India, 1961.

Family planning problems of rural India, a survey of peasantry's attitudes and motivations. *Family Planning News*, 1961, Vol. 2, No. 1: 21–22.

Introductory Comments. Session on Factors Affecting Fertility in Pre-Industrialized Countries in Particular, Physiological Factors, Food and Health Status. International Population Conference, New York, 1961, 2: 83.

The organisation and case load of family planning clinics in Greater Bombay (with K. Kuder). Proceedings of the Fourth All-India Conference on Family Planning, Hyderabad, 1961, 116–24.

Physiological factors affecting fertility in India. Proceedings of the International Population Conference, New York, 1961, Session 8, Paper 76: 342–49. Also *Pakistan Medical Journal*, 1963, Vol. 14, No. 9: 29–36.

A population survey in Mysore State, India. Family Living Studies: A Symposium. Geneva: International Labour Office, 1961.

Population trends in India. Symposium on Food Needs and Resources, Mysore, 1961. New Delhi: N.I.S.I. 1962, 1–4.
Concepts used in defining urban population and data available on its characteristics in countries of Southern Asia (with K.C. Zachariah). UNESCO Regional Seminar on Urban-Rural Differences in Southern Asia: Some Aspects and Methods of Analysis, New Delhi, 1962, 51–70.
Estimating human numbers—a must for planners and a challenge to statisticians. Presidential Address to the Statistics Section, Forty-ninth Indian Science Congress. 1962.
Growing population problems. *Times Survey of India*, 26 January 1962.
Indian fertility in a changing economic and social setting. Prepared for the use of the Indian Parliamentary and Scientific Committee. *Family Planning News*, 1962, Vol. 3, No. 10: 228–36.
The mechanisms underlying the differences in fertility patterns of Bengalee women and three socio-economic groups (with M.V. George). *Milbank Memorial Fund Quarterly*, 1962, Vol. 40, No. 1: 59–89.
Population control in India. *World Medical Journal*, 1962, 8: 402–3.
Population control. *In* B.C. Tandon (Ed.), Third five year plan and India's economic growth. Allahabad: Chaitanya Publishing House, 1962, 293–96.
Family as a unit. *Family Planning News*, 1963, Vol. 4, No. 12: 229–30. Inaugural talk in the National Programme Series: Men Behind the Figures, broadcast by All India Radio, Delhi Station on 16 October 1963.
Fertility indices from limited data. Proceedings of the Meeting of International Union for the Scientific Study of Population, Ottawa; 1963, 91–105.
K.C.K.E. Raja—an appreciation. *Journal of the International Statistical Institute*, 1963.
Policies and programmes relating to fertility, public health, family welfare and family planning. Summary statement as discussion leader. *First Asian Population Conference*, New Delhi, 1963.
Problems for research on attitudes and behaviour patterns relevant to fertility and techniques of such research. Working Paper for the Asian Population Conference (APC/WP/13), New Delhi, 1963.
The relative role of information sources in the dissemination of family planning methods in Bombay City (with Prafulla C. Bebarta). *The Journal of Family Welfare*, 1963, Vol. 9, No. 4: 5–14.
Statistical problems in evaluating the effectiveness of family planning programme. Working Paper for the Asian Population Conference (APC/WP/14), New Delhi, 1963.
Survey of family planning clinics in Greater Bombay. Seventh International Conference on Planned Parenthood, Singapore, 1963, 56–63. Also *The Journal of Family Welfare*, 1963, Vol. 9, No. 3: 30–39: and *Family Planning News*, 1963, Vol. 4, No. 6: 101–5 and 118.
Evaluating the effects of community efforts to modify family size (with M.W. Freymann). Symposium: Research Issues in Public Health and Population Change, Graduate School of Public Health, University of Pittsburg, 1964.

Evaluation of family planning programmes (with M.W. Freymann). Family Planning Communication Research Seminar, New Delhi, 1964.

Fertility trends and national policy. Fifth All India Conference on Family Planning, Patna, 1964: 18–24. Also *The Journal of Family Welfare*, 1964, Vol. 10, No. 3: 12–20; and *Maharashtra Medical Journal*, 1964, Vol. 11, No. 1: 31–39.

Preliminary report of the study of two family planning clinics in Colombo. Eleventh Annual Report of Family Planning Association of Ceylon, 1964, 41–42.

Tendencias demograficas Y control de la poblacion (Text of a lecture delivered at the Fourth World Congress of Gynaecology and Obstetrics held at Mar Del Plata). *Medicana Panamenicana*, 1964, Vol. 22, Nos. 20–21: 491–97.

Demographic research and training in the ECAFE region. Papers contributed by Indian authors to the World Population Conference, Belgrade, 30 August–10 September 1965. New Delhi: Office of the Registrar General. Also World Population Conference, Belgrade, 1965. Background Papers No. B.8, Part I/4/E/404.

Family planning through clinics: report of a survey of family planning clinics in Greater Bombay (with Katherine Kuder). Bombay: Allied Publishers, 1965.

Population programmes in ECAFE countries. Proceedings of the First Conference on Population. Population Institute of the University of the Philippines, 1965, 37–54.

Problems of research design and methods in studies of effectiveness in policy measures aimed at influencing fertility. World Population Conference, Belgrade, 1965, 260–64.

Some demographic guides to strengthen the family planning programme in India. State Family Planning Conference, Punjab, Chandigarh, 1965.

Recent trends in family planning research in India. *In* Bernard Berelson *et al.* (Eds.), *Family planning and population problems*. Chicago: The University of Chicago Press, 545–59.

Demography and the statistician. Proceedings of Seminar on Demographic Problems of Eastern India. Calcutta: Government of West Bengal, 1967.

Fertility in ECAFE countries (with J.C. Caldwell, Rapporteur). *Population change: Asia and Oceania*. Proceedings of IUSSP Regional Conference, Sydney, Australia, 1967, 65–77.

Studies of fertility in ECAFE countries. *Contributed Papers*. Session V(a) Moderator's statement, IUSSP Sydney Conference, 1967, 225–395.

Population explosion and the family. Report of the Proceedings of the 17th International Conference on the Family, New Delhi. Bombay: Family Planning Association of India, 1968, 131–34.

ECAFE program to assist fertility control. *Demography*, 1968, Vol. 5, No. 2: 651–58.

Forms of age-specific birth rates by orders of birth in an Indian community (with P.P. Talwar), *Eugenics Quarterly*, 1968, Vol. 15, No. 4: 264–72.

Comparative fertility studies. Proceedings of the IUSSP General Conference, London, 1969, 1: 377–79.

Family limitation in South and South-East Asia. Proceedings of IUSSP General Conference, London, 1969, 2: 1240–49.

National family planning programmes: aims and progress, family planning and national development. Proceedings of IPPF Conference, Bandung, 1969.

Population and Asian environments. Conference on Asian Environments, Ann Arbor, Michigan, 1971.

Some problems in determining the number of acceptors needed in a family planning programme to achieve a specified reduction in birth rate (with D.V.R. Murthy and K. Srinivasan). *Population Studies*, 1971, Vol. 25, No. 2: 303–8.

Continuation rate, 'use effectiveness' and their assessment for the diaphragm and jelly method (with Malini Karkal). *Population Studies*, 1972, Vol. 26, No. 3: 487–94.

Demographic training and research. Session XII Moderator's statement, Indian Census Centenary Seminar, New Delhi, 1972.

Future lines of research in the field of family planning in population strategy in Asia. Proceedings of the Second Asian Population Conference, Tokyo. *Asian Population Studies Series*, 1972, No. 28: 375–79.

Acceptor data in measuring the effect of family planning programmes on fertility. *In* C. Chandrasekaran and A.I. Hermalin (Eds.), *Measuring the effect of family planning programmes on fertility*. Belgium: Ordina Editions, 1975.

Measuring the effect of family planning programmes on fertility (ed. with A.I. Hermalin). OECD/IUSSP Publication. Belgium: Ordina Editions, 1975.

Overview (with A.I. Hermalin). *In* C. Chandrasekaran and A.I. Hermalin (Eds.), *Measuring the effect of family planning programmes on fertility*. Belgium: Ordina Editions, 1975, 505–54.

Role of the union in Asia. IUSSP Eleven Years of Activity 1965–76. Belgium: Liege, 1976, 5–7.

Family planning programmes and fertility changes. Proceedings of IUSSP International Population Conference, Organizer's Overview, Mexico, 1977, 373–91.

On the RFPI model for estimating family planning targets (with J.S. Parsons). *Demography India*, 1977, Vol. 2, No. 2: 112–23.

Concept of development from the demographic point of view. UNESCO Symposium International Sociological Conference, Stockholm, 1978.

Indonesia's population in the year 2000 (with S. Soeharto). *Bulletin of Indonesian Economic Studies*, 1978, Vol. 14, No. 3: 86–93.

Overview. *The methodology of measuring the impact of family planning programmes on fertility*. Manual IX, United Nations , 1979.

Ethics of controlling human fertility. *Journal of Institut d'Egypte*, 1981.

On the correlation bias in the application of Chandra-Deming method for estimating vital events. Working Paper No. 2. Cairo: Cairo Demographic Centre, 1981.

Report on the Future of the International Union for the Scientific Study of Population. Liege: IUSSP, 1981.

General summary of policy implications. World Fertility Survey Conference, 1980, WFS/ISI, London and Voorburg, 1981, Vol. 1: 469–76.

Changing concepts and data needs as methodology of evaluating family planning programmes develops. *In Evaluation of the impact of family planning programmes on fertility: sources of variance. Population Studies*, 1982, No. 76, ST/ESA/SER A/76, New York: United Nations.

Changing concepts of family planning programmes and methodological implications for evaluation. Conference Room Paper No. 9. Third Expert Group Meeting on Methods for Measuring the Impact of Family Planning Programmes on Fertility, United Nations, New York, 1982.

1981 Census count: signal for hope or despair. Karnataka State Legislators' Conference on Problems of Population and Development, Bangalore, 1982.

Fertility in Kathmandu city (with Laxman Bahadur Basnet and Mahbub Ahmad). *In Determinants of fertility in some African and Asian countries.* Chapter 27. Research Monograph Series, No. 10, 1981 seminar. Cairo: Cairo Demographic Centre, 1982.

Inaugural Address to the International Seminar on the Social Context of Demographic Change. Bangalore: Population Centre, 1982.

Meeting India's demographic challenge. Seventh IASP Conference, Tirupati, 1982.

Methodological problems in the evaluation of family planning programme impact whose programmes are integrated with other development sectors. Participant statement. Third Expert Group Meeting on Methods for Measuring the Impact of Family Planning Programmes on Fertility, United Nations, New York, 1982.

Review of developments in dual system estimation of population size and growth ed. by Karol Krotki. Edmonton: University of Alberta Press, 1978 (with Peter Gardiner). *Population Studies*, 1982, Vol. 36, No. 1.

Control of human fertility. Dr. John Matthai Endowment Lectures, University of Kerala, 9–10 February 1983, 1–16.

On two estimates of the number of events missed in a dual record system. Working Paper No. 6. Cairo: Cairo Demographic Centre, 1983.

Turning points in the fertility regulation policies of developing countries. 1982 seminar. Cairo: Cairo Demographic Centre, 1983.

Comments on 'reassessment of population trends' by Ansley J. Coale. Proceedings of IUSSP General Conference, Manila, 1981 (in press).

Biographical memoir on Late Dr. Satya Swaroop. New Delhi: Indian National Science Academy (in press).

About the Contributors

V S Badari is Assistant Director, India Population Project, Bangalore. He is interested in the evaluation of family planning programmes in South India, as well as in mortality and fertility studies.

Asha A Bhende is Professor, Family Planning at the International Institute of Population Sciences, Bombay. Her interests include demographic training and implementation of the family planning programme.

C Chandrasekaran was the first Professor of Statistics at the All India Institute of Hygiene and Public Health. He was the President of the International Union for the Scientific Study of Population. A pioneering demographer, he has many distinctions to his credit including the Chandrasekaran-Deming formula for the estimation of birth and death rates. He is a member of the International Statistical Institute.

Kingsley Davis was formerly Professor at the Department of Demography, University of California. At present, he is Professor at the Hoover Institute, Southern California. He is the most outstanding scholar in the field of Population Sciences. He has revolutionised the research on fertility through his formidable contribution, namely, Intermediate Variables.

James T Fawcett is at present Professor at the Population Centre, East-West University, Hawaii. He is a member of the American Psychological Association and Task Force on Psychology, Family Planning and Population Policy. He has written several books and papers on the psychological aspects of population.

Ronald Freedman is a senior Professor in Sociology and Population Studies at the Population Studies Centre of Michigan University. He is a pioneer in the field of empirical research on fertility behaviour. His conceptual model on Davis-Blake intermediate variables of fertility has formed the basis for fertility studies during the last two decades in several countries.

V Gowri is a faculty member at the Tata Institute of Social Sciences, Bombay. She has conducted fertility and mortality investigations, and has also worked on two micro-demographic studies. Her doctoral thesis is on the performance of *Dais* in MCH and FP Programmes.

Thomas M Harrington is a faculty member at the Southern California University, USA.

Albert I Hermalin is currently Director, Population Studies Centre, Michigan University. His chief contributions are in the fields of family planning evaluation, family size norm, and fertility.

Terence H Hull is the co-ordinator of the International Population Programme of the Australian National University. He worked for several years in Indonesia. He has done extensive research on mortality and fertility. He is an anthropologist and a social demographer.

P Krishnan is Professor of Sociology, Alberta University, Canada, and also the Head of the Population Laboratory. He has been chiefly working on the determinants of fertility behaviour. His other interests include historical demography.

R S Kurup was formerly Director, Bureau of Economics and Statistics, Kerala, and Chief of the Population Research Centre, Trivandrum. He has worked for several years in the Department of Family Welfare of the Central Ministry. His contributions are related to the evaluation of family planning and determinants of fertility.

K Mahadevan is at present Dean, Faculty of Arts and Professor of Population Studies, S.V. University, Tirupati. He was formerly a senior social scientist at the Gandhigram Institute of Health and Family Planning. His analytical framework and conceptual model on mortality are being extensively used in mortality research all over the world. He is the General Secretary and Treasurer of the Asian Forum for Development and Population Studies.

M S R Murthy, a trained biologist, is Lecturer in the Department of Population Studies, S.V. University, Tirupati. He has done research on animal and human reproduction, bio-social aspects of mortality, determinants of early menopause, abortion and population education.

D A Naidu is Reader at the Department of Population Studies, S.V. University, Tirupati. His main interests are fertility, child labour, and value of children. At present, he is involved in a study on the interrelationship of mortality and fertility.

Charles B Nam is Professor at the Florida State University, USA. He has edited the book *Population and Society* which is extensively used all over the world for teaching and research. His areas of interest are fertility and social demography.

N Krishnan Namboodiri was Chairman of the Department of Sociology at North Carolina University for several years and is currently a distinguished professor at the Ohio University. His methodological and theoretical contributions in the field of population studies are laudable.

P K B Nayar is Professor of Sociology and Dean, Faculty of Arts, at Kerala University. He has done work in the fields of sociology, population sociology, family planning and fertility. At present, he is the President of the Asian Forum for Development and Population Studies.

Pietronella van de Oèver is a faculty member at the Florida State University, USA.

S Siva Raju is Lecturer at the Tata Institute of Social Sciences, Bombay. He is a statistician-cum-demographer and has done work on the determinants of family planning behaviour and evaluation of health and family welfare programmes.

D Usha Rani is an economist and a UGC scientist in the Department of Population Studies, S.V. University, Tirupati. She has done work on the costs of rearing children in fertility behaviour.

P R Reddy is Dean, Faculty of Biological and Earth Sciences and Professor of Nutrition, S.V. University, Tirupati. In collaboration with the Department of Population Studies, she undertook three projects related to the determinants of early menopause, determinants of infant and childhood mortality, relationship of nutrition with fertility and mortality. Her interests include rural development, nutrition and population education.

P Jayarami Reddy is a population psychologist and Reader at the Department of Population Studies, S.V. University, Tirupati. His interests lie in the field of urban studies, population aging and population education.

K Srinivasan is the Director, International Institute of Population Sciences, Bombay. He has worked at the Gandhigram Institute and ESCAP for several years. His contributions are in the field of evaluation, biostatistics and family planning and mathematical demography.

Terence H. Hull is the co-ordinator of the International Population Programme of the Australian National University. He worked for several years in Indonesia. He has done extensive research on mortality and fertility. He is an anthropologist and a social demographer.

P. Krishnan is Professor of Sociology, Alberta University, Canada, and also the Head of the Population Laboratory. He has been chiefly working on the determinants of fertility behaviour. His other interests include historical demography.

P.S. Kurup is formerly Director, Bureau of Economics and Statistics, Kerala and Chief of [illegible] Research Centre, Trivandrum. He has worked for several years in the Department of Family Welfare of the Central Ministry. His contributions are related to the evaluation of family planning and determinants of fertility [illegible].

K. Mahadevan is formerly Dean, Faculty of Arts and Professor of Population Studies, S.V. University, Tirupati. He was formerly a senior social scientist at the Gandhigram Institute of Health and Family Planning. His analytical framework and conceptual model on mortality has been extensively used in mortality research all over the world. He is the General Secretary and Treasurer of the Asian Forum for Development and Population Studies [illegible].

M.S.R. Murthy, a trained biologist, is lecturer in the Department of Population Studies, S.V. University, Tirupati. He has done research on [illegible] human reproduction, biosocial aspects of mortality, determinants of early [illegible], abortion and population education.

D.A. Naidu is Reader at the Department of Population Studies, S.V. University, Tirupati. His main interests are fertility, child [illegible] and value of children. At present he is involved in a study on the interrelationship of mortality and fertility.

Charles B. Nam is Professor at the Florida State University, USA. He has edited the book *Population and Society* which is extensively used all over the world for teaching and research. His areas of interest are fertility and social demography.

N. Krishnan Namboodiri was Chairman of the Department of Sociology at North Carolina University for several years and is currently a distinguished professor at the Ohio University. His methodological and theoretical contributions in the field of population studies are invaluable.

P.K.B. Nayar is Professor of Sociology and Dean, Faculty of Arts at Kerala University. He has done work in the fields of sociology, population sociology, family planning and fertility. At present he is the President of the Asian Forum for Development and Population Studies.

Philomena van de [illegible] is a faculty member at the Florida State University, USA.